DEVELOPMENTS IN SEDIMENTOLOGY 19

SEDIMENTATION MODELS AND QUANTITATIVE STRATIGRAPHY

FURTHER TITLES IN THIS SERIES

DEVELOPMENTS IN SEDIMENTOLOGY 19

SEDIMENTATION MODELS AND QUANTITATIVE STRATIGRAPHY

BY

WALTHER SCHWARZACHER

Department of Geology,
The Queens University, Belfast (Northern Ireland)

ELSEVIER SCIENTIFIC PUBLISHING COMPANY
Amsterdam — Oxford — New York — 1975

ELSEVIER SCIENTIFIC PUBLISHING COMPANY
335 JAN VAN GALENSTRAAT, P.O. BOX 211, AMSTERDAM, THE NETHERLANDS

AMERICAN ELSEVIER PUBLISHING COMPANY, INC.
52 VANDERBILT AVENUE, NEW YORK, NEW YORK 10017

Library of Congress Card Number 74-29689

ISBN 0-444-41302-2

With 110 illustrations and 22 tables

Printed in The Netherlands

To A.B. Vistelius and
W.C. Krumbein, the
pioneers of the subject

Preface

The realization that sedimentary rocks are a record of the earth's history is possibly the most important discovery in geology. Stratigraphy, the science which studies such records, is impossible without sedimentology, yet most sedimentological problems cannot be solved without a stratigraphic framework. The main concern of this book is with this interrelationship between sedimentation and stratigraphy.

If nature were more obliging, sedimentation would be a continuous process, proceeding at a constant rate in each environment. The dating of geological events could then be achieved by taking simple measurements on the accumulated sediments. Unfortunately, reality is different and sedimentary records usually provide merely relative dates and only rarely can they provide an approximation to absolute dating. However, one can consider theoretical systems of sedimentation which are intermediate between the idealized solution which has been mentioned and the complexities of real geology. A study of such artificial systems can lead to a greatly improved understanding of geological processes in general and to the development of better methods of sedimentary stratigraphy eventually.

The theoretical approach to problems of sedimentation and stratigraphy involves the use of mathematics inevitably, some of which may be outside the average knowledge of the geologist who is trained in, let us say, calculus and statistics. Given a certain amount of patience, it should be possible however, to read large parts of the book without having more than an elementary knowledge of probability theory. The treatment of problems is "geologically oriented" throughout and the mathematical background is developed only when a good understanding of the theory is thought to be essential. Furthermore, the mathematics in the book is not original and excellent text books exist which can be consulted by any reader who wishes to go into the subject more deeply. A short list of such books is given.

The literature on the specific subject of mathematical approaches to sedimentary stratigraphy is still mercifully small and it is hoped that no important contribution has been missed, although some of the Russian and French work has probably not been as frequently quoted as it deserves.

Finally, it is my pleasant duty to thank everybody who has helped in the writing of this book. Thanks are due to Miss E. Moore of the typing pool, The Queen's University, Belfast who, together with her staff, helped considerably in preparing the manuscript. Dr. J. Graham and Mrs. J. Graham-White prepared most of the text figures. Very special thanks are due to my wife who not only attempted to translate my notes into English but also typed the first draft with one finger. I also thank all authors and journals who have permitted the reproduction of text figures.

The writing of this book has been frequently interrupted by the noise of exploding bombs and periods without gas, water and electricity. The kindness of my brothers enabled us to work and recover during prolonged visits in a more peaceful country.

Department of Geology,
The Queen's University of Belfast,
BELFAST

FURTHER READING

The geological background to this book is very wide and it would be difficult to select a reading list. However "Stratigraphy and Sedimentation" by Krumbein and Sloss (1963) gives an introduction to many of the problems which will be discussed, "Computer Applications in Stratigraphic Analysis" by Harbaugh and Merriam (1968) and "Computer Simulation in Geology" by Harbaugh and Bonham-Carter (1970) provide an introduction to the numerical treatment of stratigraphical problems.

The following might be found useful for the mathematical background. Feller (1957) as an introduction to probability theory, Cox and Miller (1965) for stochastic processes and Granger and Hatanaka (1964) for spectral analysis. The method of Laplace transforms which will be used repeatedly, is treated by Churchill (1958).

Contents

Chapter 1

Sedimentation and environment

At the beginning of this century a very sharp distinction was made between the exact and descriptive sciences. Physics, as the prototype of the exact sciences, consisted of two branches: the experimental physicist made observations which were thought to be of limitless accuracy and the theoretical physicist interpreted such data, using the impeccable logic of mathematics. The geologist, like the biologist who was designated by implication as an "inexact scientist", realized that he was dealing with much more complex problems. The observations of the geologist and the biologist, although they were as accurate as possible, were mostly non-quantitative or else the results of their experiments were highly variable. It was impossible to use the mathematical approach for deductions which could be based on such data and, therefore, their methods were called descriptive.

The introduction of statistical methods was a considerable step forward. The sedimentologists were leading in this field and were now able to use quantitative measurements for the description of rocks. Then, statistical decision methods were introduced into the interpretation of various stratigraphical and sedimentary problems. From about 1950 onwards, the new term "model" became fashionable and this term has since been used increasingly in geology. A model is a simplification of a real system to such an extent that it can be easily understood and analysed. Indeed, the simpler models can be formulated in mathematical terms.

Meanwhile, the truly exact sciences had undergone a certain revolution. Towards the turn of the century, physics discovered that it had to develop new mathematical methods in order to stay quantitative. Probability theory and mathematical statistics suddenly became the tools of theoretical physics. Furthermore, the modern physicist is quite aware that he too is dealing with models rather than with phenomena of the real world. Probabilistic models are beginning to be used in geology and it is now possible to examine the problems which were considered to be quite unsuitable for quantitative analysis only a few decades ago.

It must be clear from this account that the fundamental differences between the exact and descriptive sciences have become smaller, at the same time it would be over-optimistic to see in this trend an indication that geology is on its way to becoming an exact science. Fundamentally, nothing has changed, the geologist is still dealing with highly complex systems and he is still trying to understand processes which are out of the range of his observation. The new working methods, and particularly the model concept, permit a limited theoretical treatment of geological problems but, since one is dealing with highly simplified systems, results can only be taken as an approximation to reality.

It might appear that sedimentary stratigraphy is the most unsuitable section of geology to which mathematical methods should be applied. Other sections like geochemistry, petrology and even palaeontology, are much more concerned with measurable quantities than stratigraphy, which is largely descriptive in its outlook. In practice, stratigraphy has always used all sections of geology which could contribute in any possible way to the solution of stratigraphical problems. The mathematical approach is no exception, but in order to become useful, any mathematical methods which are employed must become fully integrated with the geological observations and theories which are used in stratigraphy. To achieve this, certain adjustments must be made on both sides. Of course the mathematical methods must be suitable for the types of problems which are to be considered and the mathematical geologist is constantly on the lookout for new methods. But the geological background itself will become slightly changed and certain sections will be emphasized, which might not be regarded as important by other geologists. Unavoidably, considerable time must be spent with the definition of geological concepts. The definitions which are used in mathematics are quite clear and unambiguous, which is not always true in geology, so that a certain amount of discussion and hair-splitting is needed to bring the two subjects together. Such an exercise can only benefit geology and indeed some geologists may see in this the main benefit which stratigraphy has received from the mathematical approach.

1.2 THE RELATIONSHIP BETWEEN OBSERVED ROCKS AND THE SEDIMENTARY ENVIRONMENT

The measured stratigraphical section is the basis of historical geology and provides the raw materials for the reconstruction of physical and biological conditions in the past. The section enables the geologist to put past events into a correct order and thus trace the development of our planet from the past to the present.

It is natural that the stratified sedimentary rocks have received most of our attention because these rocks contain the best record of the geological history.

The nature of a sediment will obviously depend to a large extent on the physical conditions under which their formation took place. The geologist expresses this by stating that the environment determines the aspect of a sediment. It is understood that within the environment certain conditions exist which lead to sedimentary processes and these in turn leave the sediment as an end product. The stratigrapher interprets sedimentary rocks by trying to reconstruct the environments under which the sediments were formed. This sequence can be shown diagrammatically as follows:

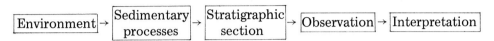

| Environment | → | Sedimentary processes | → | Stratigraphic section | → | Observation | → | Interpretation |

The whole object of sedimentary stratigraphy is to collect information which somehow has been transmitted from the past to the present. This type of problem is dealt with specifically by information theory. Consider, for example, the well-known problem of transmitting a message by wireless telegraphy. This process involves four stages which may be shown as follows:

| Message | → | Encoding | → | Transmission | → | Reception | → | Decoding |

If the system is perfect, the message will be fully received; if "noise" enters at any of the stages, a loss of information will occur. It is the aim of information theory to determine the efficiency of such a system and this is done by measuring the information loss in quantitative terms. The analogy between the wireless and the stratigraphic transmission problem is almost complete except for one important difference in practice. The wireless example is entirely man-made and is amenable to observation but the stratigraphic process involves the human operator in its second half only; the first or transmitting stages occurred in the geological past. The information loss in the second stages can be determined to a certain extent. At the best, the efficiency of the transmitting stage can be estimated only roughly.

It is obviously of great importance to know more about the information loss which is concerned with the geological process, and this can be illustrated by a deliberately naive example. Let it be accepted that there are only two types of environment, viz. sunshine or rain. These two constitute the message the stratigrapher wishes to receive. A faultless transmission system would produce, say, shale for sunshine and sand for rain. If this were true, even the most inexperienced stratigrapher could interpret a sand—shale sequence. The system would still be faultless if rain corresponded to either sand or silt, but considerable complications would arise if silt could indicate either sun or rain. In the language of communication theory, this is known as a failure of the first kind and it comes about because the symbol does not correspond uniquely to the message. This mistake is not to be confused with the so-called failure of the second kind which occurs when a symbol is identified wrongly, and would obtain if shale were wrongly interpreted as rain. The following diagram indicates all the mistakes which can be made in this simple example.

Message	Record	Identification	Interpretation

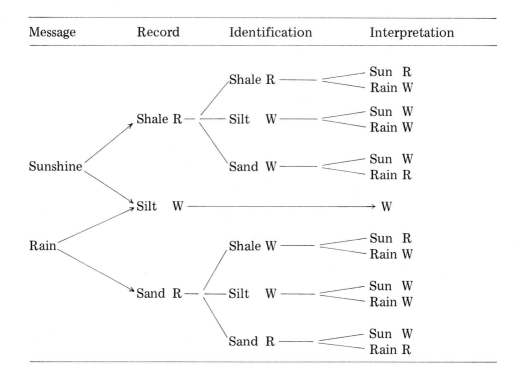

The stage of identification has been added to indicate how observational errors can enter the system. Judgment is given after each stage by indicating whether it has been right (R) or wrong (W). It may seem unfair to blame nature for producing a "wrong" record when silt is laid down in sunshine, but this judgment is made because the record is not unique and is therefore faulty within the system. It is these recording errors which are so difficult to trace and it is important to discuss the first stages of the information system in some detail.

1.3 THE ENVIRONMENT

The sedimentary environment has been defined as the complex of physical, chemical and biological conditions under which a sediment accumulates. There is nothing abstract about the term environment itself; it is simply the real world which surrounds any point on the earth or, for that matter, in the universe. Associated with the environment, there is always a three-dimensional space in which a variety of physical, chemical and biological processes occur and at each instant of time there exists a set of conditions which give a full description of the space. The various conditions which constitute the state of the environment have been named the environmental factors (Twenhofel, 1939).

Since the environment is changing continuously, it would be very uneconomical to define a new environment for each moment. In practice, many factors of the environment are quantities which represent average observations over a considerable period of time. As a typical example, one may think of climate, which is frequently used for the description of environments but it is well-known that a knowledge of the climate cannot be derived from a single set of meteorological measurements. The question of how long a time period is needed for the description of an environment will be considered in some detail later. At this stage it is sufficient to regard the environment as a dynamic system which is persistent over a sufficiently long time for certain processes to develop. Some environmental processes like sedimentation or erosion may be relatively fast; others like crustal sinking, or mountain uplift, may be comparatively slow. It will depend entirely on the chosen time-scale whether such processes are considered as part of the environment.

Accepting the environment as a dynamic system implies that the environmental factors are linked by a variety of functional relationships which can be used as additional descriptors of the environment. To take a typical example, the process of erosion may be related by some function to the local current velocity; the current may be the result of a pressure gradient in the environment, and this in turn may be related to some temperature difference and so on. The relationships in such a dynamic system are always extraordinarily complex and there is no environmental variable which is not, in one way or another, related to some other factor. Indeed, the processes themselves are part of the system and will change the factors which have caused the processes initially. In the example above, erosion is initiated by current action and once it occurs it will certainly influence the local current velocity. Such relationships between factors and response processes are known as feedback and they are a very important part of the system.

1.4 THE ENVIRONMENTAL MODEL

A moment's thought will show that the environment, as it has been defined, is so extremely complex that it is a useless concept for any practical consideration. One has to restrict the environment both in scope and scale in order to understand the environmental processes. This can be achieved by simplifying the dynamic system (which occupies the space associated with the environment) to a system which contains only those variables which are relevant to a certain geological process. Here, the main interest is in the processes of sedimentation and the relevant system is called the sedimentary system. Ecologists, palaeontologists or vulcanologists may be interested in different processes and therefore would consider different systems.

Each system represents an abstract environment. The sedimentation system is related to the so-called sedimentary environment model (Krumbein and Sloss, 1963), whereby the term model is added to indicate that this environment is in fact a simplification of the real world environment. In careless usage one often refers to the sedimentary environment model simply as the sedimentary environment.

It is now possible to define the area in which environmental studies are made so that the sedimentologist will concentrate on those areas which are relevant for the formation of sediments. For example, the first stages of sediment formation are commonly processes of weathering and erosion, in what is generally called the source area. The environment of this area consists of all the point environments which are situated within the source area. In a similar way one can define a deposition area and call the environment which reigns in this area, the deposition environment. There are many other sedimentological terms which can be used to demarcate boundaries in a geographical sense. For example, a basin which may either be an actual depression on the sea floor or a structurally-caused thickening of sediment is such a unit with definite geographical limits. Similar terms are 'drainage system', 'belt of sand dunes', 'estuaries' or 'deltas'. These geographical units are regarded as essential parts of the sedimentation system which usually comprises much wider boundaries, but in exceptional cases, it can be contained within the geographical unit under discussion. For instance, if one wishes to examine the depositional environment of a basin, the environment of each point on the sea floor has to be established. It might be sufficient to record all the physical and other conditions within the basin itself, and under such circumstances the basin would be regarded as a self-contained sedimentation system. In many other examples, important environmental factors are active outside the area of the actual geographical unit which has to be examined and so the sedimentation system must be chosen with much wider boundaries. In the study of a delta which reacts with a marine basin and which may be fed by a large river system, it may be more appropriate to consider the delta as only a part of a much larger sedimentation system. There are no theoretical limits to the size of the sedimentation system, but in practice one attempts to keep the system as small as possible and just big enough to understand the essential processes of the environment.

The erection of system boundaries is always artificial and will invariably result in some interaction between the system and its surrounding real world. This may be allowed for by defining outside components, that is, factors which influence the variables and processes within the system. Such outside factors have been called exogenous components in contrast to the endogenous components of the system (Harbaugh and Merriam, 1968). The endogenous components of the system are all linked with each other by functional relationships and their behaviour can be studied within the model. The exogenous components are in no way determined by the model and they have to be regarded as independent variables. The use of this concept enables one to make the model very much more compact and therefore more easily understood. Consider, for example, the sedimentary environment of a small lake; most of the processes which are relevant to sedimentation are contained within the lake itself and perhaps on the land by which it is surrounded. It is possible that the amount of sunlight hitting the lake surface is an important additional factor. Rather than incorporating relationships between sun and earth and possibly the solar system, it will be clearly advisable to regard the solar radiation as an exogenous component.

1.5 ORGANIZATION OF THE ENVIRONMENT

Within an environmental model, a certain amount of order can be obtained by organizing the manifold environmental factors into groups. These groups have been called environmental elements (Krumbein and Sloss, 1963). The grouping of factors is again directed towards the purpose of the model; for sedimentation and stratigraphical problems the following four elements are used.

Geometry — The geometry of the sedimentary environment includes the essential description of shape and configuration of the deposition area and its framework. Such factors as water depth, shore distance or areal extent are included in this group.

Material — The material factors include the medium of sedimentation (air, water) and the full description of any material which is involved in the sedimentation processes.

Energy — The energy factors describe the kinetic and potential energy status of the environment. Potential energy is stored as thermal energy, as pressure differentials or as differences in chemical composition. The water or air masses of the sedimentary medium and parts of the solid materials are usually in a state of movement providing kinetic energy. Current velocities, wind speeds, or wave height are energy-related factors.

Biology — In many sedimentary environments, the biological factors are very important. Organisms either provide a large amount of the sedimentary materials or they may influence processes of sedimentation indirectly by aiding or preventing erosion, by bioturbation or by creating special chemical environments.

The environmental elements like the factors are interdependent and environmental processes are often linked to more than one element. To illustrate the usage of the environmental model the following example is used.

It is assumed that a small pond which has the shape of an inverted cone is to be examined. Although the system of this pond is self-contained, there are two external factors: a constant wind which produces waves on the surface of the pond and solar radiation which penetrates the water. No sediment enters the system from the outside but plankton is produced throughout the pond and this provides a constant source of sediment. Wave action prevents the sediment from settling above a certain critical depth and sedimentation can only occur below this level. The solar radiation maintains the growth of plankton. The model may be summarized in the following systems diagram:

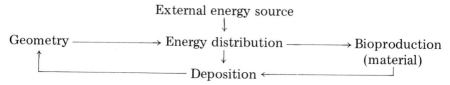

It is assumed that the distribution of light energy determines the amount of plankton which grows and that this in turn determines the amount of dead plankton which is available for deposition as sediment. Of course, the energy distribution is largely determined by the geometry of the basin. If the process continues over a sufficiently long period, enough deposition will occur to change eventually the geometry of the basin. In this case, a feed-back mechanism is present. This can be ignored over short time periods.

In order to define the environment of the pond, it is necessary to intro-duce the functional relationship between light penetration and geometry of the basin. It is known from modern observations that light energy declines exponentially with depth. It is further assumed that the wave energy gener-ated by wind on the surface decreases linearly with increasing depth. Given this information, the bottom environment of the pond can be determined and this is shown graphically on the right-hand side of Fig. 1.1. It is also possible to find the environment for any other locations and the conditions on the pond surface are illustrated on the left-hand side of the figure. The actual methods used for developing environmental data from sedimentation systems and the reversed process of finding a sedimentation system from the collected environmental information will be discussed in more detail later.

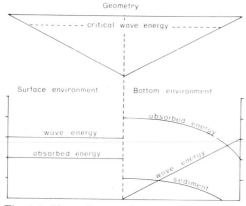

Fig. 1.1. The sedimentation environment of a small pond.

1.6 THE CLASSIFICATION OF SEDIMENTARY ENVIRONMENTS

The question of how accurately an ancient environment can be reconstructed from geological data will be discussed at a later stage. It must be obvious that the geologist has much less information on the ancient environment than the student of a modern sedimentation system. To make up for this lack of information, one has to rely heavily on observations which have been made on modern sedimentary environments. The ancient environment is compared with a modern equivalent and it is then assumed that the factors, and even more important, the relationships between the factors in the past environment, were similar to the modern analogue. To make such studies, it is sometimes useful to have a classification of environmental types.

Twenhofel (1939) initiated the systematic study of sedimentary environments and proposed the three groups of continental, mixed continental and marine types (Table 1.1).

Each group has been subdivided on a geographical basis. In recent years much more attention has been paid to modern environments and one finds a number of terms which are not necessarily geographic. Thus "high-energy environment", "low-energy environment", "reducing environment" and

TABLE 1.1

Classification of sedimentary environments (after Twenhofel, 1939)

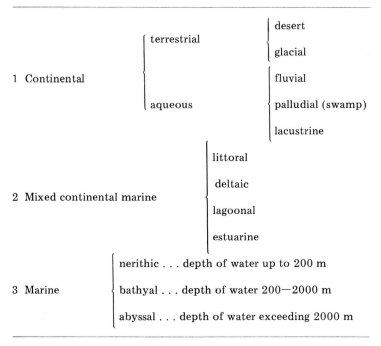

1 Continental	terrestrial	desert
		glacial
	aqueous	fluvial
		palludial (swamp)
		lacustrine
2 Mixed continental marine		littoral
		deltaic
		lagoonal
		estuarine
3 Marine		nerithic . . . depth of water up to 200 m
		bathyal . . . depth of water 200—2000 m
		abyssal . . . depth of water exceeding 2000 m

other similar terms refer to environmental factors and there is a need possibly, for developing a classification of environments which is more closely related to the sedimentary processes than to a geographical system.

1.7 SEDIMENTATION PROCESSES

The sedimentation processes are an integral part of the environmental system and they are treated separately from all other environmental processes only because they are so closely linked with the formation of sediments. One usually associates sedimentation processes with the three sequential stages of weathering, transportation and deposition. But in reality it is often found that the stages are not clearly separated. Weathering (halmyrolysis etc.) continues during the transport stage which itself is usually a series of transport, deposition, erosion stages following each other many times. Most annoying perhaps, for the quantitative geologist, is the fact that the so-called deposition process is very ill-defined.

Logically, deposition is simply the termination of transport and as such cannot be called a process. However, common usage in geology implies that the process of deposition is the building up of newly formed sediment and this may include the last stages of transport as well as the time when individual particles are already buried. Therefore, there are no definite limits to the beginning and ending of this process.

Clearly, a more precise definition is needed if the process is to be treated in mathematical terms. Any such definition must be artificial as it involves finding the precise moment in time when the sedimentary material ceases to be transported and becomes deposited sediment. Sedimentologists are well aware that the last stages of transportation are often difficult to separate from early diagenesis which in turn may go through all grades into metamorphism. There are situations where a particle may slow down considerably even before it reaches the sediment interface, but its movement continues after it has entered the sediment and as compaction progresses. One can define deposition as the passing of a particle through the interface between the sedimenting medium and the sediment, but it must be remembered that such an interface is not always physically well defined. It is necessary to define the interface which is a conceptual and often artificial boundary. In some instances the liquid limit of soil mechanics may be useful in defining such an artificial boundary. In this case, the interface is the surface which separates sediment with lower water content from sediment which has a higher water content than that of the liquid limit. But any other predetermined porosity or void-ratio could be used for such a purpose as long as it defines a definite reference level. Very similar arguments can be developed for erosion which can be treated as negative accumulation.

The sedimentation processes themselves determine the nature of the sediment which they produce to a great extent, but these processes themselves are in turn controlled by environmental factors. One can regard sedimentary processes in this sense as the link between sediment and environment. For example, it is commonly accepted that the transport of sand is related to current velocity and indeed, if specific parameters like grain size are taken, this relationship can be expressed in quantitative terms. Hjulström's well-known graph (Hjulström, 1939) shows that for each grain size a critical velocity exists at which erosion, transport and settling occurs. In addition, it was found that such velocities cannot be established with absolute precision; sometimes, experiments indicate that a relatively slow current can lift a grain of a given size and sometimes a relatively fast current is not competent to lift the same grain. The relationship between competent current which is an environmental factor and grain size in the stage of transportation is therefore not absolute. It is a relationship which can only be classified by statistics because the competence values for a given grain size fluctuate around a mean value.

In this particular example it is relatively easy to see how such fluctuations arise and why statistics enters into the relationship. Although the only forces which can erode a sand particle are determined with mathematical precision by the velocity field surrounding the grain, the relationship cannot be established because the velocities vary unpredictably in the micro-environment near the sand grain. The same is true for the forces which hold the grain in place; the resistance against erosion is determined by the weight of the grain, but also by the positions of the neighbouring grains which may be placed in a random fashion. The unpredictability of the acting forces is therefore not due to a breakdown of physical laws, but due to the complexity of the process which involves so many variables that they cannot be controlled either by experiment or theory. The situation is very similar to the classic example of "gas laws" in physics. Under given temperature and pressure conditions gas molecules undergo a thermal movement. The path of an individual molecule is determined by its energy and the vast multiplicity of collisions which it suffers with other molecules and the walls of the container. There are so many environmental factors (position of each molecule at one time) that the path of an individual molecule becomes unpredictable. The gas "laws" then must be understood as statistical laws which give the average velocity of molecules in a given volume. It is possible only because the population of molecules is so enormous that one can apply the laws with something amounting to certainty but, as it is well-known, irregularities do occur, particularly at extremely low temperatures.

The reason for substituting statistical for deterministic laws is simply that the former give a very much better description of the complex processes of a real world. For the interpretation of ancient environments the recognition of such statistical relationships has some very direct consequences. Consider the following hypothetical example: in a fictitious environment a bottom current is caused by a temperature gradient which in turn may be somehow related to a change in water depth. There may be some information about the functional relationships of these factors, but the relationships will invariably incorporate a probabilistic element. It follows directly that the sediment is not uniquely determined by the environmental factors and therefore the environment can never be reconstructed with certainty.

This uncertainty will increase obviously, with the number of stages involved in the argument. Thus it could be fairly safe to deduce a competent current from a given grain size but by the time water depth is reconstructed from the size of a sand grain, any relationship if it does in fact exist, will be very loose.

A similar loss of information occurs when an attempt is made to reconstruct the earlier stages of sedimentation. Deposition itself is the last stage and it is usually well documented by the character of the sediment it produces and fairly good criteria exist for deciding whether deposition has been mechanical, chemical or biological (Sander, 1936). However, the transportation stages are much more difficult to decipher since the later phases of transport very often eliminate the traces which were left by the earlier stages. The further removed an event is from the final deposition stage, the more difficult it becomes to find evidence of its occurrence. One is frequently left with the estimation of nothing but the total length of transport. The same applies to the information about weathering and a reconstruction of conditions in the source area is only possible if specialized minerals were formed or when some particularly characteristic rocks were decomposed.

The inability of sediments to record the environment correctly in every detail has been called at the beginning of this chapter the "recording error of sediments". It should be realized now that this error is unavoidable and, of course, no amount of sedimentological research or improvement in techniques can ever reduce this error because it has been made in the geological past.

If it is impossible to remove this error, one can at least try to estimate its importance, although this is a very difficult undertaking. Some experience could be gained by the study of modern environments since it is commonly found that the relation between environmental factors and the sediment which is produced is not at all strict. However, data are rare and too many workers regard any statistical variability as a nuisance rather than an extra bit of information. An alternative approach of estimating the error of sedimentation is far more speculative. It is based on making prior assumptions about the actual processes in the environment followed by a study of sediments which it produced. This experimental method will be discussed in more detail later.

The statistical relationships which have been introduced into the framework of the environmental variables do not, as a rule, make it any easier to understand the processes in the environment or the process of sediment formation. In justification of this step there are two points which should be considered primarily. As has been explained, the concept of statistical probability permits processes to be treated quantitatively, even when they are too complex for a deterministic treatment. Quite apart from this, it seems intuitively right to associate some uncertainty with any geological process and the idea of probabilistic elements in the environment is certainly not against the prevalent geological thinking. Other, and probably more philosophical, arguments have been used to justify the use of probability theory in geology (see for example, Mann, 1970). However, the argument of efficiency must be conceded by everybody.

1.8 OBSERVATION AND INTERPRETATION

At the beginning of this chapter the working process of the stratigrapher was compared with a communication problem. The environment was regarded as a kind of message which was to be translated into sedimentary rocks. This translation is, not without fault, achieved by the sedimentation processes. Once the sediment is formed it becomes part of the stratigraphic sequence and rests more or less undisturbed until it is discovered by the geologist. The period is equivalent to the "transmission" stage and during it the sediment undergoes all sorts of changes; at first diagenesis, later possible metamorphism and finally weathering on exposed outcrops will alter the sediment and contribute to the general information loss. Although such post-depositional processes are important they will not be discussed further at this stage. It is assumed for simplicity that any changes during this phase can be detected.

The final two stages in stratigraphy which are reception and decoding are the work of the geologist. Information is accumulated, by observing sediments in the field and laboratory, which forms the basis of interpretation. The two stages may be considered as separate but, during the actual working process, observation and interpretation very often overlap. Observations are frequently guided by a hypothesis which might be formed quite early in the process. This makes it very difficult to give a formal analysis of the observation process but some general statements can be made.

The type of information and also the quality of observation varies with different workers and with the different sediments which are studied. Generally speaking, two types of observations are made by every stratigrapher. One is based on measurements where the information is recorded by number and this set of observations is complemented by information in purely descriptive terms. Each individual observation refers to some particular feature of the stratigraphical section or rock specimen under study and the division into descriptive and measurable characteristics comes quite naturally. Features like colour, composition, faunal content and so on, are descriptive and these are known as attributes. Features like bed thickness, grain size or porosity are measurable and are known as variables.

The attribute is a property which objects either have or do not have and information about attributes are therefore qualitative statements. Variables which are measurable characters have both quality and quantity and the latter can be uniquely identified by making reference to a scale. Quantities are naturally described by numbers.

The number system might be considered as analogous to the vocabulary and it is therefore quite possible to express any statement, including descriptive information, in numbers. Sometimes this can be useful. Consider for example the following coding system defining colour: yellow = 1, green = 2, grey = 3. This leads to a so-called nominal scale (Krumbein and Graybill, 1965) which does not measure colour but simply substitutes numerals for words. This procedure is sometimes useful in book-keeping and data processing because it allows one to condense information into a much smaller space. There is a similar scale which is known as an ordinal scale and this applies when an attribute can be arranged in some order of rank which means that there are degrees to the quality which can be coded. For example, the series white—1, light grey—2, dark grey—3, black— 4 is a ranked series referring to the degree of lightness and it is possible to derive some basic statistics like mean rank and percentile measurements from such a sequence and such information could be regarded as semiquantitative.

Coding does not produce quantitative data but such data can be generated from descriptive information by determining the frequency with which certain attributes occur. As an example, the observation that a certain section contains red beds is qualitative whereas the observation that the same sequence contains five red beds is quantitative. This procedure is always possible and the modern tendency is to obtain as much quantitative information as possible and this applies particularly to the mathematically orientated stratigrapher who can handle numerical material much more easily than purely qualitative information. Certain statements like: "the middle Glencar Limestone shows strong bioturbation", cannot be quantified in this manner yet this statement has a high information content. It would be a great mistake to regard such information as in any way inferior from quantitative data and in the mathematical approach to stratigraphy such evidence must become part of the developed theory.

Any observation, and therefore any property which is derived from observations, can be either right or wrong. It is not necessary to bring in metaphysical arguments to accept an observation as right when it uniquely corresponds to the observed phenomenon. Right observations are reproducible whenever the observations are repeated. In a similar way, wrong means quite simply that a mistake has been made and, for example, a yellow sandstone has been called a green shale. It is well known that each observational act incorporates some errors. If the longest direction of a pebble is measured several times the measurements will not be identical because of errors in observation. It is equally well known that statistical methods can be designed to estimate this error and that by the simple process of repeating observations they can be made as accurate as it is required. Similar methods can be applied to the statistical properties of sediments and several text books deal specifically with geological statistical problems (e.g., Krumbein and Graybill, 1965).

Not all observational mistakes are as easily controlled as errors in measurement and there is a much more elusive mistake which can arise from incomplete observation or from not knowing the type of observation which is relevant. The sedimentary rock has been thought of as a carrier of information but it is impossible to know a priori how many messages are carried by each rock type. If some information is missed, then an incomplete and indeed wrong picture of the original environment will be gained.

It is known that sedimentological techniques are continuously improving and therefore observational powers are developed. To take but two examples, some twenty years ago it was unthought of that the temperatures of an ancient sea could be measured with any degree of accuracy; the work on isotopes in carbonates (Urey et al., 1951) suddenly suggested completely new possibilities for establishing such palaeo-temperatures. Perhaps less spectacular but quite typical is the relatively new use of the electron microscope in sedimentology. By simply looking at the surface of a sand grain it is now possible to see abrasion marks which are characteristic of various environments, wind transport and various ways of water transport can be recognized (Krinsley and Funnel, 1965). Such pieces of information were not available to earlier geologists and it is reasonable to assume that there are still many undiscovered bits of information at the present time. Once again one must come to the conclusion that the observation stage involves information loss, some of which is quite immeasurable.

A very similar situation exists when one attempts to analyse the final stage of interpretation. The special difficulties in assessing information loss during this stage can be appreciated by referring once more to the communication example. Decoding a wireless message consists in substituting a letter for each symbol. This is done by following rules which have been firmly laid down and an interpretation is wrong when any of these rules is violated. An assessment of this kind was used in judging the interpretation of the simple sunshine—rain example (see Section 1.2). None of the geological rules are as precise as the Morse code used in wireless transmissions where a symbol corresponds uniquely to a letter, but there are some geological rules. An example might be that: "red beds indicate an arid climate", or "flysch type deposits originate in geosynclines". These are statements which are generally accepted although exceptions to such rules are known to exist. Many other rules which are used are more debatable and principles used by some geologists are refuted absolutely by others. The interpretation of the stratigraphical observations cannot consequently be regarded as the mechanical application of rules which are already in existence, and very often each interpretation must be developed from first principles in each individual case. Developing an interpretation from first principles means in this context that the explanation of an observed phenomenon must be given within the framework of established laws. A variety of mistakes can be made in this process: invalid "laws" could be referred to, the explanation might not be unique, or the logic of the explanation may be faulty. Although it is difficult to assess just how many geological explanations are in error, undoubtedly some of them are.

1.9 CONCLUSIONS

The analysis of the stratigraphical working method has brought out a number of points which may be summarized.

(1) Many environmental processes, including sedimentation, incorporate random elements.

(2) The sediment is a statistical record of environmental conditions.

(3) The sedimentary record is altered during the stages of diagenesis and subsequent metamorphism.

(4) The geologist makes mistakes both during observation and interpretation.

The first two points in this summary can perhaps be clarified further. Let it be assumed that the sediment is always a perfect record of the environmental history, then point 1 states that this record will have statistical properties involving chance elements. Let it be assumed, on the other hand, that certain factors in the environment develop with perfect predictability, in which case point 2 states that the record of this environmental factor will once again involve random elements. The theoretical study of environmental and deposition processes must take account of this situation and many of the theoretical models which can be studied will therefore be based on probability theory.

Chapter 2

Sediment and time

2.1 INTRODUCTION

Stratigraphy, as an historical science, is based on the recognition that younger strata are superimposed on older rocks. This principle of superposition, which is attributed to Hutton (1726—1797), is the basis for constructing a geological time scale. By gradually building up the stratigraphical column, it is possible to bring the major events of the earth's history into a proper sequence and thus create a relative time scale.

Many attempts have been made to estimate the absolute time involved in geological history and a variety of methods have been used. For example, if the average rate of sedimentation can be determined, then time can be calculated from the total thickness of the deposits laid down in a certain formation. Using this principle, C.D. Wallcot in 1893 estimated the age of the Archaean as 10^7 years, and Sollas in 1905 came to a figure of $3.35 \cdot 10^6$ years.

Kelvin (1824—1907) attacked the problem from a geophysical point of view and calculated the cooling rates of the earth, and came to the conclusion that some $2.4 \cdot 10^7$ years ago the surface of the planet was molten. Other calculations based, for example, on the salt content of the oceans and assuming that all sodium is originally derived from dissolving igneous rocks, led Joly in 1899 to believe that the age of the oceans is some $1.5 \cdot 10^8$ years.

Radioactive dating started in 1910 when R.J. Strutt (later Lord Rayleigh) determined the age of pitchblende by the helium method. Since this time, the methods have been very much developed and a large amount of data is now available. Thanks to A. Holmes (1965) who collected and critically reviewed many of these data, we have now an absolute time scale which at least gives an age for all the systems in the stratigraphical column.

The development of an absolute time scale has been of immense importance to geology but it is not accurate and complete enough to replace the standard methods of classical stratigraphy. The geologist who works within a certain system still relies entirely on relative dating. Indeed, many practising stratigraphers regard absolute time as not very essential and absolute ages are mentioned with great care, if not suspicion.

The sedimentologist, on the other hand, is mainly concerned with the study of physical processes and time is an essential physical quantity. Time, as in our daily lives, is regarded as a continuum and many sedimentation processes involve time measurements in fractions of a second rather than millions of years. Such precise time information is hardly ever available from the stratigraphic record and the student of sediments, therefore, uses experiments, observations on modern sedimentation and theory as his main tools. In this sense, sedimentology is also based on Hutton's important work which is summarized in the principle of uniformitarianism stating that the present is the key to the past.

Like any dogmatic statement, this principle has been attacked by some geologists and too literally applied by others. The general consensus, however, is still that the same processes which make up our modern environments acted in the past quite simply because the "laws of nature" have not changed. At the same time one must recognize that our planet undergoes certain developments and it cannot be denied that the advent of man on earth, for example, had a profound influence on sedimentation. The same is probably true when grasses spread for the first time towards the end of the Mesozoic. It is probably also true that the present time, following a period of intense tectonic activity and glaciation, is not very typical of some past periods. Stoke's law, for example, was applicable in the Precambrian just as it is now, except that some of the constants have changed slightly. Theoretical sedimentology, together with actualistic observations, is therefore well justified and can be used as substitutes when information is lacking in ancient sediments.

One of the major problems which arises from sedimentary stratigraphy is the fact that the sedimentologist and stratigrapher use two different time scales. The former thinks in terms of absolute time just like the classical physicist, whereas the latter uses a much more complex scale which will now be discussed in some detail.

2.2 THE PRINCIPLE OF SUPERPOSITION

The principle of superposition appears to us self-evident and today it seems hardly necessary to give any further justification for this basic rule. It was, however, not until Hutton's work that sedimentation was clearly recognized as a rock-forming mechanism and it is only this process of piling particle on top of particle which makes the law of superposition so inevitable.

If deposition is continuous, more and more material will be accumulated and one can measure the thickness of sediment by the distance Z from a reference level to the surface of the sediment (see Fig. 2.1). Provided the rate of sedimentation R is constant, the trivial relationship:

$$Z = Rt \qquad\qquad\qquad [2.1]$$

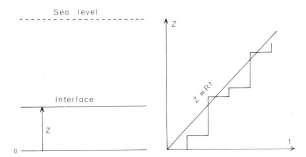

Fig. 2.1. The relationship between sedimentation and time.

obtains. The thickness scale Z is a transformation of the time scale and, even if the rate of sedimentation varies but remains positive, the principle of superposition applies and one may formally express this by:

$$t_0 \leqslant t_1 \text{ if } Z_0 \leqslant Z_1 \qquad\qquad [2.2]$$

A moment's consideration will show that this relationship breaks down if one considers sedimentation on a microscopic scale. All sediments consist of particles with a finite size; this may be molecules deposited onto a crystal face or particles up to the size of big boulders. From this it follows that the stratigraphic thickness scale cannot represent continuous time since this scale is by its very nature discrete and the sedimentation—time relationship on a microscopic scale will always have the appearance of a step function as indicated in Fig. 2.1 on the right. The relationship [2.2] i.e. the law of superposition, is not applicable for sediment thicknesses which are smaller than the size of the steps. This means in practice that the material of the sediment will determine the limits within which stratigraphical measurements are meaningful. It is well known that, in finely laminated shales, relative dating is possible within millimeters or less whereas this would be quite impossible in coarse conglomerates.

To make full use of the principle of superposition a certain knowledge of the sedimentation mechanism is necessary. For example, it is usually tacitly assumed that the sediment accumulates under gravity only, meaning that each particle settles strictly along a vertical path. Under this condition horizontal surfaces must be progressively younger. One may see this by considering the example which is shown in Fig. 2.2 where it is assumed that spherical particles of equal size have been deposited on a horizontal surface. If such particles assume close packing, one can define for any one a precise number of particles which are of necessity older than the chosen particle X. In the geometrically simplified case of close packing, all older particles are contained within a cone which slopes at $60°$ and which has the particle at its apex. One may now generate a new surface S_2 by distributing a number of

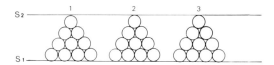

Fig. 2.2. The law of superposition involving spherical particles.

similar pyramids on the original surface S_1. Obviously every point in the new surface must be younger than the original surface S_1. However, it does not follow from the superposition principle that the new surface is of a simultaneous age and indeed the points 1, 2, 3 may have been deposited in this order, each pyramid having been formed after the previous one has been completed. Only if all the points are of exactly the same age will it be called a time plane. This situation will only arise if either sedimentation occurred with equal rates over the whole area or if sedimentation stopped after the completion of the second surface so that it became simultaneously exposed on the sea floor. Clearly, any time plane must have been at one time an interface between sediment and sedimenting medium, and if this interface can be recognized, time planes can be constructed. It is fortunate that in many sediments a permanent record of the positions of interfaces is kept in the form of bedding planes and it is such bedding planes which define the strata of Hutton's principle. A discussion of bed formation is therefore very essential in developing the stratigraphic time scale.

2.3 SEDIMENTARY BEDDING

It has been frequently stated that, although bedding in sediments is one of the most common features, surprisingly little is known about its formation.

Bedding is primarily a mechanical property of sediments and very often it is discovered in the field by hitting the rock with a hammer. The fracture of sediments into parallel slabs is due to the presence of bedding planes which produce mechanical non-homogeneous surfaces. The presence of bedding always implies anisotropy in the fabric, but not all anisotropic fabrics are bedded (Schwarzacher, 1946). The three most common types of sedimentary fabrics are the isotropic, the anisotropic—homogeneous, and the anisotropic bedded fabrics. These are shown diagrammatically in Fig. 2.3. Compo-

isotropic   anisotropic   anisotropic in-
homogeneous   homogeneous   homogeneous

Fig. 2.3. Sedimentary fabrics.

sitional changes and fabric changes very often go together and both combine to give the rock differing resistance against weathering which often accentuates the bedded nature of sediments.

The terminology which has been applied to layered rocks is slightly confusing. Otto (1938) introduced the term "sedimentation unit" which is frequently quoted as, "that thickness of sediment which was deposited under essentially constant physical conditions". Because the term "sedimentation unit" is somewhat cumbersome, the word bed has been substituted and most geologists regard bed and sedimentation unit as equivalent. Otto recognized that beds can be subdivided by shorter fluctuations which produce layers within the sedimentation unit and these he called laminae. His choice of the term lamina is somewhat unfortunate as it is clear that Otto intended the term to be purely genetical. The word laminated implies 'thinly layered', and it is difficult to think of laminae as layers which may occasionally measure as much as 25 cm in thickness.

Payne (1942) has made the thickness of the layer the essential criterium, a lamina is defined as any layer which is thinner than 1 cm and any layer which is thicker is called a stratum. Payne's definition has lost the genetical implications of Otto's terminology and it is also possible under this definition that a stratum changes into a lamina purely by local thinning.

Modern usage of bed and lamina (cf. Campbell, 1967) is more in agreement with Otto's terminology. A bed is largely determined by bedding planes which are formed by a distinct change of sedimentation which frequently comes to a complete halt at the bedding plane. Laminae, which Otto attributed to momentary fluctuations in the current velocity, may be produced by various "fluid dynamic and rheologic processes even under steady-state conditions of laboratory experimentation" (Jopling, 1964). It has been shown by Jopling (1966) that such laminae may form in extremely short times and Campbell (1967) defines a lamina as a small bed which forms rapidly and which is of more limited areal extent. It is also considered important that laminae show no internal layering and we may add that the lamina surface does not show any signs of prolonged non-deposition. The definition is somewhat vague but sufficient for most descriptive purposes; a definition which is particularly suitable for stratigraphical work will be given shortly.

It is now recognized that the thickness of layers in sedimentary rocks can provide a valuable description of sedimentary facies and for this reason a number of size scales have been developed for sedimentary bedding. In the quantitative approach, such scales will be replaced by statistics derived directly from measurements and no extra terminology is needed. One classification of external bed shapes, however, should be mentioned as it not only considers the thickness of the beds but also their lateral extent. This is the classification of Pettijohn and Potter (1964) which divides bedding forms into four groups as shown in Table 2.1.

TABLE 2.1

External form of beds (after Pettijohn and Potter, 1964)

1 Beds equal or subequal in thickness	beds laterally uniform in thickness	beds continuous
2 Beds unequal in thickness	beds laterally uniform in thickness	beds continuous
3 Beds unequal in thickness	beds laterally variable in thickness	beds continuous
4 Beds unequal in thickness	beds laterally variable in thickness	beds discontinuous

The last group in this classification contains all the types of bedding which are normally found in high-energy environments, particularly the various forms of current bedding which have been extensively studied (Allen, 1965). Much less is known about the laterally uniform beds. It is usually stated that over large distances all sedimentary bedding is lense shaped and this is a fairly safe statement since it is known that most sedimentary basins are of a finite size and therefore beds must of necessity end somewhere.

The following field relationships of bedding are frequently seen. In almost any exposure of bedded sediments, some bedding planes are more prominent and can be traced over the whole extent of the exposure or indeed sometimes over an area linking many outcrops (see Fig. 2.4). This type of bedding plane has been called a master bedding plane (Schwarzacher, 1958), a term which refers to the mapability of such a horizon and which should therefore be linked to the size of an area or outcrop in which it can be traced. Other bedding planes are seen to terminate but if the unit is traced laterally, such bedding planes sometimes reappear in exactly the same stratigraphical position and they are, therefore, truly homotaxial bedding planes (Fig. 2.4a). Often, the termination of the bedding planes is caused by uneven weathering but there are examples in which detailed petrographical analysis seems to indicate a complete disappearance of a bedding plane. Depending on the environment, it may happen that two or sometimes more bedding planes merge, in which case a single bedding plane is the time-equivalent of several bedding planes and the sediment enclosed between them (Fig. 2.4b). This is

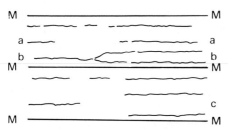

Fig. 2.4. The field relationships of bedding planes.

SW   NE

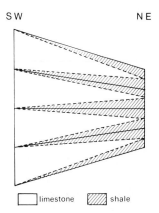

☐ limestone   ▨ shale

Fig. 2.5. The relationship between lithological boundaries and sedimentary bedding.

very much to be expected when sedimentation rates are variable over an area. The last example shown in Fig. 2.4c indicates two bedding planes which have no lateral equivalent. The situation may arise from two causes. Either, this is again a case of two merging bedding planes in which the actual merging is not recorded, or, there may be a genuine difference in the sedimentation history of the two areas.

The technical difficulties in tracing individual bedding planes over large areas are naturally quite considerable. Such work requires painstaking mapping, together with a constant awareness that changes are not only possible but are to be expected. It should not come as a surprise that some beds contained in persistent bedding planes change their facies. Examples of this can sometimes be seen in a single hand specimen but they are quite common on a regional scale. A typical case is shown in Fig. 2.5 which represents a limestone—shale series from the Lower Carboniferous of NW Ireland in which the master bedding planes have been traced over a distance of ten miles. In the SW the sequence consists of almost pure limestone and when beds are traced towards the NE they become thinner and at the same time shale progressively replaces the limestone. It can be seen clearly that the trace of the successive facies boundaries cuts the bedding planes at an acute angle and facies changes occur within a single bed.

The fact that bedding planes cut lithological boundaries has been frequently reported (cf. Vortisch, 1930) and is an indication that the migration of facies boundaries must have occurred at a different rate from that of the spreading of the bedding planes in the sediment.

The mechanism of the migration of bedding planes is best understood by considering the well-known example of current bedding. Fig. 2.6 shows a series of current bedded units which have been deposited from left to right and the numbers refer to consecutive time stages. It is clear that the surface of the sediment at time 2 consisted of the top set of unit 1, the foreset 2 and

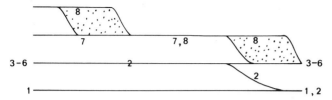

Fig. 2.6. Time planes in current bedded sediments.

the bottom set 1, 2. The tops and bottoms of the cross-bedded units originate by the merging of bedding planes. Since each of these surfaces take a
certain time to develop, the tops and bottoms must be of a composite age. In
the diagram it is assumed that the second unit was deposited after the first
unit had been completed and this is indicated by giving the top of the lowest
unit the age 3—6 which incorporates a short period of non-deposition. Sedimentological examination can often reveal such gaps and, for example, the
situation shown for the time stages 7 and 8 indicates that the two beds were
deposited almost simultaneously. In between 7 and 8 a compositional change
occurred permitting us to see that the foreset of 8 travelled a very short
distance behind the foreset of 7.

Indeed, in some environments it may happen that several layers are deposited practically at the same time and it is believed that this is particularly
common in turbidites. Under such specialized sedimentation conditions the
material can sort itself into layers while it is still being transported and when
deposition occurs several such layers may be put down practically together.
Apart from such exceptions, Fig. 2.6 shows clearly that whenever lateral
transport of sediment takes place, horizontal surfaces must be of composite
age. This is true for parallel beds also where no cross-bedding is recorded.
Such a composite surface can only become a synchronous surface when
sedimentation pauses long enough for such a surface to be exposed in its
entire extent on the sea floor.

Many observations indicate that such pauses in deposition occur frequently and, indeed, good bedding planes are usually developed when sedimentation slows down or comes to a complete standstill. Examples of bedding
which indicate a decrease in sedimentation are shown in Fig. 2.7; the types A
and B are particularly common in sequences where massive limestones alternate with shales. The laminations in the shale decrease in thickness and this
does at least suggest decreasing sedimentation rates. The decrease may be
followed by a gradual increase (as in type A) or by an abrupt change of
sedimentation (as in type B). In other rocks the evidence of non-deposition
may be seen directly on the bedding plane which may either show strong
biological encrustation or the effects of prolonged exposure to sea water
such as partial solution and phosphoritization. Very often such bedding
planes can also be covered by trails and burrows as indicated in Fig. 2.7C.

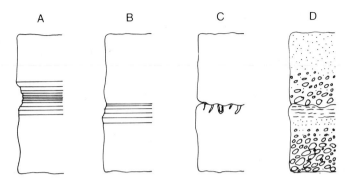

Fig. 2.7. Various types of bedding planes.

Precisely the same slowing of deposition and indeed very often slight erosion is noted in many clastic sediments as indicated in Fig. 2.7D.

Many well-known criteria exist to enable the sedimentologist to decide whether a surface has been at one time the sea floor. Shrock's classical work (1948) gives an excellent account of such primary sedimentation fabrics. If primary bedding is recognized and if there are indications that sedimentation near the bedding plane has decreased or come to a standstill, it is justifiable to regard such a surface as a genuine time plane. As a rule it is not difficult to find such time planes for such small areas as in a hand specimen or in a single outcrop. It will be much more difficult to establish the time equivalence of bedding planes over larger areas on a regional scale and this problem will be discussed at a later point.

2.4 RATES OF SEDIMENTATION

It has already been indicated that the rates of sedimentation must be known if a stratigraphic section is to be interpreted in terms of time. If such rates could be known at any point of the section, one would at least be able to measure time within the interval in which sedimentation has been continuous. However, it has just been argued that sedimentation rates varied considerably and that it is very likely that deposition has often been interrupted and in some environments even reversed. If one attempts to estimate rates of sedimentation, one can expect average values only over a considerable time period. An absolute time measurement is always involved regardless of whether such determinations are made on recent or on ancient sediments. Very often this is the biggest source of error in such estimates.

A variety of methods have been employed with modern sediments. Traps have been used to collect sediment over a year or more and sometimes annual layering or varving was recognized and used for dating. In some sediments, historical events such as eruptions of volcanoes can be identified and used as a time control. Most modern deep-sea sediments have been dated by ^{14}C methods which are relatively accurate. Most determinations of ancient sedimentation rates are based on the radioactive time scale. Average values are taken over a whole system and this of course represents a very much longer time interval than the one used on recent sediments, a fact which is important when recent and ancient rates are compared. Following a proposal of Fischer (1969), sedimentation rates can be expressed in Bubnoff units which were specially introduced for the measurement of geological time/distance rates. One Bubnoff (B) is defined as 1 micron/1 year which is equivalent to 1 mm/1,000 year or 1 m/1 million years.

A collection of modern sedimentation data is given in Table 2.2; unfortunately only a very few observations are available and the sample is still very inadequate. Nevertheless, a certain grouping in the rates may be seen and the near-shore marine sedimentation gives a rough average value of $2 \cdot 10^3$ B, the value for inland seas is approximately $2 \cdot 10^2$ B and the rates in the deep-sea are of the order of 10 B.

These measurements represent unconsolidated sediments and, if a comparison is made with ancient sediments, allowance has to be made for com-

TABLE 2.2

Modern sedimentation rates in μmm/year

Lake Vierwaldstätter[1]	3500—5000	freshwater
Lake Lunz[1]	1800	
Rhone delta	700	delta
Nile delta	660	
Clyde Sea[2] (shallow)	5000	largely terrigenous
Norwegian fjord[2]	1500	
Gulf of California[2]	1000	
Moluccas[2] (volcanic ash)	700	
Tyrrhenian Sea[2]	100—500	inland seas
Black Sea	200	
Bahamas[3]	33.8	carbonate environments
Florida Keys[3]	80	
Florida inner reef tract[3] (contaminated by terrigenous material)	220	
Globigerina ooze[2]	8—14	deep sea
Red clay[2]	7—13	

[1] Quoted after Schwarzacher (1946).
[2] Quoted after Kuenen (1950).
[3] Stockmann et al. (1967).

paction. According to Kuenen (1950) most recent sediments have water contents of between 50 and 80%, whereas ancient sediments usually have porosities of between 10 and 20%; Kuenen, therefore, recommends that the recent sedimentation rates should be divided by a factor of 2 or 3 when compared with ancient rates. Ancient sedimentation rates have not only been calculated for determining geological time but also because the amount of sedimentation can be taken as a measure of crustal activity in a certain time period. The interplay between tectonism and sedimentation will be treated in more detail later, but it is clear that, for any sedimentation to take place, two conditions must be fulfilled. There must be a source of sedimentary material and a basin into which the sediment can be dumped, thus both subsidence and mountain formation will largely determine the amount of epicontinental sedimentation. Bubnoff (1950), who examined this problem, came to the conclusion that the best measure of tectonism is the maximum thickness of sediment which has developed in a specific basin and this is known as the effective maximal thickness. Most previous workers used maximal thicknesses which were compiled from the thickest development of the sequence taken from anywhere in the basin. As Bubnoff pointed out, such a maximum thickness is never realized in a single locality and is therefore rather artificial. To the sedimentologist, it is of particular interest that Bubnoff also collected data on shelf sedimentation, and these together with the maximum and the effective maximum thicknesses are given in Table 2.3. The rates have been recalculated from Bubnoff's (1950) data using the improved Holmes (1965) time scale; all measurements are again given in microns/year (Bubnoff's).

TABLE 2.3

Sedimentation rates after Bubnoff (1950) in μmm/year

	Duration (m.y.)	Maximal America	Maximal Europe	Effective maximum	Shelf
Cambrian	100	86	55	40	15
Ordovician	60	147	77	66	25
Silurian	40	49	154	113	30
Devonian	50	78	314*	160	32
Lower Carboniferous	40	51	88	50	13
Upper Carboniferous	40	188	210	150	13
Permian	45	62	224*	100	23
Triassic	45	178	140	67	43
Jurassic	45	152	111	67	33
Cretaceous	65	354	230*	230	43
Paleogene	45	236	224	133	31
Neogene	23	533	533*	226	47
Weighted average		160.9	171.5	108.9	27.7

* Caledonian, Variscian and Alpine orogenic periods.

The data in Table 2.3 show three important points. First, it is clearly seen that sedimentation rates increase during times of special tectonic activity. The Caledonian, the Variscian and the two Alpine periods of orogenesis are indicated by an asterisk in the European column and the increase in sedimentation rates is quite striking. Secondly, it will be noted that Bubnoff's data do not seem to indicate a trend-like rise of sedimentation activity. Admittedly, the very value which is under the influence of tertiary mountain building is exceptionally high but otherwise the values fluctuate strongly but apparently not systematically. Finally, it may be noted that the modern sedimentation rates seem to be considerably higher than even the maximum sedimentation rates of the past. If one accepts that the few values for terrigenous sedimentation together with the delta values are equivalent to the geosynclinal type of sedimentation which is largely recorded by the maximum sedimentation rates, one obtains a mean value of 912 B. This excludes the obviously exceptionally high rate of the Clyde estuary. Adjusting for compaction, this average gives 304 B which is twice the amount of maximal sedimentation in older rocks.

The discrepancy between ancient and modern sedimentation rates may arise for two reasons. The present-day sedimentation is unique in the sense that it follows very close to the last ice-age and although ice-ages have occurred before, the determination of ancient rates has not been restricted to such a short period following the glacial upset. The second and probably much more effective reason is the discrepancy between the length of the observation period applied to ancient and modern sediments. It has always been assumed that even rapidly sinking basins experience periods of non-deposition and that such gaps would be incorporated into the estimates of sedimentation. On the other hand, modern sedimentation is always observed when it actually takes place and although some of the modern estimates may include sedimentation gaps, all observations come from areas of active sedimentation.

It is tempting to estimate the incompleteness of the stratigraphic record by comparing modern values with ancient ones. Such a comparison would indicate (as has been shown) that the stratigraphic record contains approximately 50% of time gaps. To show how tenuous such estimates are, one can use a different method to check the above result. Some old sediments contain stratification which has been interpreted as varves and which therefore give a direct measurement of the annual sedimentation rates. Such data as could be found are shown in Table 2.4.

The data in Table 2.4 fall into two groups. It is believed that the first group represents more or less stable shelf conditions whereas the second group represents rapidly subsiding basins. Comparing the varves with the maximum and shelf rates respectively of Table 2.3 leads to the two values of 91% and 82% time gaps in the stratigraphic record. Thus, one might say that the stratigraphic sequence records only 10—50% of the total time. Although

TABLE 2.4

Ancient sedimentation rates as deduced from fossil varves (in μmm/year)

Montery formation[1]	Tertiary	100—200
Kuban formation[2]	Tertiary	200
Black Hills formation[1]	Cretacous	200
Toledo limestone[3]	Jurassic	170
Blue Lias (shale)[1]	Permian	25—42
Blue Lias (limestone)[1]	Permian	230
Castille formation[3]	Permian	130
Loferites[4]	Triassic	1300
Gypsum[3]	Permian	1398
Silts[2]	Carboniferous	2760
Shales[2]	Silurian	2106

[1] Quoted after Duff et al. (1967).
[2] Quoted after Bubnoff (1947).
[3] Anderson and Kirkland (1960).
[4] Schwarzacher (1947).

these estimates cannot be regarded as anything but a first approximation, they do give an idea of the order of magnitude of the time gap in the stratigraphic succession. These values will vary considerably from environment to environment like the sedimentation rates themselves.

It is perhaps of some interest to record that Bubnoff (1947) interpreted the discrepancy between varve thickness and sediment accumulation as a possible indication that the astronomical year has shortened since Palaeozoic times. It is now known that the astronomical time units are indeed changeable and that the rotation of the earth around its own axis is now slower than in the past, but this affects only the length of the night—day cycle. The work of Wells (1963), who showed from a study of coral growth that the Devonian year had a length of approximately 400 days, is some indirect evidence that the time taken for the earth to rotate around the sun has not changed since the lengthening of days is in agreement with the slowing down of the earth's rotation, which itself can be measured directly. Similar indirect evidence for the constant length of years may be seen in the observation that the sunspot "cycle" has not changed in length either in the Permian or Devonian, where it has been recorded in varved sediments. Choosing the year as the geological time unit makes it, therefore, fairly safe to accept that the time standards have not changed at least since Cambrian times.

Sedimentation involves two stages which are transport and deposition. The transport consists of a lateral movement of sedimentary material and clearly the deposition of sediments will largely depend on the rate of lateral supply. In some environments, like travelling sand dunes, the actual creation of sediments proceeds much faster in a sideways direction than upwards and one suspects that such a lateral component is present in almost any kind of

TABLE 2.5

Horizontal sedimentation rates

Turbidity currents (Grand Banks)	$9.4 - 1.5 \cdot 10^{14}$
Gravel and boulder movements (experimental)	$3 \cdot 10^{12} - 3 \cdot 10^{10}$
Long shore drift	$2.4 \cdot 10^{11} - 8 \cdot 10^{10}$
Offshore drift (largely tidal)	$6.7 \cdot 10^{11}$
Advance Mississippi delta (Garden Island Bay)	$4.2 \cdot 10^{8}$
Mississippi delta (migration of main delta)	$4.5 \cdot 10^{7}$
Dunes	$1 - 2 \cdot 10^{7}$

sedimentation process. Unfortunately, very few data are available for estimating the magnitude of such rates but some observations which may be relevant have been collected in Table 2.5.

The observations which are summarized in Table 2.5 are again taken over very short time periods but one can use some geological examples to obtain at least a rough estimate of the rate with which sediment transgresses. For example, the classical interpretation of the Pennsylvanian cyclothems involves the repeated transgression and regression of marine environments. The length of such a cycle can be estimated as 10^{5} years. During this time, the sea spreads over the whole area involving distances of up to 1,000 km. As will be discussed in more detail later, it is very unlikely that a single transgressive state took longer than a single cyclothem and one obtains the rough figure of 10^{7} B for the rate of transgression. This is very likely to be a minimum figure since it is quite clear from the rock sequence that the cyclothem remained for quite an extensive period in the marine environment and that the actual transgression, therefore, proceeded much faster.

More direct observations have been made recently by C.D. Gebelein (personal communication, 1972) who dated the strand terraces of Andros Island in the Bahamas. These terraces are built out successively towards the sea and, therefore, represent a transgression of terrestrial environment which proceeds with an approximate rate of 10^{6} B.

Taking all these observations into account it seems reasonable to assume that the lateral rate of sediment migration is of the order of 10^{8} B. Comparing this with the vertical sedimentation rates we find the lateral migration to be 10^{6} times faster than maximal sedimentation or approximately 10^{7} times faster than shelf-type sedimentation. Expressed differently, this means that a lithological boundary may migrate 10 km horizontally in a rapid sedimentation environment, or 100 km in a shelf environment, whilst only 1 cm is laid down in the vertical succession.

The question of lateral transport compared with sedimentation rates is naturally of vital importance in problems of facies interpretation. Far too little is known about this as yet. It is possible that facies changes spread much slower than the figures which have been derived here, but perhaps these are the magnitudes which we should keep in mind when talking about the spreading of bedding planes.

2.5 STRATIGRAPHICAL UNITS AND TIME

In formal stratigraphy, lithostratigraphic units are ranked into bed, formation and group. The smallest of these units which is the bed corresponds to Otto's sedimentation unit. It has been argued that the bedding planes which enclose the unit are synchronous time planes in a restricted area. It is obvious that, when a bedding plane is exposed as the sea floor, it is an absolutely synchronous surface and there is no question of it migrating impercepti bly in time over the exposed surface. Such an assumption cannot be made for laminae which, although they may be traceable over the whole exposure, are often time transgressive. This difference is one of origin. While it is commonly found that a bed is followed by a short period of non-deposition the same is not necessarily true for laminae.

For stratigraphical work, the following definitions are useful. A bed is a lithostratigraphic unit which on the scale of outcrops is enclosed by synchronous time planes. It is the smallest unit which can be used for relative dating and one must avoid applying the principle of superposition to stratigraphic measurements below bed thickness. The lamina is regarded as a unit which has generally diachronous boundaries but which may be used for relative dating within the size range of thin sections and hand specimens. The principle of superposition becomes useless when applied to stratigraphic measurements below lamina thickness. The definitions do not provide for any thickness restriction of either bed or lamina and they are intended as a working terminology for practical sedimentary and stratigraphic work.

The quantum-like behaviour of beds as stratigraphical units can be accentuated by certain sedimentation processes. For example, the bed may be homogenized after deposition by diagenetic or bioturbation processes and this obviously makes it impossible to carry out any relative dating in the bed. In other situations the successive depositional surfaces may be so complex that it is impossible to reconstruct them. This is the case where a bed is formed by a series of bushy corals where the resulting time planes could hardly be analysed over more than a very restricted area. In yet other examples, so little may be known about the formation of the bed that it has to be regarded as simply a unit which cannot be divided.

Considering the bed as a unit which cannot be resolved in terms of relative dating, does not imply that the individual bed carries no time information, although this may often be the case. According to the optimistic relationship in [2.1] at the beginning of this chapter, the thicknesses of individual beds should be proportional to the time of their formation. This, of course, only applies when the sedimentation rates are constant. In real situations (as shown in Fig. 2.1) the sedimentation varies at random, most probably in the form of step functions. It is possible that the formation of a single bed is so fast that this itself may be best represented by a single step. Obviously, the thickness of such a bed carries no time information at all. The classical example for this type of process is the formation of a single-graded unit by a turbidity current. Much of the size grading occurs already during transport and the bed is laid down, as it were, ready-made and practically instantaneously.

A survey of the geological literature shows that the thickness of a single bed is hardly ever claimed to be related to the time of its formation. However, when repeated bedding is observed, one can assume that an average rate of bed formation exists. If a sufficiently long interval is observed, then the thickness of the sedimentary succession must be somehow related to time. Very little is as yet known about the minimum thickness of sediment for which such a claim can be made and detailed sedimentological analysis will be necessary in order to obtain more information about the time—thickness relationship. The problem will be treated later from a theoretical point of view and at this stage it is important to recognize that the stratigraphic record is not only made up of quantum-like steps, within which relative dating is impossible, but that there are also steps in the time recording abilities of the sediment. In exceptional circumstances, such steps can be small. For example, varved sediments allow precise timing of a single bed unit. Even more spectacularly, growth fabrics of fossils may permit the timing of a single day. Dating in such cases is achieved simply by counting the number of layers, and the thickness of the layer is exclusively determined by the environmental conditions and not by time.

It is because such ideal situations are so rare that many sedimentologists have been searching for periodicities which extend over more than one year. Such long cycles would obviously make the ideal time units in lithostratigraphy and this possibility will be discussed later.

2.6 TIME AND THE SEDIMENTARY ENVIRONMENT

It has been indicated in the first chapter that time plays an important role in the classification of environmental processes. Any environment contains variables which develop at very different rates. On the one hand there are such factors as daily temperature changes, etc., which are extremely fast. On the other hand, such factors as crustal sinking may only be noticed after a very long time. A table (Fig. 2.8) which has been prepared by Fischer (1969) shows the rates of various time—distance processes which are seen to vary from $1-10^5$ B. Other time processes which do not involve distance would show a similar spread and, if one included such factors as current velocity, sediment transport, etc., one could easily add some very much faster processes.

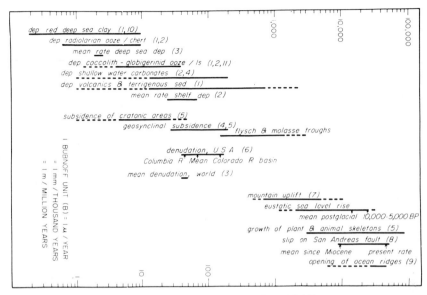

Fig. 2.8. Rates of geological processes (after Fischer, 1969).

Time processes which contribute to the environment are, however, not always continuous and it is instructive to think of an environment as of a dynamic system which contains numerous thresholds to which the system reacts when they are exceeded. Such thresholds may be purely mechanical (as in the processes which lead up to the triggering of a turbidity current), or purely biological (as in the spreading of an epidemic) or much more complex. In each case it is possible that a chance event causes the final result and if such processes happen frequently enough they can be studied in modern environments. However, it is well known that there are many phenomena which occur only a few times during a man's life. Major earthquakes, flood disasters, or volcanic eruptions are such examples and it is these sporadic events which lead to the most dramatic geological records. It is possible that, over the longer geological periods, events may occur of which we have little actualistic knowledge. To accept such a possibility is not a return to Cuvier's catastrophist philosophy. On the contrary, it follows clearly from the probabilistic nature of the environment which has been introduced in the first chapter.

The idea of environmental processes which have to take their course, once certain initial conditions or thresholds are satisfied, leads us to consider geological history as being made up of a number of completed processes which can repeat themselves several times. Thus there may have been several periods of glaciation. Geosynclinal sedimentation has occurred repeatedly in different parts of the world and the cycle of erosion, if not in the idealized sense of Davis, has operated again and again. In shorter time intervals we find

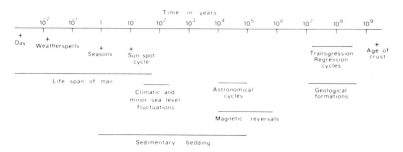

Fig. 2.9. Ages and durations of geological processes.

repeated sedimentary successions such as the well-known coal measure cycles or the phenomena of bedding. Much too little is as yet known about how long such repeated phenomena took and it is difficult to estimate the average length of such processes. To give some idea of the magnitudes involved, an attempt has been made to compile time data for geological processes in Fig. 2.9 and although these are based on examples treated in the literature they should not be taken too literally. Time values for relatively regularly repeated processes are of course more easily obtained and this is the reason why so-called cyclical phenomena are better known and why there is very little information about the more irregular events. The diagram Fig. 2.9 seems to indicate a gap between 10^5 and 10^7 years. Whether this is real or caused by our ignorance of processes which occurred during this time is still uncertain. The age and the duration of some geological phenomena and processes which have been estimated in a second group of data suggests that the spectrum of time events is continuous. It is convenient to differentiate between two groups of phenomena. These represent time-spans which are of the same order as the time involved in bed formation ranging up to approximately 10^5 years and processes which are completed between (let us say) 10^7 and 10^8 years. These two groups are responsible for what will later be referred to as the oscillating and trend components of stratigraphical sections. This division is simply a way of indicating that environmental changes occur at different rates and that the more rapid changes are superimposed on the slower development. This invariably affects sedimentary environments.

It has been seen previously that any description of environmental conditions involves the recognition of average conditions over certain time spans. In normal stratigraphical work, variables which are of the time order of bedding phenomena are as a rule averaged. As a very rough guide, one might say that the observational period which we associate with the description of an ancient environment is of the order of 10^5 years. This, however, is a generalization and many micro-stratigraphic studies have been made in which much shorter fluctuations have been considered as well as some studies which take an even broader view of stratigraphic developments (e.g., Krynine, 1942; Sloss, 1963).

2.7 SUMMARY

In this chapter some of the basic properties of the stratigraphical record have been discussed. In doing this, no reference has been made to biostratigraphic problems because biostratigraphy is a specialist's subject. It would be valuable if closer links between biostratigraphy and lithostratigraphy could eventually be established.

When considering lithostratigraphic problems, two points are of prime importance. In order to apply even the most basic stratigraphic principles, a knowledge of the sedimentation processes is required and sedimentological investigations are, therefore, the pre-requisite to any lithostratigraphic study. Secondly, any sedimentary record is discrete and there are several orders of quantum-like intervals which have different stratigraphic meaning. In ascending order it was found that the particle is the ultimate step of stratigraphical resolution. In practice, relative dating is possible by using laminae in thin sections and specimens and sedimentary beds in outcrops. Indeed this stratigraphical usage can be made the basis for defining these units. Again sedimentological knowledge is needed to ensure that such lithostratigraphic units are correctly chosen. A single bed may carry no time information and the mechanism of bed formation decides the minimum interval which can be related to time. This is a statistical sample problem which will be considered in more detail later.

Most stratigraphical records are incomplete and contain time gaps. Making some fairly pessimistic assumptions, these have been estimated to be of the order of 50—90%. Finally, it is suggested that the environment itself contains discontinuous processes, but too little is as yet known for making it possible to associate these definitely with stratigraphical units.

Chapter 3

Deterministic models

In the first chapter we have introduced the model concept as a method to simplify the sedimentary environment; it shall now be shown how this concept can be used in the quantitative approach to stratigraphic and sedimentation problems.

A model must be, by its very meaning, something which resembles reality but it is also implied that the model is somehow more easily handled than the real thing. This may be only a question of physical size, as is the case with the scale models of the railway enthusiast, or the model may be a more simplified version of something which therefore can be more easily understood. It is this idea of simplification which leads from the physical reality to the abstract formulation of a mathematical model which represents all or the most essential features by the use of meaningful symbols. In this way, the word 'model' has become almost synonymous with hypothesis and, indeed, it is used by many scientists in this sense. For example, Kendall and Buckland (1960) define a model as "a formalized expression of a theory or the causal situation which is regarded as having generated observed data". Krumbein and Graybill (1965) add that the model "provides a framework for organizing or structuring a geological study". Harbaugh and Bonham-Carter (1970) identify a so-called conceptual model with hypothesis and show how such models are continuously used in scientific methods of interpretation

3.2 DETERMINISTIC AND STOCHASTIC MODELS

When discussing the environmental model, it has been said that it consists of inter-related factors and elements which are responsible for the environmental processes. This essential make-up of an environment can be directly translated into the language of mathematics — the factors become variables and constants, the relationships are functions between variables, and the processes can be described by differential equations. In replacing a physical model like the environment by a mathematical model, one gains several advantages. Foremost is the economy and precision with which the model can be formulated; for example, the well-known problem of free fall in physics may be expressed as: "a body falling in the gravitational field of the earth is subject to the constant acceleration g", or much more compactly as:

$$\frac{d^2s}{dt^2} = g$$

[3.1]

The mathematical equation does not only contain everything which has been said in words but it can also be operated on with the greatest of ease. For example, two-fold integration of [3.1] gives the well-known relation:

$$s = \frac{gt^2}{2}$$

[3.2]

which is obviously useful if one is interested in the distance which a falling body travels in a given time. Or again, without much mental effort, one may write:

$$t = \sqrt{\frac{2s}{g}}$$

[3.3]

if one's interest is in the time which it takes for the body to travel a certain distance.

This simplicity of manipulation can be used to "operate" the model and any physical process can be followed through from the beginning to the end. For example, if the initial position of the body is known, its future positions can be predicted for any time. Position and time are by the definition of [3.1] predictable variables and, therefore, this is called a deterministic model. Any model which contains only predictable variables is deterministic and most "laws" of classical physics are of this kind.

Mathematics is not restricted to deterministic models, and processes with uncertain outcomes can be investigated by introducing so-called stochastic variables. A stochastic variable is unpredictable in the sense that it may take on any value or a defined range of values. Usually, the probabilities with which such values are taken are specified and, therefore, stochastic variables are also sometimes called probabilistic variables. The use of stochastic models will be discussed in the next chapter.

At this stage, it is essential to realize that whether a model is deterministic or stochastic is entirely a matter of definition and has nothing to do with the philosophical question of whether the universe is deterministic or not. It is, however, up to the investigator to decide if a particular problem is to be treated from a deterministic or probabilistic point of view and in this respect his philosophical attitude to the problem may be of some influence. Fortunately, it is usually quite obvious which approach will give the more useful results. Many problems in geology are more profitably treated from the stochastic point of view and the attitude is taken here that there is nothing inferior about probability models. A contrary opinion has been expressed by some geologists (cf. Pollack, 1968) who claim that the use of stochastic models is a sign of immaturity in the subject, an attitude which like the one expressed here, must be coloured by personal beliefs. At the same time, it would be foolish to neglect the deterministic approach for such reasons. It has been said (cf. Mann, 1970) that the principle of determinism and even causality may not exist in nature; this may be true, but the principles work very well in the practice of our daily life and most geologists will be quite satisfied to understand their subject in a physical rather than metaphysical world.

3.3 THE USE OF MODELS IN STRATIGRAPHIC ANALYSIS

The interpretation of stratigraphic sections is an intricate mixture of speculation and observation. There are always a number of collected facts which can be directly translated into environmental factors. Thus, if a limestone contains a well-developed marine fauna, one will not hesitate to assume that the formation took place in a marine environment; if such limestones are repeatedly followed by sandstones containing a terrestrial assemblage of fossils, one will conclude that there has been an alternating marine —continental environment. Stratigraphic analysis, however, goes beyond this obvious conclusion and will search for some cause for this alternation. There are two well-known explanations for the outlined situation — either eustatic sea level fluctuations occurred or the water level changed because of relatively local crustal movements. At this point the geologist begins to think in models; he will analyse, at least in his mind, the implications of the eustatic model and the regional tectonic model. Obviously, the first model implies simultaneous facies change on a world-wide scale, whereas, by the second model, such changes should be locally restricted. The two alternative models therefore clearly indicate the type of observation which is needed for further interpretation and such data may be already available or they may have to be collected.

This simple example illustrates that the speculative approach of forming a hypothesis or model is usually resorted to when the direct evidence of the sedimentary record cannot be interpreted any further, and one may regard the model approach as a complete reversal of the normal geological working methods. The established method is the reconstruction of events in the past, based on evidence which is studied in the present. The model constructor assumes some situation in the past and tries to develop the sequence of processes which lead to the rocks which are found in the present.

3.4 THEORETICAL VERSUS EMPIRICAL MODELS

In physics, models are either based on theoretical considerations or on experimental evidence. Let us again consider the example of free fall; the theoretical approach which has been discussed is based on the assumption of constant acceleration. If one further assumes that the fall commences at $t = 0$ then the distance travelled after a given time will be given by the expression of [3.2].

The experimentalist would proceed differently; he would actually drop a body and measure the distance travelled after a given time. Such data can be plotted and would give the same relationship as [3.2]. Fitting a curve to the observed data to obtain the distance as a function of time can be done without knowing anything about the physical significance of the variables and it is, therefore, an obvious but important step that in the experimental approach all variables are identified.

The construction of geological models proceeds on very similar lines; either, a theoretical approach is chosen and it is argued from first principles what processes have occurred after an initial state has been assumed, or, alternatively, the geological record is studied and regarded as the outcome of an experiment which took place somewhere in the geological past. The model must then be formulated in such a way that it produces a similar result. This type of model can be called an empirical model.

A purely hypothetical example may serve to illustrate the methods. Assume that the model of a volcano is to be constructed and this could be based on the few physical facts which are known. One might argue that there is a magma chamber beneath the volcano and that the material in it receives a certain influx of heat. The chamber may be connected with the surface of a plugged vent, thus pressure increases until an eruption, together with increased heat loss, occurs. Whichever mechanism is chosen for the model, one will try to keep the theory as close as possible to the processes in a real volcano and the rules by which the calculations are carried out are the laws of physics and, possibly, chemistry. One may, for example, calculate the pressure in the magma chamber by applying established principles of thermodynamics.

The same problem could be approached from a different direction. It is well known that the eruption sequences of real volcanoes contain certain regularities; indeed, it is this pattern of activity which is observed in the geological record. One could develop a mathematical model which generates a series of numbers which, in their sequence, have the same pattern as that of the eruptions. Obviously, the rules which such a model incorporates would be mathematical rules which need not have any immediate connection with the processes of the real volcano. Since such a model would be derived from an actual case history, the model would be empirical.

Either model, theoretical or empirical, has its own merits. The empirical model generally gives the better performance when it is tested against reality and empirical formulas are frequently used by scientists and engineers. Unfortunately, the reliability of an empirical model cannot be judged if it is applied to a problem which is outside the field for which experience of the model's performance exists, and it, therefore, lacks the generality which is a feature of the theoretical model. The theoretical model provides an insight into the actual mechanism of a process and, for this reason, it is very often preferable, even if it does not give such good results when applied to a specific real case.

3.5 THE OPTIMIZATION OF MODELS

It will be shown shortly that an absolute evaluation of models is impossible. A model can only fulfil certain functions and the only judgement which is possible is to decide how well it fulfils these functions. For example, a theoretical model may be formulated and its performance can be tested against some real geological process. The previously mentioned theoretical volcano model may be used to simulate a sequence of eruptions and this series may be compared with the history of an actual volcano, or, the theoretical model of free fall may be applied to some experiments and judgement may be given of how well it describes such a real situation. The comparison does not have to be always between model and reality and one might equally ask how, for example, an empirical model describes some purely theoretical process. As long as there is some standard against which the model can be compared there will be a way of assessing the performance of the model and, if necessary, a way to improve this performance. This process of finding the best model fulfilling a certain function is known as optimization.

Clearly, the development of an empirical model is nothing but an optimization procedure. The standard, in this case, is always a series of observations and the model is some analytical expression which has to be found and which is to give the best description of the data. If the model is to be deterministic, the well-known procedure of curve-fitting is used and the criterion of least squares is commonly applied as optimization principle (Bonham-Carter, 1972).

Theoretical models frequently contain a number of parameters which cannot be derived from first principles and which must be somehow evaluated. The determination of such parameters is, once again, an optimization procedure and the standard against which such a comparison is compared can be, again, some real observation.

Bonham-Carter (1972) and Harbaugh and Bonham-Carter (1970) have shown how formal, mathematical methods can be used in a search for the best model. A very simple example illustrates the procedure.

The model of free fall gave us a relationship between height of fall and

TABLE 3.1

Times of free fall with different values of g

Gravitation constant	$s = 1$ m	$s = 2$ m	$s = 3$ m	$s = 4$ m
60	0.577	0.816	1.000	1.148
70	0.534	0.756	0.926	1.068
80	0.500	0.707	0.866	1.000
90	0.471	0.666	0.816	0.943
100	0.447	0.633	0.774	0.894
110	0.426	0.602	0.738	0.853
120	0.407	0.577	0.707	0.816
130	0.392	0.555	0.679	0.784
Observed time	0.35	0.62	0.78	0.90

time, [3.2]. The model is universally applicable but, if it is to be used under terrestrial conditions, the appropriate constant g has to be found and this is done by optimizing the model with respect to a real experiment. The gravitational constant is treated as a variable and, using [3.3], the time which it takes to fall 1, 2, 3, 4 m, respectively, are calculated, assuming g takes on the values from 60 to 1300 cm sec^{-2}; such values are shown in Table 3.1. Next, the experiment has been carried out of dropping a stone through the four measured distances and timing this with a stop watch. To simulate "geological" conditions, this experiment was performed without any refinement and the times given are simple averages of five measurements each.

Each optimization procedure needs a so-called decision variable which describes the agreement between the model and the observed or otherwise assumed ideal values. In this particular example, the sum of the squared deviations between observed and theoretical values was taken as such a measure, and the variable itself is called the objective or state variable. If the objective variable is expressed as a function of the decision variable, the minimum of this function will obviously represent the point at which the model performs at its best. This particular value of the decision variable will indicate the smallest disagreement between the performance of the model and reality. The objective variable is shown as a function of g, the decision variable in Fig. 3.1 and it can be seen that there is a reasonably clear minimum at approximately $g = 1000$. It is, of course, well known that this value should be equal to 981, and considering the crudity of the experiment the result is quite good.

Optimization is not restricted to a simple variable and several model parameters can be optimized simultaneously; the methods used are fully described in the book by Harbaugh and Bonham-Carter (1970).

The adjustment of the parameters in a model leads, of course, only to an improvement in its performance and leaves its basic structure unchanged but

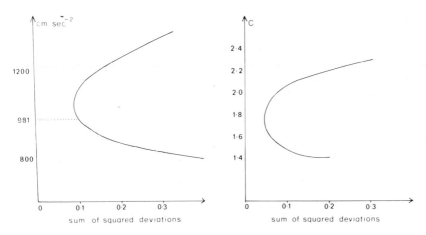

Fig. 3.1. Optimization of the free-fall experiment model.

there is no technical reason why the optimization procedure should not also be extended to the model itself and, indeed, sometimes this has been done. The consequences of such an approach may be seen in the previous example. Let us assume that we are not quite certain as to what the exponent in [3.2] should be and we shall write the equation for the free drop as:

$$s = \frac{g}{2} \, t^c \qquad\qquad\qquad [3.4]$$

where c is considered to be the new decision variable which has to be optimized against the experimental data. The procedure is exactly the same as before and the correct value of $g = 981$ is assumed. The objective variable is again the sum of squared deviations which is shown as a function of c in Fig. 3.1. In this diagram we find a minimum at the value of $c = 1.8$ and furthermore, it is seen that this new model gives an even closer approximation to the observed data than before, in other words, we have once more improved the performance of the model. This improvement, however, has been achieved at the cost of generality of the model and a different set of experiments would have given quite different results. It can be seen, therefore, that optimizing the structure of the model reduces a theoretical model into an empirical one and the advantages of the theoretical model are lost in this process.

It is always possible to make the change from a theoretical to an empirical model, but the reverse process can be much more difficult. Interpreting an empirical model involves the translation of a mathematical formula into a real physical process. In the free-fall example this would not have been very difficult, reasonably accurate experimentation would provide a c-value near to 2, and [3.2] could be obtained as an empirical model. The equation itself contains the basic assumption that the accelerating force is constant and, since one is dealing here with a very simple and fully defined process, no difficulties of interpretation should arise. On the other hand, consider a geological example. Assume some sand has been collected on a beach and let us say its grain size has been plotted as a function of the prevalent wave height. Of course, a curve can be fitted to such data but it is extremely doubtful whether the mathematical expression which generates this curve can be immediately interpreted in physical terms. At the best, the empirical model will be suggestive of some physical process which can be taken as a starting point for the development of a theoretical model. Indeed, it may be seen from this discussion that the empirical and theoretical approach can compliment each other in practice, and the development of more complex models is usually achieved by mixing empirical and theoretical methods.

3.6 THE EVALUATION OF MODELS

A model is by definition a simplification of reality and to expect a theoretical model which performs exactly like a real system would be a contradiction in terms. It is, therefore, not possible to establish the absolute correctness of a model. This discrepancy between theory and reality is much less obvious in physics, where a theory may be tested under such controlled conditions that model and experiment become almost identical. The geologist cannot, as a rule, simplify nature and perform a controlled experiment, and by necessity compares the simple model with the much more complex geological record. It has been shown that the model can be optimized against this standard, but this "best" model is not necessarily the most useful one, indeed, optimizing the free-fall experiments provided us with a model which is of much less use than the original theory.

There are, as yet, no established rules by which a model can be judged, but there are obviously two sometimes opposing qualities for which one has to look. One is performance and this can be measured against a standard. The second is generality which can only be judged in qualitative terms. Generality is, in the first instance, the property of theoretical models and is based on the complete understanding of a physical process. If the function of each model component is known, then the understanding of a geological process is complete. However, if the models are partly empirical and partly theoretical, then such models which contain more physically explained processes will be the more general ones. The generality of a model does not, of course, ensure that it is applicable to a specific case, but the knowledge of how the model works in detail will make it easier to decide whether the model is relevant or not. The stratigrapher who is largely concerned with processes which can no longer be controlled by experiments, always needs some theoretical foundation to his models and it may often be advisable to increase the theoretical content of models even if this leads to a decrease in performance.

3.7 EROSION AND DEPOSITION MODELS

Since stratigraphy is essentially historical geology, our main interest will be in models which treat the development of geological processes in time. Perhaps the most basic problem in stratigraphy concerns the rates of sedimentation; if the amount of sediment deposited per unit time was known for an observed section, one could not only construct an accurate time scale, but could also derive a great deal of information about the environment in which such rocks were formed. It has been shown in the previous chapter that we know very little about this vital problem, and any attempt to investigate sedimentation rates on a theoretical basis is very well justified.

Any geological problem has to be severely simplified before it can be treated by mathematical analysis, and the erosion—deposition problem can be reduced to such simple terms that formal mathematical treatment is hardly necessary. This means that the basic conclusions can be reached by simply "thinking through" the consequences of the assumptions which will be made; at the same time, a formal mathematical treatment will establish the procedure for developing deterministic models.

Let it be first assumed that a basin receives all its sediment from a nearby mountain range which is lowered by continuous erosion (see Fig. 3.2). Obviously the sediment in the basin will accumulate just as fast as the mountain disintegrates, and if S_t is the thickness of sediment at time t and E_t the amount of material which has been removed up to time t, we may write $S = cE$, where c is a constant expressing the relative areas of the mountain and basin districts. The height of the mountain h_t will be strictly determined

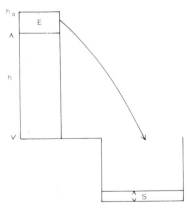

Fig. 3.2. A simple erosion—deposition model.

by the amount of eroded material which has been removed and, as can be seen from Fig. 3.2, one can write: $h_t = h_0 - E_t$ where h_0 is the original height of the mountain. Let us next assume that the amount by which the mountain decreases in height during a short time interval is proportional to the height of the mountain at the time. We may write this in the form of a simple differential equation as:

$$-\frac{dh}{dt} = \alpha h \qquad\qquad [3.5]$$

which can be directly integrated to give the solution:

$$h_t = h_0 \, e^{-\alpha t} \qquad\qquad [3.6]$$

The constant h_0 indicates that the mountain had its original height at $t = 0$ and the constant α gives the rate of disintegration. Substituting $h_0 - E$ for h in [3.5] gives:

$$\frac{dE}{dt} = \alpha(h_0 - E) \qquad\qquad [3.7]$$

which under the condition that $E = 0$ at $t = 0$ has the formal solution:

$$E = h_0(1 - e^{-\alpha t}) \qquad\qquad [3.8]$$

A typical solution of this equation is illustrated in Fig. 3.3 and, since E_t is

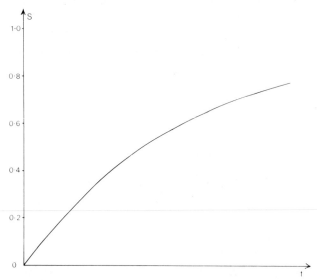

Fig. 3.3. The sediment—time relationship in the exponential model.

directly proportional to the cumulative amount of sediment in the basin, the
curve also represents the history of filling this basin. In the beginning, it can
be seen that the sedimentation rates are high and once the mountain has
been removed, no further sediment reaches the basin and the sediment fill
takes on a constant level.

This simple theory has been developed around the assumption that erosion is proportional to the height above a given base level. The justification
for making this assumption is primarily based on the fact that the potential
energy of mass increases with its height. Since the process of levelling a
mountain can be regarded as a process which tends to reduce the potential
energy the assumption of [3.5] seems reasonable. It is, however, known that
erosion is a very complex process depending on many factors of which
height is only one. The study of present-day environments has shown that
such factors as rainfall, vegetation cover, rock resistance and slope are of
very great importance. Although these factors may be correlated with height
to some extent, it is not immediately obvious how they will affect the
relationship. There are also very few observational data which would help in
the development of an empirical model. Fournier (1960) tried to estimate
the gross rates of erosion by measuring the sediment load of rivers and
relating these to such factors as average height $\bar{h}$, average slope $\bar{\phi}$ and climatic
conditions of the drainage basin. He obtained an empirical formula which
essentially states that the rate of erosion is determined by a climatic factor
C, which is derived from rainfall measurements, and the product of mean
height and slope:

$$\frac{\mathrm{d}E}{\mathrm{d}t} = C(\bar{h} \, \tan \bar{\phi} \,)^a + b \qquad\qquad [3.9]$$

a and C in this equation are constant. Fournier's formula suggests that, if one
ignores the climatic factor and keeps the slope a constant, [3.5] should be
written as:

$$-\frac{\mathrm{d}h}{\mathrm{d}t} = \alpha h^c \qquad\qquad [3.10]$$

in which c is a constant. One could now take this relationship and find the
best value of c to fit Fournier's formula but very little would be gained by
this in understanding the process and, in judging the reliability of the model.
It is, however, of some interest to see how sensitive the solution of [3.5] is
to some changes in the exponent c, and solutions for $c = 0.5, 2, 3$ are shown
in Fig. 3.4. Integration of these equations shows that the general picture of
the basin history is not very much affected by this change and the curves of
accumulating sediments will be similar. If the exponent is smaller than 1, the
erosion process would not stop when base level is reached, but, if only

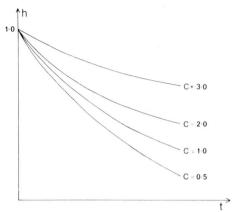

Fig. 3.4. Various solutions to [3.10].

positive heights are considered, the mountain would disappear within a finite time.

It is obvious that a better understanding of the process will only be achieved by taking additional variables into account. Such a model, which is based on the elevation and slope of the land mass, has been developed by geomorphologists.

Scheidegger (1961) argues that the development of natural slopes is largely determined by water-induced deposition and erosion. Sediment-laden water is moving over the slope and the carrying capacity, c, of this flow is determined by its speed. If the carrying capacity is exceeded, deposition will occur, but if there is less sediment than can be carried, erosion will take place. It is assumed that the slope consists of the same material as that which is carried in the water and that the speed v with which the water body moves, is proportional to the slope. The model predicts, in general terms, that erosion will take place along steep slopes and deposition will occur in the flatter parts and that a gradual levelling of the surface topography will occur.

Let us introduce the co-ordinates x and y for the horizontal and vertical directions, together with time t. Sedimentation can then be regarded as an increase in height, whereas erosion would be height loss. The increase in height is due to the loss of sediment mass carried by the water and therefore:

$$\frac{\partial y}{\partial t} \sim - \frac{\partial(vc)}{\partial x} \qquad\qquad [3.11]$$

Since the velocity is proportional to the slope and the carrying capacity is proportional to the velocity, one may write:

$$c \sim v \sim - \frac{\partial y}{\partial x} \qquad\qquad [3.12]$$

or the sedimentation:

$$\frac{\partial y}{\partial t} \sim \frac{\partial v^2}{\partial x} \sim -\frac{\partial}{\partial x}\left(\frac{\partial y}{\partial x}\right)^2 \qquad\qquad [3.13]$$

If [3.13] is written as an equality, one obtains a nonlinear, second-order, differential equation which can be linearized by introducing a new constant depending on the boundary conditions and the same constant. The linear equation for the change of height will be given by:

$$\frac{\partial y}{\partial t} = a\frac{\partial^2 y}{\partial x^2} \qquad\qquad [3.14]$$

which is the familiar equation used in physics to describe deterministic heat-flow or diffusion.

In Scheidegger's derivation, it was assumed that the water velocity, which carries the sediment, is proportional to the slope ([3.12]); if one assumes that the product vc, which is the mass flow velocity, is proportional to dy/dx, one can obtain the linear equation directly. Equation [3.14] could be expressed in words by stating that the change in height is proportional to the curvature of the topographic relief. Sharp corners will be eroded more quick-ly, flats and depression will be filled up by sedimentation.

The solution of [3.14] is given by:

$$y(x, t) = \frac{1}{2a\sqrt{\pi t}} \int_{-\infty}^{+\infty} \phi(\alpha)\exp\frac{(x-\alpha)^2}{4a^2 t}\, d\alpha \qquad\qquad [3.15]$$

in which a is a constant determining the mass flow and $\phi(\alpha)$ a function describing the initial conditions of y at $t = 0$ or in other words, the original topographical relief.

Let us consider the following, very simple, configuration as shown in Fig. 3.5. On the left-hand side is a landmass of height y. On the right is a basin of depth $y_0/2$ whereby, all heights are measured from the sea floor. The bound-ary conditions, therefore, are:

$$\phi(\alpha) = 0 \quad \text{for } \alpha > 0$$

$$\phi(\alpha) = y_0 \text{ for } \alpha \leqslant 0 \qquad\qquad [3.16]$$

Substituting [3.16] into [3.15] gives:

$$y(x, t) = \frac{y_0}{2a\sqrt{\pi t}} \int_{-y}^{0} \exp\frac{(x-\alpha)^2}{4a^2 t}\, d\alpha \qquad\qquad [3.17]$$

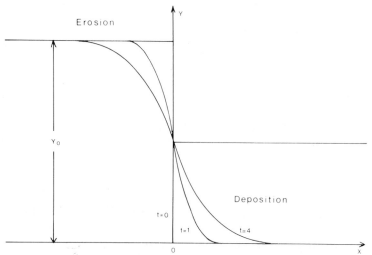

Fig. 3.5. The diffusion model of erosion and deposition.

Introducing a new variable $\xi = (x - \alpha)/2a\sqrt{t}$, we can write [3.17] as:

$$y(x,\ t) = -\frac{y_0}{\sqrt{\pi}}\left[\int_{-y}^{0} e^{-\xi^2}\ d\xi + \int_{0}^{\frac{x}{2a\sqrt{t}}} e^{-\xi^2}\ d\xi\right] \tag{3.18}$$

which, by making use of the error function $\mathrm{erf}(x) = \dfrac{2}{\sqrt{\pi}}\displaystyle\int_{0}^{x} e^{-u^2}\ du$ may be written as:

$$y(x,\ t) = \frac{y_0}{2}\left[1 - \mathrm{erf}\left(\frac{x}{2a\sqrt{t}}\right)\right] \tag{3.19}$$

The solution enables us to calculate the position of the land surface at various times, and this is shown in Fig. 3.5. Alternatively, one can calculate the cumulative sedimentation curve for any given point in the area of deposition and such a solution is shown in Fig. 3.6 for the point $x = 1$ and assuming $a = 0.5$.

The sedimentation curve, derived from the diffusion model, shows a sigmoidal flexure which is not present in the exponential decay model (Fig. 3.3) but, once past this inflexion, both models behave similarly and approach asymptotically a value of ultimate sedimentation which, in the notation of [3.19], equals $y_0/2$ corresponding to h_0 in [3.8]. If the boundary condition in the diffusion model had specified a more natural convex mountain slope and concave basin slope, sedimentation curves could be obtained which in their general aspect are identical with curves derived from the exponential model.

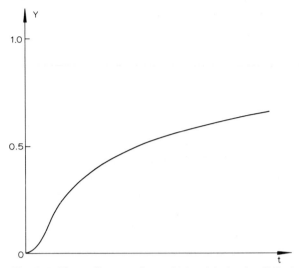

Fig. 3.6. The sediment—time relationship in the diffusion model.

There are obvious ways in which one could modify the diffusion model; for example, one might introduce the coefficient of mass flow, a, as a variable. This would reflect different resistance to erosion in the land areas and different modes of transportation in the basin. However, such elaborations add little to the general picture and at this stage a short discussion of the stratigraphical implications of the erosion—sedimentation models is indicated.

The theories show the not unexpected fact that the formation of an upland area will produce sedimentation which, at first, is rapid and then more or less exponentially declines until the source of sediment is flattened out. The present-day geology shows, as is well known, both types of mountain areas, the systems which are relatively young and which are active suppliers of sediments together with the mature areas which have reached a level where erosion is no longer active. However, there is also clear evidence that this process of mountain levelling is not continuous; indeed, it is well-known that many of the now mature areas have been reactivated by repeated tectonic uplift and the history of sedimentation must reflect such interruptions.

Obviously, if the periods of uplift continue all the time, erosion will never stop and the supply of sediment will be more or less constant over an average period. This situation will be discussed in the following section.

3.8 SEDIMENTATION IN SUBSIDING BASINS

If one considers sedimentation throughout geological time intervals, one comes to the conclusion that it must have been a more or less continuous process. It is true that there have been periods of low- and high-rate sedimentation, but the overall rates show neither an increase or decline (see Chapter 2). A constant supply of sedimentary material raises questions which are different from the ones treated in the previous section. If one considers an isolated basin receiving such a continuous input, two fundamental situations are possible. Either, the basin is filled to capacity and any excess material by-passes the basin to be deposited somewhere else, or, the basin is actively sinking and, therefore, creates new space for the sediment all the time. These are extreme situations and any combination may occur. Active sinking can be combined with some material being removed to other areas and, indeed, it is possible that filling and sinking are exactly in balance, which would, of course, lead to a constant water-depth in the basin. It is this latter case which has received a good deal of attention from geologists because this situation seems to apply to many areas of shallow-water deposition in the past. To explain such a balance, geologists rightly assumed that there must be a causal connection between deposition and crustal subsidence; at first it was generally accepted that it was the weight of the sediment which depresses the basin and, thus, creates its own space. However, Barrel (1917) discussed this problem in some detail and came to precisely the opposite conclusion that it is not the sedimentation which causes subsidence, but the subsidence which controls the sedimentation. Barrel assumed that each environment has a critical depth above which deposition is impossible — this he called the wave base or base level; once this level is reached, any added material is removed by erosion. If crustal sinking takes place, more room is created and sedimentation will again reach its base level. Barrel's arguments read to-day still very much up-to-date, and his ideas are incorporated in most modern sedimentation models. For example, Sloss (1962) bases a general stratigraphical model on the principles of Barrel by stating that the shape, or external geometry, of a body of sedimentary rocks varies as a function of the quantity Q of material supplied to the depositional site, the subsidence R, the dispersal D and the nature of the material, M; formally expressed as:

$$\text{shape} = f(Q, R, D, M) \tag{3.20}$$

If one ignores possible compaction, the simple geometrical relationship between water-depth W_t and Q, R, D obtains:

$$W_t = R_t + D_t - Q_t \tag{3.21}$$

The variables are all functions of time, but one may simplify the model by assuming that the rate of crustal sinking and the rate of sediment supply are constant:

$$\frac{dR}{dt} = R' \text{ and } \frac{dQ}{dt} = Q' \tag{3.22}$$

The rate of dispersal depends, according to Barrel's arguments, on the water-depth in the basin and, as a first approximation, one may assume this to be a linear decrease, let us say:

$$\frac{\mathrm{d}D}{\mathrm{d}t} = D_0 - aW_t \qquad [3.23]$$

where D_0 is the dispersal at sea-level and a is an environmental constant. Equation [3.21] may now be written as:

$$\frac{\mathrm{d}W}{\mathrm{d}t} = C - aW_t \qquad [3.24]$$

which, again, is the exponential decay equation which, under the condition that $W = W_0$, the initial water-depth, at $t = 0$ has the solution:

$$W_t = \frac{C}{a} + \left(W_0 - \frac{C}{a} \right) e^{-at} \qquad [3.25]$$

Setting $\mathrm{d}W/\mathrm{d}t = 0$, meaning that a constant water-depth is achieved, it is found that the value of this equilibrium depth $W_e = C/a$, the same can be seen by setting t in [3.25] equal to infinity. The constant $C = R' + D_0 - Q'$ must be chosen positive to ensure that deposition does not take place above sea-level, but, when this is assured, an equilibrium depth will develop under any circumstances. It is obvious from [3.21] that the variables in these equations differ from each other only by their signs and an increase in subsidence, for example, has precisely the same effect as a decrease of Q, the material supplied. Furthermore, the amount of sediment S, which is actually deposited is, of course, the difference between input Q and output D:

$$S = Q - D \qquad [3.26]$$

S, being opposite in its effect to water-depth, would enter [3.17] with a negative sign.

It is now quite possible that sedimentation rates vary with water-depth, even if there is no erosion at all. To obtain an equilibrium water-depth, one has to make assumptions about this dependence. Two models can be considered in this context:

$$(a) \frac{\mathrm{d}S}{\mathrm{d}t} = Q_0 \, e^{-kW}$$

$$(b) \frac{\mathrm{d}S}{\mathrm{d}t} = \frac{Q_0}{k} \left(1 - e^{-kW} \right) \qquad [3.27]$$

These models are based on the assumption that the production of precipitated material is proportional to the organic activity within the water. It is
known that the distribution of modern plankton is largely determined by the
amount of light penetration which controls the photosynthetic rate of the
phytoplankton. The relationship [3.27a] is a fair approximation of energy
reaching a certain water-depth, and if, for example, lime production is proportional to the organic activity, then the rate of sedimentation will exponentially decrease from a maximum Q_0 at the surface, to the value given by
[3.27a]. The constant k is related to the energy absorption in the water, and
is again, an environmental constant. Model [3.27a] will therefore describe
conditions on the sea floor and sedimentation caused by benthonic activity
such as algal or reef developments, should follow this model. To obtain the
amount of precipitation contributed by the whole column, it is necessary to
integrate [3.27a] over the water-depth and this leads to [3.27b]. This model
would be appropriate if it is taken that the entire plankton population
contributes to the sedimentation.

The differential equations of both models can be integrated without difficulty but lead to transcendental equations and, therefore, numerical solutions are given in Fig. 3.7. Setting $dW/dx = 0$ the benthos model is found to
have an equilibrium water-depth $W_e = (1/k)\ln(Q_0/R')$ which, under the assumed conditions, is unstable. The slightest amount of organic overproduc-
tion will lead to a filling of the basin, and any increase in subsidence will lead
to a rapid decline of organic activity. Under more realistic conditions, one
would assume that the maximum of organic activity is situated a short
distance below the surface of the water, and any benthonic activity would
build up to this level and then remain constant. The equilibrium depth W_e,
therefore, is, in this particular case, a critical depth below which reef forma-

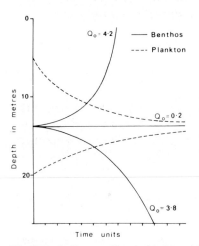

Fig. 3.7. Sedimentation in basins with constant subsidence.

tion is impossible — once this depth is slightly exceeded, a base level will be established which depends only on the environmental conditions near the surface.

The plankton model leads to an equilibrium water-depth W_e = $(1/k)\ln[Q_0/(Q_0 - R'k)]$ which will establish itself over a wide range of conditions. As long as $Q_0 - R'k > 0$, meaning, as long as sedimentation is capable of compensating for subsidence, the base level will be reached.

The two models of [3.27] are of some interest as they indicate the type of sedimentation processes which can lead to the formation of a base level without involving any lateral transport at all; it would be possible, of course, to combine such models with the dispersion model and this might lead to a more realistic picture.

3.9 SEDIMENTATION RATES AND BASIN CONTROL

The models which have been discussed in the previous sections concentrate on the two main factors which determine the absolute amount of deposited sediment, that is, the supply of sedimentary material and the availability of space for depositing the material. The space is determined by the geometry of the basin which, at any point may change either by crustal movement, eustatic sea-level changes and sedimentation. For the time being, tectonic movements and eustatic changes are regarded as equivalent.

Particularly if one is working with such simplified models, one is easily tempted to attribute sedimentation to a single, all-determining factor. Barrell wrote: "every foot of sediment was supposed to be the cause, leading, as an effect, to a depression of the basin of just one foot ..." and this was not acceptable. Barrell clearly avoided drawing the opposite conclusion that: every foot of depression of the basin leads to precisely just one foot of sediment being deposited; in his opinion, the base level has been sufficiently variable throughout time to allow a certain amount of sedimentation control which is independent of the tectonic control within the basin. It is obviously of great stratigraphical importance to examine this interplay between tectonic basinal control and external factors in some detail. The sedimentation models, simple as they are, allow us at least qualitatively to explore some of the relationships. Various combinations can be set up and numerical solutions can be obtained from a digital computer (Harbaugh and Bonham-Carter, 1970). The simple differential equations which have been used as sedimentation models, are, however, also very suitable for analogue computers and since we are only interested in the general picture, no expensive equipment or any special knowledge of analogue computing is needed. Indeed, a variety of instructive experiments can be performed with very small machines, like the Heath Kit educational model EC-1U. Two examples illustrate this.

Consider the following situation: A mountain is situated near a basin in which base-level control is exerted by dispersion which decreases linearly with water-depth. Assuming that the input Q equals the erosion E one can write according to [3.8]:

$$\frac{dQ}{dt} = -h_0 e^{-at} \qquad\qquad [3.28]$$

Assuming no crustal movement, i.e., $R' = 0$, [3.28] and [3.23] change into [3.21] and we may write:

$$\frac{dW}{dt} = D_0 - aW_t + h_0\, e^{-at} \qquad\qquad [3.29]$$

By setting the equation equal to zero and letting t go to infinity, we find the base level $W_e = D_0/a$. Any sedimentation below this critical depth is not affected by base level control and, if the source consists of a relatively low mountain, the base level may not be reached at all since the basin is only partly filled. The sedimentation curve is the exponential, as curve 1 in Fig. 3.8. Using higher initial inputs corresponding to higher source areas will lead, at first, to an excess of sedimentation, but dispersal sets in as soon as the critical height W_e is reached and this will eventually remove all sediment above the base level. If one regards curves 2 and 3 in Fig. 3.8 as successive uplifts, each supplying a bulge of sediment which has been dispersed, no record of this history will be preserved and the sequence contains a diastem

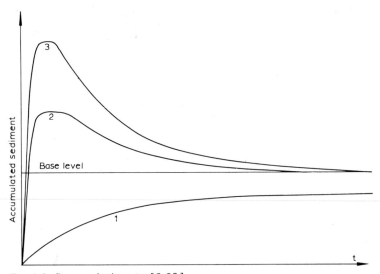

Fig. 3.8. Some solutions to [3.29].

at base level. At this point, further sedimentation is only possible if subsidence takes place and sedimentation will strictly reflect tectonic movements as long as it takes place exactly at base level. Such a situation is very unlikely. Accepting constant subsidence and a precise balance between sediment supply and crustal movement, one would also have to assume that the sedimentary material is homogeneous and consists of, let us say, one grain size only. Clearly in mixed sediments each grain size group, to mention but one sedimentary characteristic, will have its own base level and these cannot all be in equilibrium with the sinking movement. A strict control of sedimentation by tectonism is more likely if subsidence is interrupted and step-like, each step followed by long periods of no crustal activity. Under such circumstances, the basin will rapidly fill to capacity and stay for long time intervals exactly at base level. The resulting sequence of sediments would be frequently interrupted by diastems, and the thickness of sediment between diastems would indicate the amount of subsidence. Just how reliable a record of tectonism can be produced by sedimentation is an important question which the geologist has to decide; the scale of movement is obviously an important factor. Not many sedimentologists would accept that the increase of water-depth by, let us say, one inch would lead to a clear cut change, from non-deposition to deposition. Obviously crustal movement of only one inch cannot be recorded, but this type of argument must be individually developed for each sedimentation environment.

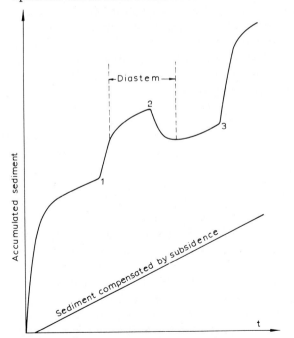

Fig. 3.9. Sedimentation with variable sediment supply.

If it is accepted that a precise control of sedimentation by subsidence is unlikely, one still has to admit the overall importance of the tectonic element in sedimentation. The second analogue computed example of a linear dispersal model, Fig. 3.9, illustrates how base-level control is effective, even if there is continuous subsidence and how, in addition, fluctuations in the supply will affect the cumulative sedimentation curve. The example shows the effect of increasing the original constant supply at point 1 in Fig. 3.9 by a certain amount and is followed by a decrease of Q at point 2, to be followed again by an increase in point 3. Although the decreased sedimentation is still well above the amount of crustal movement which has to be compensated, a diastem develops between 2 and 3 because the erosive stage has removed part of the record. Obviously the model cannot differentiate between changes in dispersal rates or changes in supply, and the same type of record would result if either occurred.

The models which have been discussed in this chapter are specifically intended as models of sedimentation trends and in stratigraphic sections their application to geological problems will be further discussed. Similar principles could be employed to develop two- and three-dimensional models which describe sedimentation conditions in depositional basins. Several attempts have been made in this direction and Harbaugh (1966) designed a model to simulate the quite complex processes involved in the formation of marine carbonate banks. Sedimentation in this model was partly due to organic activity and partly due to terrigenous clastic sediment which was introduced into the basin. The carbonate sedimentation, caused by several organism communities, was determined by the interference of these communities amongst each other and their sensitivity to depth and influx of clastic sediment. The Harbaugh model, which was the first of its kind, incorporated some stochastic elements and its main purpose is to demonstrate the development of facies pattern, but the interplay between crustal movement, water-depth and sedimentation is an integral part of the model.

A somewhat simpler model was developed by Harbaugh and Bonham-Carter (1970) as a computerized extension of the general shelf deposit model by Sloss (1962). This model deals with clastic sedimentation introduced into a shallow sea. It was assumed that deposition declined exponentially with distance from a source point. A critical water-depth, above which all sediment was removed, was taken to represent a base level below which sedimentation proceeded being strictly controlled by the amount of sediment supplied. The solution of this problem would be quite trivial but for the fact that if sedimentation reaches base level, the source point migrates forward and is situated at the brink of the foreset slope as can be seen from Fig. 3.10, a delta-like deposit builds out into the basin and three stages of development are shown in the figure. The computerized version of this model permits experiments with different grain sizes which each have different base levels and it is also possible to investigate the effect of crustal movements.

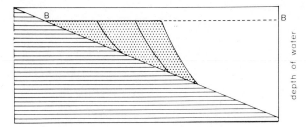

Fig. 3.10. A simple model of delta sedimentation.

The sedimentary basin model of Bonham-Carter and Harbaugh like the carbonate model is useful for studying facies variations marginal to a shelf. More specialized environments like evaporite basins (Briggs and Pollack, 1967) and deltas (Bonham-Carter and Sutherland, 1968) have also been investigated and an excellent summary of these is found in the book by Harbaugh and Bonham-Carter (1970).

Chapter 4

Stochastic models

4.1 INTRODUCTION

In the preceding chapter it was shown how models can be used to describe physical phenomena, and how simplified physical systems can be expressed in mathematical terms. The deterministic models, which were used in Chapter 3, were based on the assumption that each state of a physical system can be derived from the initial state or the boundary conditions. Stochastic models are indicated when a physical system appears to change according to probabilistic laws. Probabilistic and stochastic mean precisely the same thing; the term "stochastic process", which will be frequently used, is commonly reserved for the mathematical abstraction, or model, of a probabilistic system.

Perhaps it is self-evident that stochastic models, or processes, are most suitably employed when the system under investigation incorporates stochastic variables; such systems should not be confused with deterministic processes which have a statistical error superimposed. This difference may be seen from the following example. At the beginning of Chapter 2, it was stated that the fundamental relationship between time and thickness of sediment may be expressed as $Z = Rt$, where R is the rate of sedimentation. Suppose this equation is to be applied to the formation of a sandstone. The particles are finite and the discontinuous nature of the stratigraphic record can be modelled by assuming that time itself is discrete. If the sand grains are of approximately equal size, one may write:

$$Z(t) = R(t) + \epsilon(t) \quad (t = 1, 2, 3 \ldots) \tag{4.1}$$

This means that, at each time increment, the sediment grows by a step which is a multiple of the grain diameter. The random variable $\epsilon(t)$ has been added to make the model more realistic, and this could represent some variation in grain size. The effect of such an added random variable is precisely the same as an error of measurement; it is well known that this can be eliminated by simple regression methods. The error is, therefore, superimposed and the model is essentially deterministic. Next, consider a stochastic model of this sedimentation process. Once again, discontinuous time is assumed but at any consecutive moment, one of the two events may occur: either, a sand grain is deposited, let us say with probability p, or, a sand grain is eroded and this occurs with probability q. The question now is not, what is the thickness of the sediment at time t? It is, how great is the probability of $Z(t)$ taking on a specific value, say z? The solution to the problem will, therefore, be written symbolically as:

$$P\{Z(t) = z\} = P_z(t) \qquad\qquad [4.2]$$

This reads in words as: the probability that the variable $Z(t)$ takes on the specific value z at time t, is equal to a number $P_z(t)$. The actual solution will be given shortly, but it is clear that this model incorporates the probabilistic element into the system and, therefore, it cannot be eliminated as in the previous example. Indeed, the probabilistic nature of the solution is part of the answer which is expected from such a stochastic model.

To understand stochastic models a certain grasp of the probability concept is necessary. The probabilities which will be dealt with here, are physical or statistical probabilities, that is, they do not refer to judgments but to the possible outcomes of real or conceptional experiments (Feller, 1957). Thus, if one takes the statement, "Cox is probably a greater artist than De Wint", then one deals with a philosophical probability which refers to a judgment and which cannot be tested by any experiment. The use of the word 'probable' is legitimate but it is used in a different sense from using probability in a statement like, "the probability of obtaining a one with a single throw of a dice is 1/6". This is a statistical probability and it can be tested either by a real or conceptional experiment. The essential pre-requisite of stating such a probability is to know all the possible outcomes of the experiment with which it is associated; in the example which was quoted, we recognize six possible outcomes because we assume that the dice can only land on one of its six sides. The complete collection of the possible outcomes is technically known as the 'sample space' and this must be rigidly defined. It is, of course, possible that, in the above experiment, the dice comes to rest on one of its corners, but this position is not recognized as part of the sample space and it is, therefore, ignored. It is possible to simplify real nature by defining the sample space in this strict manner and by doing this, give the conceptional experiments the essential property of models.

Any result which is obtained in an experiment is called an 'event' and this can be linked to one or more sample points. For example, the event of obtaining 6 in the dice throwing experiment is associated with a single sample point, but the event of obtaining an even number is associated with three sample points that is the sample points which mark the results 2, 4 and 6. It is useful to think of the sample space as of a plane covered with sample points and subareas like A and B in Fig. 4.1 which surround the points which cover the events A and B. The probability of a particular event is the frequency with which it occurred divided by the number of trials, that is, the total number of experiments which have been performed. For example, if a

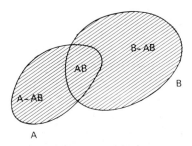

Fig. 4.1. The relationship of two sets of events A and B.

dice is thrown a hundred times, a 6 may have turned up say sixteen times. The empirical probability of obtaining a 6 is, therefore, 16/100 = 0.160. The same result could have been obtained by the following argument: the sample space of this problem contains 6 points, each event has the same probability; therefore, obtaining a 6 has the probability 1/6 = 0.166. It is common practice to express probabilities in decimal fractions; for the probability 1/4 one writes 0.25. Clearly, probability = 1 means certainty or that the event must occur, and in contrast probability = 0 means that the event is impossible.

Probability theory has its own set of basic rules and operations like any other branch of mathematics, and they will be reviewed briefly in the following section.

4.2 THE BASIC ELEMENTS OF PROBABILITY THEORY

The rules by which probability theory operates can be directly derived from the concept of a sample space. This space is thought to contain a number of points E_1, E_2.... each of which represents a possible outcome of an experiment. Associated with each point is a number, called its probability, which is written as $P\{E_i\}$. The numbers are all non-negative and such that:

$$P\{E_1\} + P\{E_2\} + ... = 1 \qquad [4.3]$$

The equation [4.3] follows directly from the definition of a sample space, because, if one takes the empirical frequency ratios r_i/n, in which r_i is the frequency, with which result i turned up in n experiments, and if the sample space contains S points, then obviously:

$$\sum_1^s r_i/n = 1$$

Consider next, two mutually exclusive events, A or B. The probability of either A or B occurring can be symbolically written as:

$$P\{A + B\} = P\{A\} + P\{B\}$$ [4.4]

This equation is also known as the addition law for mutually exclusive events. It may be applied to the dice-throwing experiments to determine the probability of obtaining either a one or a two. This is equal to $1/6 + 1/6 = 1/3$. Similarly, one may determine the probability of obtaining an even number in a single throw. An even number can be provided by three events and the probability, therefore, is equal to $3/6 = 0.5$.

If two events A, B are not mutually exclusive, one can use the symbol $\cup$ which stands for either A or B, or both occurring together; (see Fig. 4.1) in such a case, sample points of A partly overlap the sample points of B. If the sample points of two events fall into the same region of a sample space, one indicates this by the symbolic product $P\{AB\}$ and this is represented by the shaded area in Fig. 4.1. Thus, to obtain the union of two events, one adds the probability of A and B and subtracts the areas of overlap which ensures that the points which fall into this region are not counted twice, that is, amongst the A's and the B's. The probability of either A or B or both occurring together is then written as:

$$P\{A \cup B\} = P\{A\} + P\{B\} - P\{AB\}$$ [4.5]

It is often important to find the probability value which is associated with the product (AB) and the way in which this is done is best seen by using an example. Consider a sample of sand containing N grains which are either rounded (A) or angular, and feldspathic (B) or non-feldspathic. There are N_A rounded grains and N_B feldspathic grains but we are interested in the probability of obtaining a rounded grain which is, at the same time, feldspathic. Now:

$$P\{A\} = \frac{N_A}{N}, \ P\{B\} = \frac{N_B}{N}$$

and the number of grains which are both rounded and feldspathic may be written as N_{AB}. To obtain the probability of choosing a rounded grain which belongs to the sub-population of feldspathic grains, one has to form the ratio N_{AB}/N_B. A new symbol is needed for such probabilities which refer to sub-populations and which are called 'conditional probabilities'. One writes $P\{A|B\}$ which reads as, "the probability of event A provided the event B occurs simultaneously". One can write, therefore:

$$P\{A|B\} = \frac{N_{AB}}{N_B} = \frac{P\{AB\}}{P\{B\}}$$

from which one obtains:

$$P\{AB\} = P\{B\} \cdot P\{A|B\}$$ [4.6]

The equation [4.6] is also known as the multiplication law of probabilities. An important special case is given when $P\{A|B\} = P\{A\}$ and when it is said that the event A is stochastically independent from B and the probability multiplication reduces to:

$$P\{AB\} = P\{A\} \cdot P\{B\}$$ [4.7]

It is unlikely that this condition of independence applies to the example of rounded feldspathic grains, but a typical example of stochastically independent events is given by throwing two true dice simultaneously. It is well known that the chance of obtaining two sixes at one throw is given by $1/6 \cdot 1/6 = 1/36$.

To give a complete specification of a probability problem, one uses the probability density distribution which is simply a listing of all probabilities in the sample space. If the sample space is continuous, then the random variable, let us say X, is continuous and the probability that it takes on a specific value is given by the probability density function $f(x)$ which may be written as:

$$f(x) = \lim_{\Delta x \to 0} \frac{P\{x < X \leqslant x + \Delta x\}}{\Delta x}$$ [4.8]

and this has the property (similar to [4.3]) that:

$$\int_{-\infty}^{\infty} f(x)\, dx = 1$$ [4.9]

Very often it is more convenient to work with the cumulative probability distribution which is defined as:

$$F(x) = \int_{-\infty}^{x} f(u)\, du$$ [4.10]

This summary of the methods used in probability theory is of necessity very condensed, and reference should be made to a standard textbook on the subject (cf. Feller, 1957).

4.3 THE SIMPLE RANDOM WALK

One of the simplest stochastic processes which has several direct applications in geology is known as the random walk. This is the problem: a particle or point moves in a horizontal direction which may represent time and at discrete intervals, that is, at times $n(n = 0, 1, 2...)$ this particle performs a jump J_n. The jump may be either in an upward or a downward direction. It is assumed that each step is constant, being either +1 or −1. The position of the particle is measured on the vertical axis Z and is given by:

$$Z_n = Z_{n-1} + J_n \qquad\qquad [4.11]$$

The decision about whether the particle moves upwards or downwards is made independently at each time interval and depends on a random experiment. One may, for example, imagine that before each jump, a coin is tossed and that heads implies a move in the upward direction and tails a move in the downward direction. In a more general case, one may say that:

$$P\{J_n = +1\} = p, \quad P\{J_n = -1\} = q \qquad\qquad [4.12]$$

Such a sequence of independent trials, of which each can have only two outcomes, are known as Bernoulli trials. The independence is assessed by the condition that whether the particle moves upwards or downwards at any time, in no way depends on how it moved during the preceding jump. An actual observed sequence of such a random walk is called a 'realization', and an example of this is shown in Fig. 4.2. Here it is assumed that $p = 0.7$ and $q = 0.3$ and since the probability for moving upwards is larger than the probability for moving downwards, one would expect a steady upwards trend of the particle path.

The random walk can be adapted as a model for a variety of physical processes which include one-dimensional Brownian movement, the fortune

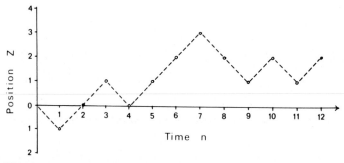

Fig. 4.2. A simple random walk.

of a gambler or an insurance risk. In the introduction it was noted that the random walk is also one of the simplest sedimentation models. In this case, the jump is associated with the deposition or erosion of a sedimentation unit, which is assumed to be a sand grain for the sake of argument. For simplicity, the diameters of all the sand grains are assumed to be unity. The possibility that neither deposition nor erosion occurs at any stage is allowed for in order to make the model slightly more general, and one can add to the condition [4.12] that:

$$P\{J_n = 0\} = 1 - p - q$$

A condition like this which allows for non-deposition or erosion does imply that $p + q \leqslant 1$. The sequence of trials is still independent although the decision of how the jump is performed is no longer a Bernoulli trial. It is intuitively clear that, as long as $p > q$, deposition will prevail and the path of the random walk will indicate the gradually accumulating sediment. If sedimentation starts at time $n = 0$, Z_n can be identified with the thickness of deposited sediment at time n.

The first problem to be considered is the thickness of the sediment after a certain time has elapsed. If sedimentation starts at the origin $n = 0$, then we have:

$$Z_n = \sum_{r=1}^{n} J_r$$

The thickness k at time n can take all the possible values of $k = 0, \pm 1, \pm 2, ..., \pm n$. In order to reach the position k, the random walk must have gone through r_1 positive jumps, r_2 negative jumps and r_3 zero jumps. Therefore, the probability that $Z_n = k$ is given by the summation of the multinomial probabilities:

$$P\{Z_n = k\} = \sum \frac{n!}{r_1! \, r_2! \, r_3!} \, p^{r_1} q^{r_2} (1 - p - q)^{r_3} \qquad [4.13]$$

The summation in [4.13] has to be taken over all values of r_1, r_2, r_3 which satisfy the simultaneous equalities:

$$r_1 - r_2 = k, \quad r_3 = n - r_1 - r_2$$

It may be noted that if $p + q = 1$, the distribution [4.13] reduces to the binomial but it must be observed that, in this case, even k's can only be reached when n is even and odd k's can only be reached when n is odd.

The statistical properties of the calculated distribution can be derived from the generating function of jump J_r and most textbooks on theoretical statistics will explain the procedure. (See for example, Feller, 1957; Alexander, 1961). If one writes for the generating function $G(s)$ of a single jump:

$$G(s) = ps + (1 - p - q) + qs^{-1}$$

then the generating function of Z_n is the n-fold convolution (see section 4.7) of $G(s)$ with itself. Hence:

$$E(s^{Z_n}) = [G(s)]^n$$

It follows that the mean or expectation for which we use the symbol E and variance of Z_n may be written as:

$$E(Z_n) = n(p - q), \text{var}(Z_n) = n[p + q - (p - q)^2] \qquad [4.14]$$

If n is large, [4.13] can be approximated by a normal distribution (see Cox and Miller, 1965). Writing μ and σ^2 for the expectation and variance we have:

$$P\{j \leqslant Z_n \leqslant k\} \simeq \Phi\left(\frac{k + c - \mu}{\sigma/n\sqrt{n}}\right) - \Phi\left(\frac{j - c - \mu}{\sigma/n\sqrt{n}}\right) \qquad [4.15]$$

where:

$$\Phi(y) = \frac{1}{\sqrt{2\pi}} \int_{-\infty}^{y} e^{-x^2/2} \, dx$$

and $c = 1$ when $p + q = 1$; $c = \frac{1}{2}$ when $p + q < 1$.

Examples of such random walks have been simulated on the computer to illustrate the outcome of this analysis, and Fig. 4.3 shows a single realization which has been stopped after twenty steps. If the experiment is repeated many times, then one will obtain a frequency curve of final thicknesses which should come close to the distribution of [4.15]. A comparison of the calculated curve with five hundred simulated examples shows that this is indeed the case and this is shown in Fig. 4.3.

The bed thickness distribution, shown in Fig. 4.3, develops when random-walk sedimentation operates through repeated intervals of exactly the same length of time — a condition which one might expect in varved sediments. One may now investigate the different problem of how long it takes to produce a layer of a certain thickness. Since this, too, is a stochastic prob-

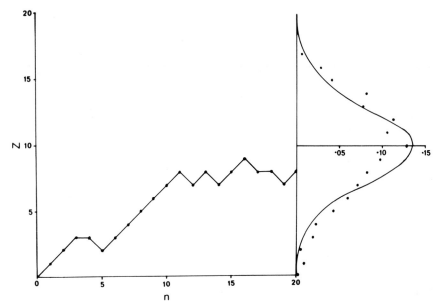

Fig. 4.3. A random walk consisting of twenty steps.

lem, the answer will again be a probability distribution which, in this case, refers to the time which it takes to produce a layer of a certain thickness.

In the language of random-walk analysis, a predetermined bed thickness is called an 'absorbing barrier' because the point which carries out the random walk stops, or is absorbed, as soon as it hits the barrier for the first time. The problem is well known from the analysis of diffusion processes and the following solution follows the arguments of Cox and Miller (1965, p. 36).

Let it be assumed that the particle starts at the origin and that we are interested in finding the time it takes for the particle to reach the absorbing state at $Z_n = a$, which is somewhere in the positive field of the random walk. If $p > q$, this will happen with certainty if there is no limitation on the time. Thus, it is also certain that the particle will move from the original zero position to position +1. The time N_1 which it takes to do this will be a random variable which has the probability generating function $F_1(s)$. Since each jump is independent of position and time, the jumps from state 1 to state 2, state 2 to state 3 and so on, will each occur at times which have the same distribution as N_1. The generating function for the time it takes to reach state a can, therefore, be written as the a-fold convolution of $F_1(s)$:

$$F_a(s) = [F_1(s)]^a \qquad [4.16]$$

The generating function $F_1(s)$ can be evaluated from the positions which the particle can take when leaving the origin for the first time.

There are three possibilities. Firstly, the particle may go directly from state 0 into state 1 and this will happen with probability p. Secondly, the particle may remain in state 0 with probability $1 - p - q$ when it will take the time N_1 to move into state 1. Thirdly, the particle may go into state -1, with the probability q and it will take two steps before it reaches state 1. These three probabilities can be summed to obtain the unconditional probability of the particle moving from zero to one and, in terms of the generating function, one can write:

$$F_1(s) = ps + (1 - p - q)_s \, F_1(s) + qs[F_1(s)]^2 \qquad [4.17]$$

This quadratic equation has two solutions with the roots λ_1, λ_2, which satisfy the condition $F_1(1) = 1$:

$$F_1(s) = \frac{p}{q} \, \lambda_2(s) = [\lambda_1(s)]^{-1}$$

and from [4.16] we find, therefore, the general solution:

$$F_a(s) = \left(\frac{p}{q}\right)^a [\lambda_{2(s)}]^a \qquad [4.18]$$

Differentiating [4.17] and putting $s = 1$ gives:

$$F'_1(1) = p + (1 - p - q) + (1 - p - q) \, F'_1(1) + q + 2qF'_1(1) = \frac{1}{p - q}$$

$$[4.19]$$

Calling N_a the time it takes for the particle to reach the boundary a, one can write for the expectation of N_a (from [4.16] and [4.19]):

$$E(N_a) = aF'_1(1) = \frac{a}{p - q} \quad \text{for } p > q$$

$$= \infty \quad \text{for } p = q \qquad [4.20]$$

The variance can be found in a similar way provided that absorption is certain, i.e., $p > q$:

$$\text{var}(N_a) = \frac{a[p + q - (p - q)^2]}{(p - q)^3}$$

Denoting the mean and variance of a single jump by $\mu = (p - q)$ and $\sigma^2 = p + q - (p - q)^2$ one can also write:

$$E(N_a) = \frac{a}{\mu}, \quad \mathrm{var}(N_a) = a\frac{\sigma^3}{\mu^3}$$ [4.21]

The distribution of N_a will again tend towards the normal when a is large because the probabilities of absorption in time N_a are the sum of independent random variables.

It is interesting to compare the results of [4.21] with the results of [4.14]; it is obvious that the variance of the time process, which leads to the formation of layers with equal thickness, is relatively much larger than the variance of thicknesses which are controlled by constant time intervals. This is the basis of an important rule regarding cyclic sedimentation which will be developed later.

4.4 THE RANDOM WALK AS A MODEL OF BED FORMATION

The two alternative ways of utilizing the random-walk principle for obtaining a bed-thickness distribution do not extend to the problem of bed formation itself, as in both examples, one has to formulate some outside influence which determines the bed thickness. In the first problem it was specified that sedimentation is interrupted after equal time intervals and similarly, in the second case, sedimentation was interrupted after a certain thickness was reached. The mechanism which determines the bed formation in these models is clearly something which is independent of the random walk. It is, therefore, of some interest to examine models which incorporate the mechanism of bed formation.

It has been stated in Chapter 2 that one of the features of bed formation is either a period of non-sedimentation, or a period of erosion. The random walk, however, is a model which produces alternating steps of deposition and non-deposition. One can associate the formation of a bed with each step of deposition that follows a period of non-sedimentation. A particularly simple example is given by the assumption that deposition takes place with probability p and non-deposition with probability q. The probabilities of obtaining 1, 2, 3... steps of deposition which are to be followed by one step of non-deposition are, pq, ppq, $pppq$.... This multiplication is valid because each step is independent from every other step. One can write for the probability of obtaining a certain bed thickness T:

$$P\{T = n\} = p^n q$$ [4.22]

Equation [4.22] is known as the geometric distribution. Summing T_n ($n = 1$, 2... ∞) and remembering that $q = 1 - p$, gives unity and the expression is, therefore, a valid probability density distribution with the mean and variance of:

$$E(T) = \frac{p}{q} \text{ and } \mathrm{var}(T) = \frac{p}{q^2} \tag{4.23}$$

The most remarkable property of the geometric distribution is the complete lack of memory which it implies. If T_n is the thickness of an uninterrupted sediment at time n, then this accumulated thickness has no effect on deciding whether deposition will continue at time $n + 1$ or not. Or to put it differently, if one identifies T_n with the life span of a stratum, then the strata have the peculiar property of showing no signs of aging which means that a layer of considerable thickness has just as much chance of continuing to grow as a very thin or young layer. The lack of memory is connected with the Markov property of stochastic processes which will be discussed in the next chapter.

4.5 THE KOLMOGOROV MODEL OF BED FORMATION

In the model which led to the geometrical distribution, it was assumed that beds are caused by periods of non-deposition. It has already been noted that this sedimentation standstill can often be replaced by active erosion. It is immediately evident that the thickness of the underlying bed will be reduced if erosion occurs and that the bed might be completely removed if the amount of erosion was large enough. What, then, is the thickness distribution of the remaining beds or that part of the beds which is preserved in the stratigraphic column? This problem was first met by A.B. Vistelius who handed it to the Russian mathematician A.N. Kolmogorov who provided a general solution (Kolmogorov, 1951). In preparing to understand Kolmogorov's work, it is useful to consider the problem at first as applied to a discrete random walk. By doing this it is possible to follow Kolmogorov's arguments in quite an elementary fashion. Some important results about the geometrical thickness distribution can be obtained at the same time.

Assume that a random walk process is operating and that with probability p a unit step is deposited, and with probability q a unit step is eroded. We shall say that a new bed is formed when deposition succeeds erosion. The principle of the model is shown in Fig. 4.4, and it may be noted that the first stratum S_1 is partly eroded by the second bed-formation process which itself leaves no record behind. Calling the amount of deposition in each bed ζ, and the erosion η, a new random variable δ may be defined as:

$$\delta_n = \eta_n - \zeta_n \tag{4.24}$$

The distribution of δ_n is easily obtained considering that:

$$P\{\delta_n = 0\} = pqp + p^2 q^2 p + \ldots = \frac{p^2 q}{1 - pq} = a$$

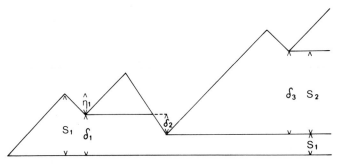

Fig. 4.4. Kolmogorov's model of bed formation.

and:

$$P\{\delta_n = r\} = ap^r \text{ for } \delta > 0$$

$$aq^r \text{ for } \delta < 0 \qquad\qquad [4.25]$$

From a study of Fig. 4.4 it will be clear that the thickness T_n of stratum S_n will be given by δ_n if no subsequent erosion occurs, and this may be expressed as:

$$P\{T_n = x\} = \delta_x \text{ provided that} \qquad\qquad \delta_{n+1} > 0$$

$$\delta_{n+1} + \delta_{n+2} > 0$$

$$\delta_{n+1} + \delta_{n+2} + \delta_{n+3} > 0$$

$$\cdot$$

$$\cdot$$

$$\cdot \qquad\qquad [4.26]$$

From [4.25] it may be seen that:

$$P\{\delta_1 + \delta_2 + ...\delta_r > 0\} = a\frac{p}{q}\left(\frac{2a}{1-pq}\right)^{r-1}$$

and summing the terms with $r = 1,2...\infty$ gives the probability that S_n is not eroded:

$$P(\text{no erosion of } S_n) = a\frac{p}{q}\frac{p^2q}{p^2q - 2a^2} \qquad\qquad [4.27]$$

If the above condition is not fulfilled, that is, if erosion does occur, then the thickness of the bed will be given by:

$$P\{T_n = x\} = \delta_n + \delta_{n+1} \qquad\qquad \text{if} \qquad\qquad \delta_{n+1} < 0$$

$$= \delta_n + \delta_{n+1} + \delta_{n+2} \quad \text{if} \quad (\delta_{n+1} + \delta_{n+2}) < 0$$

$$= \delta_n + \delta_{n+1} + ...\delta_{n+r} \; \text{if} \; (\delta_{n+1} + ...\delta_{n+r}) < 0 \qquad\qquad [4.28]$$

Since each of these are mutually exclusive events they can be added, and combining [4.26] and [4.28], one may write for the probability of obtaining a certain thickness:

$$P\{T_n = x\} = p^x a^2 \left[\frac{p}{q(1-c)} + \frac{q}{cp(1-c^2)} \right] \qquad\qquad [4.29]$$

where $c = 2a/(1-pq)$.

The result of [4.29] is particularly interesting because it demonstrates that the process of erosion does not modify the geometrical nature of the thickness distribution. As in the more general case, to which we will turn next, a truncated frequency curve is obtained. However, since the original distribution without erosion is exponential, the truncated curve must be exponential also. One can, therefore, conclude that if a sequence of beds is determined by independent random events it will keep its random character (if additional factors, which may modify this series, are also independent random variables). This is clearly the case in the example above since subsequent stages of deposition and erosion are assumed to be independent.

It will be shown in the next sections that the discrete random walk can be replaced by continuous random variables and that, in general, bed-thickness distributions are continuous and not step-like as has been accepted so far. Kolmogorov's general solution makes the following assumptions about the variable δ_n which is defined by [4.24]. As before, δ_n is independent but it has a probability density distribution $g(x)$ which can be written as:

$$P\{\delta_n < x\} = G(x), \; G(x) = \int_{-y}^{x} g(x)\, dx$$

The distribution is continuous and has a positive expectation. The latter assures that the sum:

$$\zeta_n^{(r)} = \delta_n + \delta_{n+1} + ...\delta_{n+r}$$

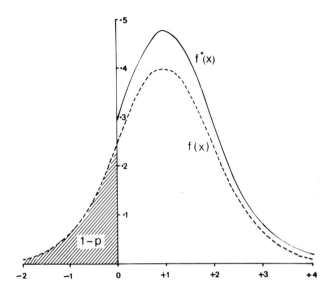

Fig. 4.5. The bed-thickness distribution of the Kolmogorov model.

goes towards infinity as $r \to \infty$; in other words, this makes certain that deposition predominates over erosion. If this is the case, then there must be a lower limit:

$$\phi_n = \lim \inf [\zeta_n^{(0)}, \zeta_n^{(1)}, \dots \zeta_n^{(r)}]$$

which will be reached for a finite number r.

Obviously, if $\phi_n \leqslant 0$, the stratum n will be fully eroded and when $\phi_n > 0$ a final thickness ϕ_n will be preserved. Therefore, it is necessary to determine the probability $p = P(\phi_n > 0)$ and the conditional probability distribution of ϕ_n, let us say $f(x)/p$, under the hypothesis that $\phi_n > 0$. The bed-thickness distribution $f^*(x)$ can then be obtained by truncating as follows:

$$f^*(x) = f(x)/p \text{ for } x > 0$$

$$f^*(x) = 0 \qquad \text{for } x < 0 \qquad\qquad [4.30]$$

Kolmogorov shows by the arguments which were used in [4.26] and [4.28] that:

$$f(x) = pg(x) + \int_{-y}^{0} g(x - y) f(y) dy$$

This integral equation can be solved numerically and Kolmogorov has shown that if one assumes $g(x)$ to be normally distributed, $f(x)$ also will be very close to a normal distribution. Kolmogorov's laborious calculations have not been repeated, but the relationship between a normally distributed $f(x)$ and a resulting bed-thickness distribution $f^*(x)$ can be seen from Fig. 4.5, where it has been assumed that $f(x) = \phi(x - a)$, $a = 1$ and $\Phi(y)$ is the normal distribution. The truncation was carried out according to [4.30] at $p = 0.8413$. The shaded area in Fig. 4.5 on the left is equal to $1 - p$ and indicates the proportion of beds which have formed but were eroded subsequently. Mizutani and Hattori (1972) call this truncation probability p the 'Kolmogorov coefficient' and have determined its numerical value for a variety of observed bed-thickness distributions. Like Kolmogorov, Mizutani and Hattori believe that this erosion—deposition model explains the asymmetric nature of many observed bed-thickness distributions. Other explanations will be considered later.

4.6 THE POISSON PROCESS

The random models which have been discussed so far are based on discrete processes in which variables like bed thickness, time or the amount of sediment, were treated as integers. We will now derive some stochastic models which refer to continuous variables.

Consider at first the following problem. Points are to be scattered on an infinitely long line in such a way that the average number of points per unit length is equal to a constant, let us say ρ. The value ρ, which may be called the point density, is the only value or parameter which must be known to find the distribution of the points. Assume, at first, that the line is of finite length t and that a total number of N points is to be distributed. One can now calculate the number of points which will fall into a finite interval τ which is completely surrounded by t. Dropping each point constitutes a Bernoulli trial and the point either falls into the interval τ with probability τ/t, or it falls inside t but outside τ with probability $1 - \tau/t$. If n points fall into the counting interval τ, one can speak of n successes and the sequence of successes and failures can, therefore, be permutated in $\binom{N}{n}$ different ways and, similar to [4.13], one may write for the probability of obtaining n points:

$$P_n = \binom{N}{n} (\tau/t)^n (1 - \tau/t)^{N-n} \quad (n = 0, 1, 2...)$$

$$= \frac{1}{n!} [N(N-1)(N-2)...(N-n+1)] \left(\frac{N}{t}\right)^n \left(\frac{\tau}{N}\right)^n \left(1 - \frac{N}{t}\frac{\tau}{N}\right)^{N-n}$$

$$= \frac{\tau^n}{n!} \left[1\left(1 - \frac{1}{N}\right)\left(1 - \frac{2}{N}\right)...\left(1 - \frac{N-n+1}{N}\right)\right]\left(\frac{N}{t}\right)^n \left(1 - \frac{N}{t}\frac{\tau}{N}\right)^{N-n}$$

If N and t approach infinity in such a way that the ratio N/t approaches the constant point density ρ, one can write:

$$P_n = \frac{T^n}{n!} \rho^n \left(1 - \frac{\rho\tau}{N}\right)^N$$

which, by making use of the exponential limit, can be written as:

$$P_n = \frac{T^n}{n!} \rho^n e^{-\rho\tau} \qquad\qquad [4.31]$$

The distribution of [4.31] is the well-known Poisson distribution with the property that its mean equals its variance and:

$$E(n) = \text{var}(n) = \rho$$

A function which is frequently used in theoretical statistics is the gamma function, defined as:

$$\Gamma(z) = \int_0^\infty e^{-t} t^{z-1} \, dt$$

The same integral can be split into two, a lower and an upper part, in such a way that:

$$\gamma(n, x) + \Gamma(n, x) = \Gamma(n)$$

The lower part is called the 'incomplete gamma function', defined as:

$$\gamma(n, x) = \int_0^x e^{-t} t^{n-1} \, dt$$

It is often useful to express the Poisson distribution in terms of the gamma function and it can be shown that the cumulative Poisson distribution may be written as:

$$P_n = \sum_{i=0}^n \frac{e^{-\rho}}{i!} \rho i = \frac{\Gamma(n+1, \rho)}{\Gamma(n+1)} \quad (n = 0, 1, 2...) \qquad [4.32]$$

A simple example will illustrate the use of the Poisson distribution. Let us take the following situation. A geologist has recorded the positions of bedding planes in a stratigraphic section, and such records may be represented as points along a line. If the bedding planes occur completely at random, then counts of beds per unit interval should follow the Poisson distribution. Indeed, the hypothesis of randomness can be tested by comparing actual counts with the calculated Poisson distribution. The only condition which must be fulfilled is that the average density of beds should not vary over the length of the section. To estimate the density ρ, one has only to take the total number of beds and divide it by the length of the section expressed in the units of the counting interval. For example, in a 30 m long section, 90 beds are recorded. Therefore, one would expect 3 beds per 1 m interval and $\rho = 3$. The Poisson distribution for this example is given in Table 4.1 in which n is the number of beds observed in a 1-m interval and P_n is the probability with which this is to be expected. Very complete tables of the Poisson distribution are found in Pearson (1948).

The Poisson distribution is, of course, again a discrete distribution as it refers to integral values of n. In order to obtain the actual bed-thickness distribution, it is necessary to investigate the spacings between the randomly scattered points.

It follows from [4.31] that the probability of no point falling into a small interval Δt on the line is given by:

$$P(\text{no point in } \Delta t) = e^{-\rho \Delta t}$$

$$= 1 - \rho \Delta t + \tfrac{1}{2}(\rho \Delta t)^2 \ldots$$

$$= 1 - \rho \Delta t + 0(\Delta t) \qquad [4.33]$$

in which the symbol $0(\Delta t)$ is used to indicate a function which tends more rapidly towards zero than Δt. For example, any function containing the factor $(\Delta t)^2$ will be indicated in this way. Similar to [4.33] we have:

TABLE 4.1

Poisson distribution for $\rho = 3.0$

n	P_n	n	P_n
0	0.0498	6	0.0504
1	0.1494	7	0.0216
2	0.2240	8	0.0081
3	0.2240	9	0.0027
4	0.1680	10	0.0008
5	0.1008		

$$P(1 \text{ point in } \Delta t) = \rho \Delta t e^{-\rho \Delta t}$$

$$= \rho \Delta t (1 - \rho \Delta t + \tfrac{1}{2}(\rho \Delta t)^2 \ldots$$

$$= \rho \Delta t + 0(\Delta t) \qquad [4.34]$$

and also:

$$P(\text{more than 1 point in } \Delta t) = \sum_{n=1}^{y} (1/n!) \, e^{-\rho \Delta t} (\rho \Delta t)^n$$

$$= 0(\Delta t) \qquad [4.35]$$

Let us now choose some origin t_0 along the line, t, which may or may not coincide with a point. According to the definition of independent randomness, anything which will happen after t_0 is quite independent from what happened before or at t_0. In particular, if $t_0 + z$ is the distance after which the first point following t_0 will be observed, then z is a random variable which does not depend on whether a point occurred at t_0 or not. To obtain the probability distribution of z one can argue as follows. Let $P(x) = P\{z > x\}$ and assume that $\Delta x > 0$, then:

$$P(x + \Delta x) = P\{z > x + \Delta x\}$$

$$= P\{z > x \text{ and no point has fallen into the interval } (t_0 + x + \Delta x)\}$$

$$[4.36]$$

Since what happens in the interval $t_0 + x + \Delta x$ is independent of what occurred in the interval $t_0 + x$, one can write [4.36] using [4.33] as:

$$P(x + \Delta x) = P(x)[1 - \rho \Delta x + 0(\Delta x)]$$

Taking the limit as $\Delta t \to 0$ and differentiating gives:

$$P'(x) = -\rho P(x)$$

Thus, $P(x) = P(0)e^{-\rho x}$, but since $P(0) = P\{z > 0\} = 1$, we find $P(x) = e^{-\rho x}$. The distribution function of z is, therefore, $1 - e^{-\rho x}$ which has the probability density function of:

$$f(x) = \rho e^{-\rho x} \quad (x > 0) \qquad [4.37]$$

which is called the exponential distribution with parameter ρ. Strictly speaking, this distribution is known as the forward recurrence time, that is, the time of an event after an arbitrarily chosen point. In the special case of an independent random sequence, these arbitrary points can be the point events along the line and [4.37], therefore, gives the distribution of distances between scattered points. Applied to the previous example of randomly arranged bedding planes in a stratigraphic section, this function is, of course, the bed-thickness distribution. It is a continuous distribution which has the mean and variance given by:

$$E(x) = \frac{1}{\rho}, \quad \text{var}(x) = \frac{1}{\rho^2} \qquad\qquad [4.38]$$

Any stochastic process in which the events are spaced according to the negative exponential distribution and in which the counting of events over a given interval leads to the Poisson distribution, is known as a Poisson process. This process is a very important model for random arrangements of events either in space or time. Further generalizations of the Poisson process will be discussed later.

4.7 OTHER CONTINUOUS DISTRIBUTIONS

One of the problems which will be met with repeatedly is to find the distribution of the sum of two or more independent random variables. For example, one may be interested in finding the thickness distribution of units when two or more beds are added together to form a composite unit. To be more explicit, one may be dealing with an alternating sequence of sand and clay beds, each having their own thickness distribution, let us say $f(x_1)$ and $f(x_2)$. The question is: what is the thickness distribution of the combined sand—clay units?

Consider, at first, two integral valued random variables X and Y which are independent and which have probability distributions $P\{X=j\} = a_j$; and $P\{Y=j\} = b_j$. The probability of the event $(X=k, Y=k)$ is, therefore, given by $a_k b_k$. The sum $X + Y = S$ is a new independent random variable which has the distribution $P\{S=r\} = c_r$ and the event of $(S=r)$ can arise in one of the following ways:

$$(X = 0, Y = r), (X = 1, Y = r{-}1) \dots (X = r, Y = 0)$$

The distribution of $P\{s = r\}$ is, therefore, given by:

$$c_r = a_0 b_r + a_1 b_{r-1} + a_2 b_{r-2} + \dots a_r b_0 \qquad\qquad [4.39]$$

The operation which leads to c_r and which is defined by [4.39] is known as convolution and is symbolically written as:

$$\{c_k\} = \{a_k\} * \{b_k\}$$

If one replaces the discrete probability distributions by some continuous density functions, say $f(x)$ and $g(y)$, one can follow similar arguments and write for the event $H(u) = P\{z \leqslant u\} = P\{x + y \leqslant u\}$ which has the density function:

$$h(u) = \int_0^u f(x)\, g(u - y)\, dy \qquad\qquad [4.40]$$

Equation [4.40] is called the convolution integral and is precisely equivalent to [4.39]. With it one may, for example, investigate the distribution of the sum of two consecutive gaps between points in a Poisson process. In this case:

$$f(x) = g(y) = \rho e^{-\rho x}$$

and the convolution of $f(x)$ with itself can be written according to [4.40] as:

$$f(x) * f(x) = f^{*2}(x) = \int_0^x \rho e^{-\rho t}\, \rho e^{-\rho(x - t)}\, dt = \rho^2 x e^{-\rho x}$$

Extending this by induction, one may find the sum of 2, 3 ... k gaps and their distribution which is the k-fold convolution of $f(x)$:

$$f^{*k}(x) = \frac{\rho(\rho x)^{k-1}}{\Gamma(k)}\, e^{-\rho x} \qquad\qquad [4.41]$$

The distribution of [4.41] is known as the gamma distribution, in the special case where k is an integer. The gamma function may be replaced by a factorial and be written as:

$$f(x) = \frac{\rho(\rho x)^{k-1}}{(k-1)!}\, e^{-\rho x}$$

which is also called the special Erlangian distribution. The mean and variance of both distributions are $E(x) = k/\rho$ and $\mathrm{var}(x) = \sqrt{k}/\rho$. It can be seen from [4.41] that if $k = 1$ the gamma distribution becomes the negative exponential distribution but if $k > 1$, $f(x)$ will always be zero at the origin and obtain a single maximum at $x = (k - 1)/\rho$. With relatively low k-values, the curves show strong positive skewness but become increasingly symmetrical with increasing k. Indeed, the gamma distribution tends towards the normal distribution when the mean k/ρ is kept fixed while k and ρ approach infinity.

The result of [4.41] can be very easily obtained by making use of the moment generating function. Generating functions are widely used in theoretical statistics and probability theory, and an introduction to the methods is found in many textbooks (cf. Feller, 1957). The moment generating function $\phi(s) = E(e^{sx})$ of a density distribution $f(x)$ has the special property that it involves the moments m_n as coefficients:

$$\phi(s) = E(e^{sx}) = \int e^{sx} f(x)dx$$

$$= \int \sum \frac{(sx)^n}{n!} f(x)dx$$

$$= \sum \frac{s^n}{n!} \int x^n f(x)dx$$

$$= \sum \frac{s^n}{n!} m_n \qquad\qquad\qquad [4.42]$$

It may be seen from [4.42] that:

$$\phi(s) = f(-s), \quad f(s) = L[f(x)] \qquad\qquad\qquad [4.43]$$

where $L[f(x)]$ stands for Laplace transform of $f(x)$ (see e.g., Churchill, 1958). Since the convolution of $f(x) * f(x)$ is simply the inverse of the Laplace transform multiplied with itself, one can derive the gamma distribution by writing the Laplace transform of $f(x)$:

$$L[f(x)] = L(\rho e^{-\rho x}) = \frac{\rho}{s + \rho} \qquad\qquad\qquad [4.44]$$

To obtain the sum of k gaps, this is raised to the power of k:

$$\left(\frac{\rho}{s + \rho}\right)^k \text{ which has the inverse } \rho^k \frac{x^{k-1}}{\Gamma(k)} e^{-\rho x}$$

which is the identical result to [4.41].

With this method, it is easy to solve the problem mentioned at the beginning of this section. This is to determine the thickness distribution of sand—clay layers whereby it is assumed that both types are independently exponentially distributed with densities ρ_1 and ρ_2. The Laplace transform of the convolution is given by:

$$L[f(x_1) * f(x_2)] = \frac{\rho_1 \rho_2}{(s + \rho_1)(s + \rho_2)}$$

which has the inverse:

$$f(x) = \frac{\rho_1 \rho_2}{\rho_2 - \rho_1} [\exp(-\rho_1 x) - \exp(-\rho_2 x)] \qquad [4.45]$$

Differentiating the moment generating function ([4.43]) and setting $s = 0$, gives the moments of the distribution and, in particular, the mean and variance are:

$$E[f(x)] = \frac{\rho_1 + \rho_2}{\rho_1 \rho_2}, \quad \text{var}[f(x)] = \frac{\rho_1 \rho_2 (\rho_1 + \rho_2)^2}{(\rho_1 \rho_2)^2}$$

To illustrate this example, a thickness distribution has been calculated for the values $\rho_1 = 0.5$ and $\rho_2 = 0.8$ and this is shown in Fig. 4.6. The assumed densities imply a mean thickness of 2 for the sand layers and 1.25 for the clay layers. The mean thickness of the combined layers is simply the sum, 3.25.

Some reference should be made also to the normal and lognormal distributions which are very familiar to the geologist. In particular, the normal

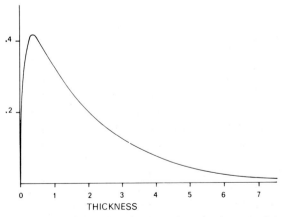

Fig. 4.6. Thickness distribution of sand—clay couplets.

distribution occurs almost universally not only in theory but also in practice, and it is found that many stochastic variables are normally distributed. This ubiquity of the normal distribution is very closely connected with the central limit theorem of theoretical statistics. This states that the sum or the mean of a number of n samples will be very nearly normally distributed irrespective of the distribution of the variable itself, provided that this distribution has a finite variance, and provided that n is not too small. Thus, if one takes n samples from an exponentially distributed random variable, the means of these samples will be approximately normally distributed.

The central limit theorem explains why the gamma distribution approaches the normal distribution when k increases and, similarly, why the binomial can be approximated by the same distribution when n is large; use has been made of this in [4.14].

The situation is very similar with the lognormal distribution. If it is considered that the variables are more "naturally" measured by their logarithms than by a linear scale, then the central limit theorem will lead to the lognormal distribution. There is an extensive literature which argues either in favour of or against the use of lognormal distributions (cf. Krumbein and Graybill, 1965), but it is clear that the logarithmic transformation has often been used only because it can make skewed geological data more symmetrical and approximately normal. In this context, it is noteworthy that it is quite difficult to differentiate statistically between the lognormal distribution and the gamma distribution unless the observed data are very complete. The main differences occur in the tails of the distributions. The density of the gamma distribution is of the order $e^{-\rho x}$ as x goes to infinity, whereas the lognormal distribution will be of the order $x^{-(\log x)/2\sigma}$ and the tail of the lognormal distribution is, therefore, much longer. All the simple models used in this approach lead to the gamma distribution, but it is quite possible that the lognormal distribution comes into its own right when more advanced methods are used.

4.8 CONTINUOUS VARIABLE MODELS AND QUEUES

With the use of continuous probability-distribution functions, a variety of sedimentation models can be constructed which are derived from the previously introduced random walk. As a first improvement, one might consider making the time axis continuous. This allows steps to occur at random intervals which could be exponentially distributed. One may further assume a continuous step distribution which, depending on the mechanism which is involved, can be anything from exponentially to normally distributed. Such models will be discussed later (Chapter 6). For the present, some new ideas which can lead to sedimentation models will be introduced.

Much of the early work on stochastic processes has been connected with practical problems, and one particular process which has received a great deal of study is the statistical behaviour of queues. In principle, a queue forms when customers arrive at some sort of service station waiting to be served. Since each customer has to spend a certain time in service, delays can arise and a queue may build up. Generally speaking, the arrival times will be controlled by some probability distribution, let us say $f(t)$, and the service time will be determined by some other distribution, $g(t)$. In this model, arrivals are statistically independent of the service, but a service only commences when there is a customer available. Several stochastic variables can be studied. For example, one is interested in the number, $n(t)$, of customers waiting to be served, including the one being served. The number $n(t)$ is known as the state of the queue and there will be a probability function, $P(n,t)$, which gives the probability that the queue is in state n ($n = 0, 1, 2...$) at time t.

The relevance of this model may be seen immediately if one substitutes "sedimentary particle" or some sedimentation units for the term customer and a waiting period between deposition and erosion, for the service time. The model, therefore, permits the study of a system in which deposition is transient; let us say, a basin which takes in sediment on the one side and loses it on the other. The deposition of sediment in the basin is temporary and a random variable. In less abstract terms, one could think of a sand bar which builds up as soon as sediment is available. To make the model really simple, one may assume that the sediment is supplied in units, one of which is used for the barrier, but the remainder is stored in the basin like a queue.

One may next make the following assumptions. The influx is Poisson distributed with density α and the time needed for clearance has the same distribution with density β. Obviously, the mean time spent in the clearance procedure (service time) is $1/\beta$ and in time t_0, precisely βt_0 units can be cleared. Therefore, if $\alpha > \beta$, the queue would increase indefinitely and more and more sediment would build up in the basin. On the other hand, if $\alpha < \beta$, the system would be in state 0 for approximately $1 - \alpha/\beta$ of the time. One can introduce the ratio $\alpha/\beta = \rho$ and clearly if $\rho > 1$, infinite growth occurs but when $\rho < 1$, one will expect some sort of equilibrium to establish itself. If the system has operated for a sufficiently long time, the probability distribution of the number of customers in the queue will be independent of time and of the initial state of the queue. It is relatively simple to show that, when equilibrium conditions have been established, the distribution for the number of customers in the queue can be written as:

$$P\{\text{state of queue} = n\} = (1 - \rho)\,\rho^n$$

where $\rho = \alpha/\beta$. The equation shows that, in the long run, the number of customers in the queue, or the number of sedimentation units in the basin, will be exponentially distributed.

The model is of some geological interest if one includes in the system some outside influence which is not connected with sedimentation but which has the property of preserving whatever sediment is present in the basin. For example, one might postulate crustal subsidence which occurs at intervals and which creates a situation in which the process can start all over again. Providing that such interruptions occur at long time intervals compared with the sedimentation processes, then the amount of sediment in the basin will have no relation to time spans between the event which is being recorded. Clearly, whether the periods of subsidence are equally spaced or quite irregular, the result will be the same: an exponentially distributed bed-thickness record. This is a model for sedimentary by-pass (see Chapter 3), and although it may be difficult to formulate an actual mechanism which can preserve sediment without interfering with the sedimentation system, the example illustrates some of the problems which will be encountered when an attempt is made to link deposited sediment with the time events which it has recorded.

The queueing models can be elaborated in several directions; both arrival- and service-time patterns can follow different probability distributions and it is possible to formulate queues in which the state space is continuous. One of these models will be discussed at a later stage.

4.9 THE USE OF STOCHASTIC MODELS IN STRATIGRAPHY

It is useful, at this stage, to review briefly the stochastic models which have been introduced from their geological aspects. Before this is done, however, there is one general point which ought to be considered.

When stochastic models were first applied to practical problems such as econometrics, meteorology and other subjects which, like stratigraphy, deal with the development of systems which are affected by random variables, it was hoped that observational series could be interpreted in terms of such models. In other words, it was hoped that once an appropriate stochastic model was found, this would provide a unique answer to the question of how such a series of observed events was generated. This expectation was never fulfilled. It soon became evident that a great variety of different stochastic models can lead to identical results and that it is, as a rule, not possible to conclude from the model that the mechanism which produced the results must have had one particular specific structure. With stratigraphic problems, the situation is very similar. It would be quite impossible to analyse an observed section by fitting a stochastic model, and to expect it to explain how the sequence has formed, but it is possible to deduce theoretically what kind of sequence would develop if certain premises are fulfilled. Thus, in geological research, it becomes important to consider each problem in two stages. The theoretical model provides the conditions under which the geological process might have developed, but the geologist has to judge whether such conditions are acceptable. A simple example may illustrate this procedure.

Almost any geologist has used the phrase "there is no pattern" in the distribution, occurrence, or whatever has been investigated at the time. By "no pattern" it is, of course, not implied that such occurrences have no arrangement at all, but that they are distributed in such a way that there is no detectable relationship between the distance from one occurrence to the next. The distances between occurrences or events which have this property are called "independent random variables" if the terminology of probability theory is used. It has been shown in [4.22] and [4.37] that the gap distribution between such independent random events is given by either the geometrical or negative exponential distribution, depending on whether the distances are measured in discrete or continuous units. An arrangement of random points along a straight line can be described by a Poisson process, which is, therefore, an ideal model for the phrase "no pattern", meaning random arrangement; the Poisson process can be developed easily for two- and more-dimensional spaces. For example, one would expect the distribution of relatively rare minerals in a thin section to follow approximately a two-dimensional Poisson distribution, as long as it is reasonable to assume that there is no interaction between neighbouring grains. If geological reasons are against such a model and if, for example, the mineral was formed by some sort of diffusion process, another model must be formulated, even if the observations seem to fit the theory.

The use of stochastic models in stratigraphic research is very well illustrated further by some work of Mizutani and Hattori (1972). These authors measured a large number of bed thicknesses in the Pliocene flysch-type sequence of the Horinouchi Formation (Japan). This series consists of alternating layers of sandstone and shales. Measurements are not only given for the individual sandstone and shale beds, but also for the combined couples. Mizutani and Hattori showed that the bed-thickness distributions of both sandstones and shales have roughly the appearance of truncated normal distributions and, furthermore, they demonstrated that the thicknesses of successive beds are not correlated with each other, showing that the amount of sedimentation and possibly erosion of successive beds may be regarded as independent random variables. Based on these observations, they propose that a Kolmogorov-type process of random erosion and deposition is responsible for the bed formation and by fitting a normal distribution to the now incomplete data they reconstruct the original distribution $f(x)$ (see [4.30]). The distribution $f(x)$ would be observed if negative deposition could be recorded. To the geologist it is of interest that this approach provides an estimate of the amount of sediment which has been by-passed at one particular locality of the basin.

Purely for argument's sake, one could construct a second model which, as will be seen, fits the observations but has otherwise nothing in common with the Kolmogorov model. Looking at the thickness distribution of sandstone beds, one is struck by how closely these follow the negative exponential distribution and, although the maximum is not in the smallest class interval of 0—2 cm, one could easily argue that this is because of bias in measuring. The shale thicknesses are somewhat less well represented by the exponential distribution but they still fit the data reasonably well. Using the locality Kagegawa A of Mizutani and Hattori, the exponentials fitted to the sands and shales had the parameters ρ_1 = 0.1136 and ρ_2 = 0.1428. The new hypothesis, then, is that both sand and shale thicknesses show no pattern at all, i.e., are independent random variables. One may next calculate the thickness distribution of the sand—shale couplets by making use of [4.41]. This distribution, together with the bed-thickness distribution, based on the Kolmogorov model and the observed data, is shown in Fig. 4.7. It may be noted that, although the relatively large number of 358 bed thicknesses has been used in this example, the empirical thickness—frequency curve is still far from smooth and such a variability is unfortunately common to many geological data. One can hardly say that one of the two models provides a better fit than the other, but one can state that in general terms both give a reasonable representation of the data.

The real value in constructing a stochastic model lies not in providing a curve of best fit to the observed data, but in pin-pointing certain geological conditions which must have been obtained if the mechanism which is portrayed in the model really operated. If alternative models can be provided, so much the better, because in this case a decision may be possible after a renewed geological examination of the evidence.

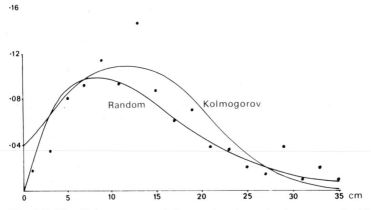

Fig. 4.7. The Kolmogorov and the random hypothesis compared with the Kagegawa data of Mizutani and Hattori (1972).

In the specific case of the Kagegawa sand—shale couplets, there is one discriminating feature between the two hypotheses which should be amenable to geological tests. The Kolmogorov hypothesis implies that there is only one erosional break at the roof of each bed since it is this erosional break which determines the bedding plane. The second hypothesis does not imply a major break along the bedding plane, but as has been shown, each bed could contain a number of erosional stages which petrographic examination should be able to demonstrate.

Mizutani and Hattori have employed the Kolmogorov principle for the formation of the individual sand and shale beds as well as for that of the couplets. This does not seem to be altogether consistent, but in fairness to the authors, it is stressed that the second interpretation which was developed here is only given as an example and it is quite impossible to give judgment without the rest of the geological evidence. On a speculative basis, one could develop the bed formation theories by accepting the hypotheses of truncated shale and sand beds but by regarding the thickness distribution of couplets as the sum (convolution) of such random variables. Alternatively, and this seems more attractive, the sand could be regarded as exponentially distributed and the shale may be truncated, but not necessarily derived, from a normal distribution.

In this chapter, particular emphasis was given to the Poisson process because the geologist finds himself very often in the position where he knows absolutely nothing about the process he has to investigate. Under such circumstances, it is often useful to make the minimum assumptions which could lead to a model of the process and this is the main principle behind the sedimentation models developed in this chapter. In sedimentary stratigraphy the main interest is in the relationship between sedimentation and time. Here the minimum assumption which can be made is that time and sedimentation are not connected and, therefore, both represent independent random variables. This leads automatically to the random-walk models used for describing the accumulation of sediment. These models are in complete contrast to the deterministic models of the previous chapter which, within the framework of simplified models, made the strongest possible assumption of complete dependence between time and sedimentation. In this respect, one can regard the independent random models and the deterministic models as two limits to the range of possible explanations of sedimentation phenomena. It is also clear that, unless a stratigraphical sequence or a series of sedimentation events can be reduced to a Poisson process, the factors involved in the formation must be in some way dependent and any such functional relationship, even if it is probabilistic, needs some geological explanation. The next two chapters will be devoted specifically to models which involve such probabilistic relationships.

Chapter 5

Markov chains

5.1 INTRODUCTION

Most of the random processes which were introduced in the previous chapter were based on the principle of independent trials. For instance, the simple random walk is based on a Bernoulli trial which is an experiment with two possible outcomes, success or failure. Associated with the Bernoulli trial are two definite probabilities which are in no way affected by how often the experiment is repeated nor by what happens during earlier or later trials.

If one tries to find examples of such chance events the choice is very limited, because most of the examples which may be quoted belong to the artificial world of gambling rather than reality. A typical real-world situation is to ask somebody for tomorrow's weather. Almost invariably, a person giving such a judgement will first assess today's weather. This is because experience has taught us that, although there is a probability element in the alternation between fine and rainy days, the weather pattern is not an independent sequence. Tomorrow's weather is clearly influenced by what happened today. In a similar way, anyone may have an accident at any time and most insurance firms know the precise probability of this happening, but it is also known that the chance of an individual having an accident depends not only on what he is doing at the time but on his past history. Sometimes trivial circumstances like a bad night's sleep may effectively increase the probability of an accident occurring and so such probabilities cannot be regarded as independent.

Many stratigraphical data are of a similar sequential nature and are somehow linked in a historical sense. If one studies a section representing, let it be assumed, a transgressive sequence, one can usually predict the lithologies which will be met with eventually. Such predictions, however, are not absolute and one might, therefore, regard the occurrence of lithological horizons as random events which are partially determined by preceding events. Formally, one can say that a random variable, in this case the lithology, takes on various values as time progresses. The values of the random variable are discrete and they are, therefore, referred to as "the states of the system". If a sequence consists of three rock types, such as sandstone, shale and limestone, then these are the three states of the system. It is useful to visualize the sequence of events, in this case the appearance of the various states, as a function of time but this is not essential and any other dimension, such as the height in the stratigraphic section, can be substituted for time.

5.2 THE MARKOV PROCESS

There exists a very wide range of stochastic processes in which a random variable depends on its past history. The simplest of these is known as a Markov process.

In a Markov process, a random variable X_n which describes the state of the system at time n, depends only on the state X_m which occurred at time m, provided that the process operated at arbitrary times ... $1 < m < n$; this may be formally written as:

$$P\{X_n = x|X_m = y, X_1 = z \ ...\} = P\{X_n = x|X_m = y\} \ , \ (1 < m < n) \qquad [5.1]$$

The equation clearly indicates that it is only the situation at time m which contributes to our knowledge of what will happen at time n and any knowledge of earlier states is irrelevant. Since all Markov processes have the same structure, [5.1] is known as the Markov property.

A Markov chain is a Markov process which operates at discrete time intervals and within a discrete state space. A discrete random variable X_0, X_1, ... may take on any value h, ... j, k ... which is a part of the state system and the conditional distribution of X_n given X_0, X_1 ... X_{n-1}, depends only on the value of X_{n-1}:

$$P\{X_n = k|X_0 = h, \ ... \ X_{n-1} = j\} = P\{X_n = k|X_{n-1} = j\} \qquad [5.2]$$

This clearly satisfies the Markov condition of [5.1].

The Markov chain is well illustrated by considering the successive positions of a particle undergoing a random walk. It is assumed that a point moves from its origin $z = 0$, at time $t = 0$ in either positive or negative unit jumps. The states which the particle can occupy during the successive times $0, 1, \ ... \ n$ are given by $z \ ... \ -1, 0, +1 \ ...$ and at time n it will be in state z_n. It is clear from the definition of the random walk that at time $n + 1$ the particle can only occupy the positions z_{n+1} or z_{n-1}. Conversely, if the position z is observed at time n this could only have been reached by an upward step, $J = +1$, from position $z - 1$ at time $n - 1$, or by a downwards step, $J = -1$, from position $z + 1$. In order to give the position of the particle, conditional probabilities must be introduced and one may write:

$$P\{Z_n = z\} = P\{J_n = +1|Z_{n-1} = z - 1\}$$

$$= P\{J_n = -1|Z_{n-1} = z + 1\} \qquad [5.3]$$

This notation is somewhat awkward and it is useful to introduce a probability which refers specifically to the transition from one state to another. For example, the probability of going from state z to state $z + 1$ is written as

$P_{z,z+1}$. If a transition occurs from state j to state k, the transition probability can be written as p_{jk}. Clearly, if a system consists of n states, then it will possess $n \times n$ different transition probabilities and if the number of states is finite, all transition probabilities can be arranged in an $n \times n$ matrix. In the specific example of the random walk there is an infinite number of states and consequently the transition matrix is infinite. Remembering that the probabilities for positive and negative jumps in a random walk are:

$$P\{J = +1\} = p \quad P\{J = -1\} = q \tag{5.4}$$

one may write all transition probabilities as:

$$
\mathbf{P} = \quad
\begin{array}{c}
 \\
 \\
-2 \\
-1 \\
0 \\
+1 \\
+2 \\
 \\
 \\
\end{array}
\begin{array}{c}
 \quad -1 \quad 0 \quad +1 \quad +2
\end{array}
\left[
\begin{array}{ccccccccc}
\cdot & \cdot & \cdot & \cdot & \cdot & \cdot & \cdot & \cdot & \cdot \\
\cdot & \cdot & \cdot & \cdot & \cdot & \cdot & \cdot & \cdot & \cdot \\
\cdot & \cdot & 0 & p & 0 & 0 & 0 & \cdot & \cdot \\
\cdot & \cdot & q & 0 & p & 0 & 0 & \cdot & \cdot \\
\cdot & \cdot & 0 & q & 0 & p & 0 & \cdot & \cdot \\
\cdot & \cdot & 0 & 0 & q & 0 & p & \cdot & \cdot \\
\cdot & \cdot & 0 & 0 & 0 & q & 0 & \cdot & \cdot \\
\cdot & \cdot & \cdot & \cdot & \cdot & \cdot & \cdot & \cdot & \cdot \\
\cdot & \cdot & \cdot & \cdot & \cdot & \cdot & \cdot & \cdot & \cdot \\
\end{array}
\right] \tag{5.5}
$$

where $\mathbf{P}$ is called the transition-probability matrix of the process. The matrix notation permits one to find any transition probability p_{jk} without difficulty; the index which refers to the rows of the matrix is used to indicate the state of the system at time $n - 1$ and the index for the columns indicates the state at time n. For example, if the system is in state +1 at time n, one will find that in time $n + 1$ it is either in state 0, which will happen with probability q, or with probability p in state +2. All other probabilities are zero because it is impossible for this system to perform anything but unit jumps.

Most of the transition probability matrices which are used in geological problems are finite and, as a typical example, one might consider the possible transitions of three rock types (sandstone, shale and limestone) which have been deposited in an imaginary section. These may be written as a

3 × 3 matrix as follows:

$$
\mathbf{P} = (p_{jk}) = \begin{array}{c} \\ \text{sandstone} \\ \\ \text{shale} \\ \\ \text{limestone} \end{array} \begin{array}{ccc} \text{sandstone} & \text{shale} & \text{limestone} \\ \begin{bmatrix} p_{11} & p_{12} & p_{13} \\ p_{21} & p_{22} & p_{23} \\ p_{31} & p_{32} & p_{33} \end{bmatrix} \end{array} \qquad [5.6]
$$

Again, the rows denote a definite lithology and the columns give the composition (state) of the stratum resting immediately above a chosen lithology; p_{jk} gives the probability with which a particular change of lithology will occur. For example, if a bed of shale is observed, then the probability that this is succeeded by a limestone is equal to p_{23}. Clearly, since the p_{jk}'s are probabilities they must be non-negative numbers and, furthermore, if the matrix is to give a full description of the process, each row must add up to unity. Assuming that a shale has been laid down, this can only be followed by either a sandstone with probability p_{21}, or a shale p_{22}, or a limestone p_{23}. One of these exclusive events must happen and, therefore, $p_{21} + p_{22} + p_{23} = 1$. A matrix in which the rows add up to unity and which consists of non-negative elements is called a stochastic or Markov matrix.

The special nature of the Markov chain may be best seen by comparing its stochastic matrix with that of an independent random process. In the three-lithology example, a random model can be provided by an urn which contains a large number of differently marked balls of three types which are in the same proportion as the rock types in the section. For example, assume that the proportions of sandstone:shale:limestone occur in an observed section as 5:3:2. An urn containing 1000 balls will, therefore, contain 500 balls marked sandstone and so on. If one chooses a lithology at random from this urn it will be, with probability 0.5, a sandstone, and this is in no way influenced by what happened at earlier or later trials. One can, therefore, write the stochastic matrix as:

$$
\mathbf{P} = \begin{array}{c} \\ \text{sandstone} \\ \\ \text{shale} \\ \\ \text{limestone} \end{array} \begin{array}{ccc} \text{sandstone} & \text{shale} & \text{limestone} \\ \begin{bmatrix} 0.5 & 0.3 & 0.2 \\ 0.5 & 0.3 & 0.2 \\ 0.5 & 0.3 & 0.2 \end{bmatrix} \end{array} \qquad [5.7]
$$

which has the distinguishing feature that each column consists of a constant probability which indicates at once that the choice of a lithology is independent of the preceding choice.

In contrast to this, consider the matrix:

	sandstone	shale	limestone
sandstone	0.70	0.20	0.10
P = shale| 0.16	0.50	0.34	[5.8]
limestone	0.50	0.25	0.25

Here all the rows are different although, as will be shown shortly, the overall proportions of sandstone:shale:limestone are precisely the same as in [5.7].

The matrix [5.8] represents a typical Markov chain. This too can be represented by an urn model but, in this case, it is necessary to have at least as many urns as states. Each urn is marked with the row number it represents and it contains balls which are marked according to the proportions in this particular row. For example, to represent matrix [5.8], urn number 1 would contain sandstones : shales : limestones in the proportions 7 : 2 : 1. Let it be assumed that the first bed in the sequence is a sandstone; one turns, therefore, first to urn number 1. A draw from this urn may result in a limestone and one, therefore, turns to urn number 3 for the next draw, and so on. It is sometimes useful to start such an experiment from an initial probability distribution $p^{(0)}$ which is a vector giving the probabilities with which the first state is drawn. If such a process is simulated, an additional urn containing the initial distribution is needed. In the case where it is specified that the first bed is to be a sandstone, $p^{(0)} = (1, 0, 0)$, and urn 0, therefore, contains nothing but sandstones.

5.3 THE N-STEP TRANSITION PROBABILITY

The Markov chain for discrete states in discrete times has been defined by [5.2] as:

$$p_{jk}^{(1)} = P\{X_{m+1} = k | X_m = j\}$$ [5.9]

Here it is assumed that the process operates at consecutive times $m < m+1$ ... and this is indicated by the superscript (1) in [5.9]. Instead of assuming a unit step, one may write:

$$p_{jk}^{(n)} = P\{X_{m+n} = k | X_m = j\} \quad (m, n = 1, 2, ...)$$ [5.10]

which for the times $m_1 < m_2 < ... m < m + n$, is a proper Markov chain moving in steps of n. If $n = 1$, one talks of one-step transition probabilities and the index (1) is usually omitted. A very important group of Markov chains is those processes in which the probability in [5.10] depends only on n but not on m. Such chains are called homogeneous, and they possess stationary transition probabilities which means that the individual p_{jk}'s are constant throughout the sequence.

If a chain is homogeneous, then it is possible to derive the higher transition probabilities from the one-step transition matrix. This can be seen as follows, writing for the one-step matrix of a three-state chain:

$$\mathbf{P} = \begin{bmatrix} p_{11} & p_{12} & p_{13} \\ p_{21} & p_{22} & p_{23} \\ p_{31} & p_{32} & p_{33} \end{bmatrix}$$

We obtain the probability that the system is in any of the possible states at time 1 by multiplying the initial conditions $p^{(0)}$ with $\mathbf{P}$, resulting in a row vector $p^{(1)}$. For example, if it is certain that the system has been in state 1 at time 0, then $p_1^{(0)} = (1, 0, 0)$ and $p_1^{(1)} = (p_{11}, p_{12}, p_{13})$. Next, we are interested in $p_1^{(2)}$ which gives the probabilities that the system is in any of the possible states at time 2, providing that it has been in state 1 at time 0. This is obviously obtained by multiplying $p_1^{(1)}$ by $\mathbf{P}$ and we obtain:

$$p_1^{(2)} = p_1^{(1)} \mathbf{P} = (p_{11}^2 + p_{12}p_{21} + p_{13}p_{31}, p_{11}p_{12} + p_{12}p_{22} + p_{13}p_{32}, ...)$$

which is the first row of the matrix product $\mathbf{P} \times \mathbf{P} = \mathbf{P}^{(2)}$. The state of the system may now be predicted for any time n from the relationship:

$$p^{(n)} = p^{(n-1)} \mathbf{P} \tag{5.11}$$

which on iteration gives:

$$p^{(n)} = p^{(n-2)} \mathbf{P}^2 = ... p^{(0)} \mathbf{P}^n . \tag{5.12}$$

The matrix $\mathbf{P}^n$ consists of the elements $p_{ik}^{(n)}$ which are called the n-step transition probabilities of $\mathbf{P}$:

$$p_{jk}^{(n)} = P\{ \text{state } k \text{ at time } n | \text{state } j \text{ at time } 0 \}$$

Applied to the example of the sandstone, shale, limestone sequences $p_{jk}^{(n)}$, ($n = 1, 2, ...$) means that, given a lithology j has been observed, then the 1st, 2nd ... beds to follow will have lithology k. For example, if a sandstone is met with, then the probability that the second bed after this sandstone is a shale is given by $p_{12}^{(2)}$. As a numerical example, take matrix [5.8] and multiplying this with itself we find:

$$\mathbf{P}^2 = \begin{bmatrix} 0.57 & 0.27 & 0.16 \\ 0.36 & 0.37 & 0.27 \\ 0.52 & 0.29 & 0.19 \end{bmatrix}, \quad \mathbf{P}^4 = \begin{bmatrix} 0.50 & 0.30 & 0.20 \\ 0.48 & 0.31 & 0.21 \\ 0.50 & 0.30 & 0.20 \end{bmatrix}$$

$$\mathbf{P}^6 = \begin{bmatrix} 0.50 & 0.30 & 0.20 \\ 0.50 & 0.30 & 0.20 \\ 0.50 & 0.30 & 0.20 \end{bmatrix}$$

From this calculation it may be suspected that, if powering is carried out sufficiently often, the matrix may reach an equilibrium which will not change on further powering. If one writes for such a stable distribution the vector $\Pi = p^{(n)} (n \to \infty)$, then [5.11] can be written as:

$$\Pi = \Pi P \text{ or } \Pi(\mathbf{I} - \mathbf{P}) = 0 \tag{5.13}$$

which represents a system of simultaneous equations with unique solutions when it is taken into account that all elements of the vector Π must add up to unity. The consequences of a stable distribution may be best seen by setting the initial distribution $p^{(0)} = \Pi$ in which case:

$$p^{(1)} = \Pi P = \Pi, \, p^{(2)} = \Pi P = \Pi, \, \dots \, p^{(n)} = \Pi P = \Pi \tag{5.14}$$

meaning that the distribution of Π is stationary and does not change with time. Now, from the property of matrix products it may be seen that if $\mathbf{P}$ has identical rows then:

$$\begin{bmatrix} p_1 & p_2 & p_3 \\ p_1 & p_2 & p_3 \\ p_1 & p_2 & p_3 \end{bmatrix} \times \begin{bmatrix} p_1 & p_2 & p_3 \\ p_1 & p_2 & p_3 \\ p_1 & p_2 & p_3 \end{bmatrix} = \begin{bmatrix} p_1 & p_2 & p_3 \\ p_1 & p_2 & p_3 \\ p_1 & p_2 & p_3 \end{bmatrix}$$

when $p_1 + p_2 + p_3 = 1.0$.

Such a matrix, therefore, does not change on further powering. Taking into account [5.14], it may be seen, therefore, that:

$$\mathbf{P}^{(n)}_{n \to \infty} = \begin{bmatrix} \pi_1 & \pi_2 & \pi_3 \ldots \pi_r \\ \pi_1 & \pi_2 & \pi_3 \ldots \pi_r \\ \pi_1 & \pi_2 & \pi_3 \ldots \pi_r \end{bmatrix} \qquad [5.15]$$

where $(\pi_1 + \pi_2 + \ldots \pi_r) = 1.0$.

The matrix $\mathbf{P}^n$ will tend to a matrix in which the stable probability vector is repeated in each row of the matrix. It was seen in [5.7] that such a matrix is typical of an independent random process and it was found that the row vector Π is determined by the unconditional probabilities of the individual states. In a stratigraphic section, therefore, the proportions of the various lithologies will determine the stable probability vector and vice versa the proportions of the lithologies can be calculated from the transition matrix by finding the stable probability vector. If electronic computers are used, then the simplest method of finding the stable vector Π is by powering the matrix $\mathbf{P}$ as indicated in [5.15], although the same result could be obtained from [5.13], a method which will be discussed later.

The result of [5.15] is of particular importance in understanding the structure of Markov chains. Imagine a stratigrapher climbing a section which is assumed to be Markovian in its bed-by-bed arrangement. Once a certain bed is reached and identified, it will be possible to predict the next bed with reasonable certainty. It may be possible to predict some 2, 3, n, beds ahead but, if n is large, this prediction becomes less and less certain until it is only determined by the relative proportion of the beds present in the sequence. At this stage, the unknown section ahead can be regarded as an independent random arrangement of beds. The predictability of the sequence depends entirely on how the transition matrix $\mathbf{P}$ approaches stability when powered and whether all Markov matrices converge towards a stable vector and how this convergence proceeds, must be investigated.

5.4 RECURRENCE AND CLASSIFICATION OF STATES

Supposing that a Markov chain is in state j at time 0, what are the probabilities that the process re-enters state j at time n, without having entered this state previously? Such a probability is called a recurrence probability and it may be formally defined as:

$$f_{jj}^{(n)} = P\{X_r \neq j, r = 1, 2 \ldots n - 1; X_n = j | X_0 = j\} \qquad [5.16]$$

Obviously, the one-step recurrence probability $f_{jj}^{(1)} = p_{jj}$ and the higher probabilities may be obtained from the recurrence relation:

$$f_{jj}^{(n)} = p_{jj}^{(n)} - f_{jj}^{(1)} p^{(n-1)} - \ldots f_{jj}^{(n-1)} p_{jj} \qquad [5.17]$$

The probability that state j will be eventually re-entered is given by the sum:

$$f_j = \sum_{n=1}^{\infty} f_{jj}^{(n)} \qquad [5.18]$$

and this must be 1.0. If it is certain that state j will be entered again, in this case j is called a recurrent state. If it is uncertain that j, once it has occurred, is ever entered again then $f_j < 1.0$ and state j is called a transient state. In any recurrent state, $f_{jj}^{(n)}$ ($n = 1, 2,\ldots$) is a probability distribution which has for its mean:

$$\mu_j = \sum_{n=1}^{\infty} n\, f_{jj}^{(n)} \qquad [5.19]$$

If the mean recurrence time μ_j is finite, then state j is called positive-recurrent. If μ_j is infinite, then j is called a null-recurrent state.

In a special group of Markov chains a recurrent state j can only be re-entered at times $\tau, 2\tau, 3\tau\ldots$ where τ is an integer greater than 1. In this case, j is called a periodic state with period τ. In any such periodic chain, the probability $p_{jj}^{(n)}$ is zero, unless n is a multiple of τ. States which are not periodic are called aperiodic and states which are at the same time aperiodic and positive-recurrent are called ergodic.

Instead of calculating the return of a state to itself, one can use the same arguments to evaluate the probability that a state j changes for the first time into k after n steps. This is written as $f_{jk}^{(n)}$ and is called the first passage probability of states j and k. Again, the probability that the passage does occur is given by:

$$f_{jk} = \sum_{n=1}^{\infty} f_{jk}^{(n)} \qquad [5.20]$$

and if it turns out that this happens with certainty, then there exists a mean passage time:

$$\mu_{jk} = \sum_{n=1}^{\infty} n\, f_{jk}^{(n)} \qquad [5.21]$$

The classification of states in Markov chains is based on the recurrence probability $f_{jj}^{(n)}$ but this is linked by [5.17] to the transition probabilities $p_{jj}^{(n)}$ and it is, therefore, possible to define the various states by the limiting behaviour of $p_{jj}^{(n)}$ and $p_{jk}^{(n)}$. For a derivation of this classification, reference should be made to textbooks (cf. Karlin, 1966) and the following summary is given from Cox and Miller (1965), p. 96.

(1) Let k be an arbitrary fixed state. Then:
(a) k is transient if and only if the series $\sum_{n=1}^{\infty} p_{kk}^{(n)}$ is convergent, in which case $\sum_{n=1}^{\infty} p_{jk}^{(n)}$ is convergent for each j; and
(b) k is recurrent if and only if the series $\sum_{n=1}^{\infty} p_{kk}^{(n)}$ is divergent, in which case $\sum_{n=1}^{\infty} p_{jk}^{(n)}$ is divergent for each j.
(2) Let k be a recurrent state. Then:
(a) if k is aperiodic:

$$\lim_{n \to \infty} p_{kk}^{(n)} = 1 \left/ \sum_{n=1}^{\infty} n f_{kk}^{(n)} \right.$$

[5.22]

and if k is positive recurrent:

$$\lim_{n \to \infty} p_{kk}^{(n)} = \Pi_k > 0$$

[5.23]

and:

$$\lim_{n \to \infty} p_{jk}^{(n)} = \lim_{n \to \infty} p_{kk}^{(n)}$$

[5.24]

(b) if k is a periodic state with period τ:

$$\lim_{n \to \infty} p_{kk}^{(n)} = \tau \left/ \sum_{n=1}^{\infty} n f_{kk}^{(n)} \right.$$

[5.25]

These results are known as the basic limit theorem for Markov chains.

A general Markov chain may contain any state from the above classification. However, there exists a theorem showing that an arbitrary chain can always be decomposed into two sets, one set which contains all recurrent states and a second set consisting of the transient states. It is possible that the recurrent states consist of more than one closed set, which means submatrices from which there is no escape. For example, the matrix:

$$P = \begin{bmatrix} \begin{bmatrix} 0.2 & 0.8 \\ 0.3 & 0.7 \end{bmatrix} & 0 & 0 \\ & 0 & 0 \\ 0 & 0 & \begin{bmatrix} 0.6 & 0.4 \\ 0.5 & 0.5 \end{bmatrix} \\ 0 & 0 & \end{bmatrix}$$

consists of two closed sets because it is impossible to enter state 3 or 4 from either state 1 or 2. As an example of a finite Markov chain with two closed sets of recurrent states and one set of transient states, consider the 6 × 6 matrix in which all non-zero transition probabilities are marked by an asterisk:

$$P = \begin{bmatrix} * & * & * & * & * & * \\ 0 & * & 0 & 0 & 0 & * \\ * & * & * & * & * & * \\ 0 & 0 & 0 & * & * & * \\ 0 & 0 & 0 & * & * & 0 \\ 0 & * & 0 & 0 & 0 & * \end{bmatrix}$$

Matrix **P** is not easily interpreted in the presented form. However, by relabelling the states s in the following way, $s_1 = s_6{}'$, $s_2 = s_1{}'$, $s_3 = s_5{}'$, $s_4 = s_3{}'$, $s_5 = s_4{}'$, $s_6 = s_1{}'$; we obtain the much more intelligible matrix:

$$P = \begin{bmatrix} * & * & 0 & 0 & 0 & 0 \\ * & * & 0 & 0 & 0 & 0 \\ 0 & 0 & * & * & 0 & 0 \\ 0 & 0 & * & * & 0 & 0 \\ * & * & * & * & * & * \\ * & * & * & * & * & * \end{bmatrix}$$

This new matrix can be seen to consist of five matrices $\mathbf{R}_1$, $\mathbf{R}_2$, $\mathbf{T}$, $\mathbf{A}$, $\mathbf{B}$, in the following way:

$$\mathbf{P} = \begin{bmatrix} \mathbf{R}_1 & 0 & 0 \\ 0 & \mathbf{R}_2 & 0 \\ \mathbf{A} & \mathbf{B} & \mathbf{T} \end{bmatrix} \qquad [5.26]$$

Clearly, $\mathbf{R}_1$ and $\mathbf{R}_2$ represent two closed sets with recurrent states. Matrix $\mathbf{T}$ comprises the transient states and the matrices $\mathbf{A}$ and $\mathbf{B}$ represent the transitions from the transient states into the two closed sets.

In the five submatrices, any state belonging to the submatrix can communicate with any other state belonging to the same submatrix. A transition matrix which has this property of communication is called irreducible and all matrices can be decomposed into such irreducible chains.

5.5 THE EIGENVALUES OF A MARKOV MATRIX

It has been seen in the previous section that one of the most efficient ways of examining a Markov matrix is to investigate the higher transition probabilities given by $\mathbf{P}^n$ and, in particular, the behaviour of $\mathbf{P}^n$ as n goes to infinity. The powering of transition matrices can be achieved by numerical methods, and in particular with the aid of computers, but it is often useful to find a more general expression for $\mathbf{P}^n$. It is known from matrix algebra that if an $n \times n$ matrix has n distinct eigenvalues it can be diagonalized as:

$$\mathbf{P} = \mathbf{U}\Lambda\mathbf{V} \qquad [5.27]$$

In this, $\mathbf{U}$ is a matrix consisting of the right eigenvectors and $\mathbf{V}$, which is the transposed inverse of $\mathbf{U}$, is the matrix of the left eigenvectors. The diagonal matrix:

$$\Lambda = \begin{bmatrix} \lambda_1 & 0 & 0 & . & . \\ 0 & \lambda_2 & 0 & . & . \\ 0 & 0 & \lambda_3 & . & . \\ . & . & . & . & \lambda_n \end{bmatrix} \qquad [5.28]$$

contains the n eigenvalues $\lambda_1 ... \lambda_n$. The diagonalized form of $\mathbf{P}$ as shown in [5.27] is known as the spectral representation of $\mathbf{P}$. Because $\mathbf{V}$ is the transposed inverse of U, the product $VU = I$ is the identity matrix. For example: $P^2 = U\Lambda V \cdot U\Lambda V = U\Lambda^2 V$, and in general:

$$P^n = U\Lambda^n V \qquad [5.29]$$

The spectral representation, however, is not only a convenient way to calculate the higher transition probabilities, but it can be used to obtain direct information about the Markov chains.

It is immediately clear that any stochastic matrix must possess an eigenvalue $\lambda = 1.0$, because if $[1]$ is a column vector of 1's then $P[1] = [1]$. This means that $[1]$ is a valid eigenvector with $\lambda = 1.0$ as the corresponding eigenvalue. By making use of the Frobenius theorem of non-negative matrices, it can be shown that no eigenvalue can exceed 1 and, indeed, the multiplicity of the eigenvalue 1 equals the number of recurrent sets in the original matrix P. Consequently, if a transition matrix has only one eigenvalue of 1 then it belongs to the class of positive-recurrent aperiodic chains, or more briefly, the Markov process is ergodic. It has been shown by the limit theorem ([5.23] and [5.24]) that $p_{jk}{}^n$ $(n \to \infty) = \pi_k$ and, because of [5.13], Π is a possible left eigenvector corresponding to the unit eigenvalue. Thus, an ergodic Markov chain must have a spectral representation of the following form:

$$P = \begin{bmatrix} 1 & u_{12} & u_{13} & . \\ 1 & u_{21} & . & . \\ 1 & . & . & . \\ . & . & . & . \end{bmatrix} \begin{bmatrix} \lambda_1 = 1 & 0 & 0 & . \\ 0 & \lambda_2 < 1 & 0 & . \\ 0 & 0 & \lambda_3 < 1 & . \\ . & . & . & . \end{bmatrix} \begin{bmatrix} \pi_1 & \pi_2 & \pi_3 & . \\ v_{21} & v_{22} & . & . \\ v_{31} & . & . & . \\ . & . & . & . \end{bmatrix}$$

$$[5.30]$$

The first right eigenvector is always $[1]$, the first eigenvalue is unity and the first left eigenvector is Π.

In calculating the matrix P^n one can write for the first element:

$$p_{11}^{(n)} = u_{11}\lambda_1^n v_{11} + u_{12} v_{21}\lambda^n + u_{13} v_{21}\lambda_3^n + \ldots \qquad [5.31]$$

Since the first term in this sum is π_1 and all λ's except the first are smaller than one, then the series converges towards π_1 as n goes to infinity. Sometimes, it is possible that the remaining eigenvalues are either negative or complex but the modulus $|\lambda| < 1$ and convergence is always certain. If the matrix contains any negative or complex eigenvalues, then the higher-order transition probabilities will approach the stable vector by a series of damped oscillations. This is seen by using again the first element as an example. If λ_2 and λ_3 are complex, then:

$$p_{11}^{(n)} = \pi_1 + C_1|\lambda_2|^n(\cos n\,\delta_1 + \sin \delta_1) + C_2|\lambda_3|^n(\cos n\,\delta_2 + \sin \delta_2) + \ldots$$

$$[5.32]$$

Here, the C's are complex constants arising from $\mathbf{U}$ and $\mathbf{V}$ and $|\lambda|$ is the modulus of λ which determines the damping of the oscillations. This oscillating behaviour of some stratigraphic chains is of particular interest in the study of cyclic sedimentation and will be returned to later.

A further detailed analysis of the transition matrix in its spectral representation (cf. Karlin, 1966) gives the following important result. If a Markov chain is irreducible and has more than one eigenvalue of modulus 1, then it must be periodic and the number of eigenvalues with modulus 1 corresponds to the period τ. It is self-evident, but often ignored in the analysis of geological data, that the length of period τ cannot exceed the number of states in the transition matrix (Schwarzacher, 1969). In general, a periodic chain can be brought into the following form:

$$\mathbf{P} = \begin{bmatrix} 0 & \mathbf{P}_1 & 0 & . & . \\ 0 & 0 & \mathbf{P}_2 & . & . \\ . & . & . & . & . \\ \mathbf{P}_\tau & 0 & 0 & . & . \end{bmatrix}$$

$$[5.33]$$

in which the matrices $\mathbf{P}_1$ to $\mathbf{P}_\tau$ are square matrices with unity row sums. If the chain has the period τ, then there are precisely τ submatrices. In the simplest case of a periodic chain, $\mathbf{P}_1 \ldots \mathbf{P}_\tau$ are 1×1 and matrix $\mathbf{P}$ takes on the form:

$$\mathbf{P} = \begin{bmatrix} 0 & 1 & 0 & 0 & . & . \\ 0 & 0 & 1 & 0 & . & . \\ . & . & . & . & . & . \\ 1 & 0 & 0 & 0 & . & . \end{bmatrix}$$

$$[5.34]$$

Clearly in this case one is dealing with a deterministic process.

Representing the transition matrix as in [5.31] and [5.32], shows that each individual transition probability can be understood as the sum of r

components, where r is the number of states in the process. Exactly the same transformation is used in multivariate analysis, where it is known as principal-component analysis. The components are linearly independent and it is the aim of the multivariate analysis to find some physical meaning for the components. It is tempting to expect similar results from an eigenvalue analysis of a transition matrix, but the interpretation of such components is difficult in practice. The only exception, of course, is the component corresponding to the unit eigenvalue which, as a stable probability vector, has a very real geological meaning. Even if it is often impossible to attribute a meaning to the components, the method can, nevertheless, be used to find a simplified model of the process. Such a simplification may be based on the principle that the absolute of the eigenvalue $|\lambda_n| < 1.0$ determines the rate of convergence for the elements $p_{jk}{}^{(n)}$. If such eigenvalues are very small, they do not contribute much and the simplified model may be based on a reduced matrix which eliminates the states corresponding to small eigenvalues.

5.6 HIGHER-ORDER CHAINS

The Markov chain has by definition (see [5.2]) a single-step dependence which means that the state at time n depends only on the state which was occupied at time $n-1$. However, if real stratigraphic sequences are analysed, it is frequently found that they are realisations of processes which depend on more than one of the preceding steps. For example, a chain may have the following structure:

$$P\{X_n = k | X_0 = h \ ..., \ X_{n-2} = i, \ X_{n-1} = j\} \qquad\qquad [5.35]$$

indicating that X_n depends not only on X_{n-1} but also on X_{n-2}. Such a higher-order chain can be transformed into a single-step dependent Markov chain by redefining the state space. Let X_n be a two-state process with the property:

$$P\{X_n = k | X_{n-2} = j, \ X_{n-1} = j\} \neq P\{X_n = k | X_{n-2} = k, \ X_{n-1} = j\}$$

The event $X_n = 1$, meaning that the system is in state 1, could follow either event $X_{n-2} = 1$, $X_{n-1} = 1$; or $X_{n-2} = 1$, $X_{n-1} = 2$; or $X_{n-2} = 2$, $X_{n-1} = 1$; or $X_{n-2} = 2$, $X_{n-1} = 2$. The four different situations before time n show that there are really four states which can lead to event $X_n = 1$ and these four states are written as, 11, 12, 21, and 22. Using this expanded state space, one can rewrite the transition probability matrix:

$$
\mathbf{P} = \begin{array}{c} \\ 11 \\ 12 \\ 21 \\ 22 \end{array}
\begin{array}{cccc}
11 & 12 & 21 & 22 \\
\left[\begin{array}{cccc}
* & * & 0 & 0 \\
0 & 0 & * & * \\
* & * & 0 & 0 \\
0 & 0 & * & *
\end{array}\right]
\end{array}
\qquad [5.36]
$$

In this matrix, rows are labelled according to the state which occurred at times $n-2$ and $n-1$. Columns are labelled according to the states at times $n-1$ and n. The expanded matrix has again the Markov property and refers strictly to one-step transition probabilities.

If such higher-order dependencies are not recognized or (as some workers have recommended) they are deliberately ignored, then some very misleading results can be obtained. There are statistical tests, which will be discussed in the next section, which can help to recognize higher-order dependencies.

5.7 THE ESTIMATION OF TRANSITION PROBABILITY MATRICES AND TESTS FOR THE MARKOV PROPERTY

The transition probability matrix can be estimated without difficulty if an observational sequence of identified states is available. It is only necessary to

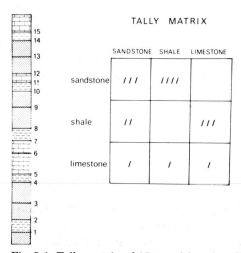

Fig. 5.1. Tally matrix of 15 transitions in a fictitious section.

TABLE 5.1

Tally matrix and estimated transition probability matrix for 309 transitions, from the Chester Formation (after Krumbein, 1967)

Tally matrix						Transition probability matrix		
	sand	shale	lime	totals		sand	shale	lime
sand	59	18	2	79	sand	0.74	0.23	0.03
shale	14	86	41	141	shale	0.10	0.61	0.29
lime	4	34	51	89	lime	0.05	0.38	0.57
totals	77	138	94	309				

obtain a count of the various types of transitions which occur in the series. If the system consists of s states, then a total of s^2 different transitions is possible, which can be written into an $s \times s$ matrix N with the elements n_{ij}. Fig. 5.1 illustrates the counting procedure for a three-state system of a fictitious section which consists of beds of sandstone, shale and limestone. Each bed is taken as one step in the Markov chain and the transitions are counted by proceeding from the lowest to the highest bed. Krumbein (1967) has called the transition frequency matrix N the tally matrix and clearly, summing the individual tallies n_{ij} either by rows or by columns must give the total number of observed transitions. For example, in a section of the Chester Formation (Upper Mississippian, Illinois) Krumbein (1967) found 309 lithological transitions and his results are given as a tally matrix in Table 5.1. The section was classified into three states: sandstone, shale and limestone and it was assumed that transitions occurred at intervals of eight feet. The row sums $\Sigma_{j=1}^{s} n_{ij} = n_i$ and the column sums $\Sigma_{i=1}^{s} n_{ij} = n_j$, individually add up to $n = 309$ which is the total number of observations. The transition probabilities p_{ij} may be estimated from these data by forming:

$$\hat{p}_{ij} = n_{ij}/n_i \qquad\qquad [5.37]$$

and the transition probability matrix corresponding to the tally matrix is shown on the right-hand side of Table 5.1. Dividing each tally n_{ij} by the row sum of the tally matrix automatically gives row sums which are equal to 1.0.

The statistical treatment of Markov chains may be based on the following considerations which have been more fully developed by Billingsley (1961). Assume that an observational series $(x_1 ... x_n)$ is indeed the realization of a Markov process which is determined by the transition probabilities p_{ij}. Assume that each of the steps $(x_1 ... x_n)$ has taken on a specific integer value which denotes one of s states, let us say the series $(a_1 ... a_n)$; then n_{ij} is the number of cases for which $a_m = i$ and $a_{m+1} = j$. The probability of obtaining the observed sequence is given by:

$$p_{a_1}p_{a_1 a_2} \cdots p_{a_n a_{n+1}} = pa_1 \prod_{ij} p_{ij}^{n\,ij} \tag{5.38}$$

Using Whittle's formula (see Billingsley, 1961) it can be shown that the probability of the series $(x_1 \dots x_n)$ having the observed tally matrix $\mathbf{N}$ is given by:

$$\prod_i \frac{n_i!}{\prod_j n_{ij}!} \prod_{ij} p_{ij}^{n_{ij}} \tag{5.39}$$

The expression [5.39] is formally the same as for the probability of obtaining s frequency counts $(n_{i1} \dots n_{is})$ in s independent samples from a multinomial population with cell probabilities $(p_{i1} \dots p_{is})$. Corresponding to the multinomial case, one can use the statistic:

$$\sum_j \frac{(n_{ij} - n_i p_{ij})^2}{n_i p_{ij}} \tag{5.40}$$

which is asymptotically χ^2-distributed. The summation in [5.40] must be restricted to those indices j for which $p_{ij} > 0$. If the number of these is d_i, then the number of degrees of freedom for the χ^2 distribution is $d_i - 1$. Furthermore, the statistics for each state are independent, and their sum:

$$\sum_{ij} \frac{(n_{ij} - n_i p_{ij})^2}{n_i p_{ij}} \tag{5.41}$$

has asymptotically, a χ^2-distribution with $d - s$ degrees of freedom. Here, $d = \Sigma_i d_i$ is the number of positive entries in the transition matrix $\mathbf{P}$.

The statistic of [5.41] can be used to test whether an observed tally count $\mathbf{N}$ could have originated from a Markov chain with the probabilities p_{ij}. An obvious application is to test whether an observed sequence could have been

TABLE 5.2

The independent random matrices for the Chester Formation

Tally matrix					Transition probability matrix		
	sand	shale	lime	totals		sand	shale	lime
sand	19.7	35.3	24.0	79	sand	0.2335	0.4480	0.3184
shale	35.1	63.0	42.9	141	shale	0.2335	0.4480	0.3184
lime	22.2	39.7	27.1	89	lime	0.2335	0.4480	0.3184

sampled from an independent random process. For this, it is necessary to find the appropriate random matrix, that is, a matrix which has the stable probability vector as rows for each state. It has been seen earlier ([5.15]) that this can be achieved by powering matrix $\mathbf{P}$ and $\mathbf{P}^n$ with $n = 43$ of the Chester Formation data (Krumbein, 1967) is given in Table 5.2. Alternatively, the tally matrix can be directly randomized (Selley, 1970). If one could shuffle all the lithologies concerned, then the expected number of transitions E, from lithology i to lithology j is given by:

$$E_{ij} = \frac{n_i n_j}{n}, \quad n = \sum_i n_i = \sum_j n_j \tag{5.42}$$

Applied to the Chester data, one finds for the first element of the randomized tally matrix $77 \times 79/309 = 19.7$. If this value is divided by the row sum of the tally matrix, one obtains again the stable probability for sandstone. Small discrepancies between the values are due to rounding errors.

In applying, next, the χ^2-test for independence one has to form the sum of [5.41], which leads to a $\chi^2 = 153.8$. Since there are no zero-elements in the random matrix, this is to be entered with six degrees of freedom and the hypothesis of independence can be rejected with confidence.

The following test, which is specifically designed to examine the order of a Markov chain, is due to Anderson and Goodman (1957). This test assumes that all entries into the transition matrix are positive. Consider, at the start, a second-order chain (cf. [5.35]). The probability $p_{ijk}(t)$ is the probability that the system is in state k at time t, provided that it has been in state j at time $t-1$ and in state i at time $t-2$. If the transition probabilities are stationary, one may obtain an estimate of p_{ijk} from:

$$\hat{p}_{ijk} = n_{ijk} \left/ \sum_{l=1}^{s} n_{ijl} = \sum_{t=2}^{T} n_{ijk}(t) \right/ \sum_{t=2}^{T} n_{ij}(t-1) \tag{5.43}$$

One may next test the null hypothesis that the chain is first-order, against the alternative that it is second-order. The null hypothesis is that $p_{1jk} = p_{2jk} = \ldots p_{ijk} = \ldots p_{sjk} = p_{jk}$ for a fixed j and $k = 1\ldots s$. The likelihood-ratio criterion for testing this hypothesis is:

$$\lambda = \prod_{ijk} (\hat{p}_{jk}/\hat{p}_{ijk})^{n_{ijk}} \tag{5.44}$$

which is more easily computed in the form:

$$-2 \log \lambda = 2 \sum_{ijk} n_{ijk} \log (\hat{p}_{ijk}/\hat{p}_{jk}) \tag{5.45}$$

Under the null hypothesis, $-2 \log \lambda$ has an asymptotic χ^2-distribution with $s(s-1)^2$ degrees of freedom. The same test was used by Krumbein (1967) for the null hypothesis that the sample comes from a zero-order Markov process (meaning an independent random process) with the alternative that it is derived from a first-order Markov process. Using the statistic [5.45], one can write:

$$- 2 \log \lambda = \sum_{ij} n_{ij} \log (\hat{p}_{ij}/\hat{p}_j) \qquad\qquad [5.46]$$

where $\hat{p}_j = n_j/n$, n being the total number of transitions. As before, $-2 \log \lambda$ is χ^2-distributed with $(s-1)^2$ degrees of freedom. The same test can be generalized for Markov chains of any order and the λ statistic can be usefully employed for investigating the probability structure of chains which may have quite complicated dependency structures (Schwarzacher, 1968).

To prove the Markov property of an observational series, one needs usually more than one test. If one has no a-priori knowledge of the process, then it is obviously best to proceed systematically and decide first the alternative between zero-order and first-order process, first-order and second-order, and so on. There is very little hardship in testing for higher-order dependencies, since the process can be automated on computers and it is certainly quite insufficient to establish only that an observational series is not a random sequence. This latter case is so unlikely in a stratigraphic sequence that one hardly has to go to the trouble of estimating transition matrices in order to disprove it. Anderson and Goodman (1957) have shown that the χ^2-tests ([5.41]) and the likelihood-ratio tests ([5.44]) are mathematically equivalent. The latter is, perhaps, slightly easier to compute, but it requires that all entries p_{ij} are positive and for this reason only some workers have preferred the χ^2-test.

Vistelius and Faas (1965) chose a slightly different approach to the testing problem and their procedure is as follows. At first, maximum-likelihood estimates of the one-step transition matrix $\hat{P}_1$ are obtained in the normal way. Secondly, an estimate of the two-step transition probabilities $\hat{P}_2$ are obtained by counting transition in steps of two. Now, if a series conforms to first-order Markov chain process, then:

$$\mathbf{P}_1^2 = \mathbf{P}_{T1}(t-2, t-1)\, \mathbf{P}_{T2}(t-1, t), \quad \text{provided } \mathbf{P}_{T1} = \mathbf{P}_{T2}$$

This means that the transition probability at times $t-2$ to $t-1$ is the same as the transition probability at times $t-1$ to t. If a series is second or higher order, then $\mathbf{P}_{T1} \neq \mathbf{P}_{T2}$, and the two-step probability $\mathbf{P}^2$ (which is still given by the product of the two matrices $\mathbf{P}_{T1}$ and $\mathbf{P}_{T2}$) will obviously be different from the two-step probability of the first-order process. Therefore, to test for the first-order Markov property, Vistelius compares the calculated $\hat{\mathbf{P}}^2$

with the observed $\hat{P}_2$ matrix. The comparison is carried out by χ^2's using [5.41]. This method of testing is probably more direct than the previously discussed likelihood criteria. However, if relatively high orders of dependency are suspected, then high powers of $\hat{P}$ have to be calculated and, because this is only an estimate, the errors are multiplied and the method soon becomes unreliable. For this reason, Vistelius did not carry the tests beyond the third-order matrix.

Possibly the most important test to be made before applying the Markov model to an observed stratigraphic section, is to investigate the stationarity of the transition probabilities. The stationarity implies that the transition probabilities are constant throughout the observed section, an assumption which from experience is very rarely justified. It is possible, as will be discussed later, to make adjustments for such changes, but in order to do so, it is necessary to recognize the situation. The only way in which stationarity in a single section can be tested is to subdivide a relatively long section into a number of subsections. Each subsection yields an estimate of the transition probability matrix which can be compared with the others by the χ^2-test (Vistelius and Faas, 1965). A reasonable idea of stationarity may be obtained by comparing at first the upper and lower half of a section, then quartering the section and repeating the same procedure. This can be done as often as the data permit.

5.8 STRUCTURING STRATIGRAPHIC SECTIONS FOR A MARKOV-CHAIN MODEL

To apply the Markov chain model or any other discrete probability model, two conditions must be fulfilled. It must be possible to define the section in terms of discrete states and it is further necessary to subdivide the sequence into steps which, ideally, should be related to some geological process. This essential classification process will be called structuring the section.

The definition of states does not usually cause any difficulty. In most geological studies, lithologies are identified with states and the section is simply regarded as a sequence of different rock types. However, other descriptive characters of the sediments could be used and continuous variables can be classified into discrete groups to provide additional states. Once the classification of the section into states has been decided, it must be rigidly adhered to. Obviously, the classification must be complete in the sense that there must be no rock type which is not included in the classification system.

It is sometimes useful to experiment with different classification schemes and it is possible that different combinations of sedimentary characters give results which may be helpful with the geological interpretation. To illustrate this, one may consider quite a trivial example. A section from a coal measure sequence could be represented at its simplest by three states of, let us say, sand, coal and shale. Both sand and shale in this classification may contain characters which enable one to separate a fourth group of seat earth or underclay. The shales could possibly be subdivided into fresh water and marine sediments, and the classification could be made more and more detailed until quite a complicated sequence of rock types results. By starting with a simple model and by gradually increasing the complexity of the model through the redefinition of states, it is sometimes possible to understand systems which, in direct analysis, would be incomprehensible. The grouping of descriptive characters into states and the classification of the observed data according to such states can be automated (Davis and Cocke, 1972), and the transition probability matrices of measured sections can be calculated by using very simple computer programmes.

Whilst it is comparatively easy to carry out the classification of observed sections into states, it is sometimes much more difficult to define the steps or intervals at which the transitions occur. The definition of a Markov chain does not require that the steps between the transitions are of equal length nor is it necessary to know anything about the time spent in a certain state, but the transitions must occur in a consecutive order. For this reason, it is convenient to use the time co-ordinate with its unidirectional increase when discussing the general properties of Markov chains. In stratigraphical analysis, time, however, is very often replaced by a vertical distance. An arbitrary marker horizon is taken as the zero position and any observation is recorded at a vertical distance above the marker; this distance is referred to as thickness. Therefore, any successive transition must have a greater thickness than its predecessor. This being the case, the steps can be numbered as a consecutive sequence.

Many sedimentary sections are naturally subdivided into steps by the occurrence of bedding planes and the single sedimentary bed is, therefore, a possible choice in investigating the transition pattern of sedimentary sections. If such an investigation is directed towards the problems of bed formation (cf. Vistelius, 1949) then beds are the logical choice for step units. In such an approach, the individual thickness of the bed is quite irrelevant and the sequence of consecutively numbered beds does not necessarily relate to the time taken up by its formation.

Other stratigraphic investigations are more concerned with finding a stochastic model which not only describes the sequence of lithologies but also their thicknesses. To obtain such a model, the section can be subdivided into artificial steps of equal or otherwise known thickness intervals (Krumbein, 1967). Most commonly, sections are sampled at fixed intervals ranging from a few centimetres up to several metres; at each sampling spot, the lithology is identified and classified into a state system. The section which is structured in this manner is again a sequence of consecutively numbered states. Because the choice of the step interval is quite arbitrary, a variety of different transition probabilities can be obtained from the same section and the geological implications of such a procedure must be discussed in some detail later.

Finally, some workers have defined a step as any thickness of sediment which has an identical lithology. Misleadingly, the term "embedded Markov chain" has also been used to describe this type of structuring and the method will be discussed more fully.

5.9 STRUCTURING BY THE EQUAL-INTERVAL METHOD

As has been repeatedly stressed in the previous chapters, a model can be regarded either as a purely descriptive tool or as something which is genetically related to the process which it describes. The Markov-chain model is no exception and it can be applied in a descriptive way to any sequence of events regardless of whether such a series has been generated by a Markov process or not. For such descriptive purposes, coding of lithological states at equal thickness intervals seems an appropriate choice and the method needs no further justification other than that simulation experiments based on the model should generate sections which are similar to the observed one. If, however, the Markov chain is regarded as a genetical process model then the step structure must relate to some real physical process and a subdivision into equal intervals can only be justified under two conditions. Either it is assumed that sedimentation is discontinuous, whereby, at possibly quite irregular time intervals, a constant thickness of sediment is deposited (see Fig. 5.2A), or alternatively, it is assumed that sedimentation is continuous and constant, but at equal time intervals, some environmental process decides whether sedimentation should continue or whether a transition should occur (see Fig. 5.2B). The third possibility, that both the time interval and the sedimentation rates are variable but always combine in such a way that they result in steps of precisely equal thickness, is extremely unlikely and can be ignored (see Fig. 5.2C).

In the precise form in which they have just been stated, neither of the two hypotheses A and B are very attractive to the geologist, but if the precision is somewhat relaxed, then both situations are feasible. For example, the inherent quantum-like nature of sedimentation (see chapter 2) may be the basis

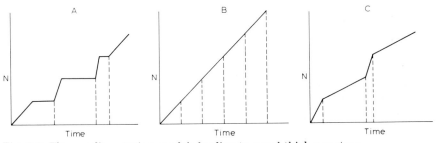

Fig. 5.2. Three sedimentation models leading to equal thickness steps.

for sedimentary steps and one may assume that a repetition of the same sedimentation process produces roughly the same quantity of sediment leading to a step-like accumulation. On the other hand, the argument may be applied to the time dimension. Many sedimentation processes take some time for their completion and this time will be roughly the same for a repetition of the process. Sediment formation, therefore, is determined by a series of discontinuous incidents which take a finite time and, in this way, justify the discrete structuring which is implied in the Markov-chain model. If, as is to be expected, there is some variation in the sedimentation rates or in the length of time between the incidents leading to the natural steps, then this should ideally be adjusted by changing the vertical interval at which the lithological states are sampled. In order to realize the effects which arise from neglecting this, one may consider the following problem. Accepting the basic model of constant sedimentation and a Markov process which operates at equal time intervals implies that a constant thickness of sediment is deposited at each step. However, it is well known that different lithologies have usually different sedimentation rates, which means that, during an identical time interval, different thicknesses of sediment are deposited when the lithology has changed. For example, a three-state system in which all the states are deposited with equal rates, will produce a sequence with the tally matrix N and the associated probability matrix P. Next consider the case in which state A is deposited α times faster than the remaining states. This means for every unit thickness of lithology A in the previous matrix N, α units have to be substituted which then yields the new tally matrix N^*. The first row sum which counts the transitions from state A will be α times larger than that of the original matrix N and, therefore:

$$n_{11}^* + n_{12}^* + n_{13}^* = \alpha n_1 \qquad\qquad [5.47]$$

The number of transitions from i to j, for $i \neq j$, is not affected and, therefore, only the leading diagonal of N^* will be different from that of N. The matrix P^* of the transformed process can be directly derived from P by:

$$p_{ii}^* = 1 - \sum_j \frac{p_{ij}}{\alpha_i} \text{ and } p_{ij}^* = \frac{p_{ij}}{\alpha_i} \text{ , } i \neq j \qquad\qquad [5.48]$$

a transformation which is to be carried out row by row.

Changing the step interval is equivalent to a uniform change in the sedimentation rates. A reduction of the step size which corresponds to increased sedimentation rates always leads to an increase of the diagonal elements p^*_{ii} which, in the limiting case, will become 1.0. Conversely, if the step interval is increased, the transition probabilities in the leading diagonal decrease. If α_i

in [5.48] is taken in such a way that:

$$\sum_j \frac{p_{ij}}{\alpha_i} = 1 \text{ for } j \neq i \qquad [5.49]$$

then the diagonal values become zero.

The practical question arises as to what step interval should be chosen when a section is approximated by a Markov-chain model. This problem was examined in an empirical fashion by Krumbein and Scherer (1970) who analysed well log data from the Oficina Formation of eastern Venezuela. The study was carried out in the following way. Transition probability matrices were computed after taking samples at 2, 3, 5 and 10-feet intervals. Each of the estimated matrices was used to simulate sections. An equal length of simulated run was compared with the actual section. Criteria like the total number of lithologic units, the average thicknesses of the units, and the overall proportion of lithologies in the section, were used to establish how close the simulated sections came to the observed one. The interpolated interval which gave the best representation was chosen for the final simulation studies (Scherer, 1968).

The problem of choosing a step interval is, however, very ill-defined if the model is intended to serve as a genetical process model. The necessary condition for a solution in this case is that the section is, indeed, a sequence which was generated by a Markov process, a hypothesis which cannot be tested unless the step structure is known. If the hypothesis is accepted as true, then there exists only one step size which will give the best representation. The logical test criterium to find the step interval is the stable probability vector which can be compared with the overall lithological composition of the sections. If a section is not the result of a Markov process which operated at discrete step intervals, then the interpretation of the transition probability matrix becomes wide open to speculations. For example, one may assume that the process which determined the environmental history which led to the formation of a certain rock sequence was in fact an independent random process. To use a specific example, three environments which were suitable for producing, say, sandstone (A), shale (B) and limestone (C) may have occurred with equal likelihood; this process would generate a sequence with the transition matrix:

$$
\mathbf{P} = \begin{array}{c} A \\ B \\ C \end{array}
\begin{bmatrix}
\frac{1}{3} & \frac{1}{3} & \frac{1}{3} \\
\frac{1}{3} & \frac{1}{3} & \frac{1}{3} \\
\frac{1}{3} & \frac{1}{3} & \frac{1}{3}
\end{bmatrix}
$$

Now, let it be assumed that the sedimentation rates A:B:C were 3:2:1. The section which is produced in this way would, if structured by the equal-interval method, yield the transition probability matrix $\mathbf{P}^*$ which, from [5.48] can be calculated as:

$$
\mathbf{P}^* = \begin{matrix} A \\ B \\ C \end{matrix} \begin{bmatrix} \frac{1}{3} & \frac{1}{3} & \frac{1}{3} \\ \frac{1}{6} & \frac{4}{6} & \frac{1}{6} \\ \frac{1}{9} & \frac{1}{9} & \frac{7}{9} \end{bmatrix}
$$

The matrix $\mathbf{P}^*$ has completely lost the character of an independent random process and, if the original time intervals were exponentially distributed (see Chapter 4), the sequence would represent a first-order Markov chain (Schwarzacher, 1967).

In a similar way, purely deterministic schemes can be designed which also will lead to a sedimentation record which has Markov-chain properties resulting only from variations in the sedimentation rates. It is essential that the geological assumptions under which an observed section is structured into equal intervals are clearly stated. In most cases, additional geological evidence to justify this method of structuring is very necessary.

5.10 STRUCTURING BY LITHOLOGICAL STATES

A third method of structuring which has been used subdivides the section into thickness intervals which have a uniform lithological composition. Each lithological unit, regardless of its thickness, forms a step in the Markov chain.

Since it is assumed that, once a state has been entered, the next state must be different, no return to the same state is possible. The leading diagonal of the transition matrix must have zero-elements. Matrices of this type can never result from an independent random process and it is, therefore, meaningless to test against such a hypothesis as was possible with the previous data.

Attempts have been made to design tests which are supposed to show that observed lithological sequences are not the outcome of random arrangements. However, such tests incorporate a-priori assumptions about the ordering of lithologies. It is best to discuss a specific example in order to clarify this point. Gingerich (1969) examined a sequence of alluvial sediments from the Paleocene in Wyoming (U.S.A.) and classified the lithology into four states, viz. sandstone, mudstone, lignite and limestone. The tally matrix N and the estimated transition probability matrix **P** are given in Table 5.3. In this study, a total of 180 lithological units have been observed and this

TABLE 5.3

Tally matrix and transition probability matrix for 180 lithological units from alluvial sediments (after Gingerich, 1969)

Tally matrix						Transition probability matrix				
					n_i					
	sandstone	0	37	3	2	42	sandstone	0.00	0.88	0.07	0.05
N =	mudstone	21	0	41	14	76	mudstone	0.28	0.00	0.54	0.18
	lignite	20	23	0	0	45	P = lignite	0.44	0.56	0.00	0.00
	limestone	1	14	1	0	16	limestone	0.06	0.88	0.06	0.00
	Total					179 + 1 sandstone					

includes the lowest unit which happens to be a sandstone. Gingerich's arguments must have been similar to the following. Out of the 180 observed units, 43 are sandstones, i.e., a sandstone could be succeeded by $180 - 43 = 137$ different units, of which 76 are mudstone. The chance that the unit which follows a sandstone is, indeed, a mudstone, is calculated as $76/137 \simeq 0.55$. In a similar way, the chance transitions into any other lithology may be calculated and summarized in the matrix E^* (see Table 5.4). Gingerich, and, in a similar way, Doveton (1971) call this an independent trial matrix. This term is somewhat misleading, because it was specified in calculating this matrix that transitions from one state into an identical state are impossible. This is a situation that cannot be realized by any random experiment. If, for example, all the lithologies were mixed in an urn and drawn out at random, one will obtain the transition matrix E (see Table 5.4). The matrix has the stable probability vector as rows which are identical, as should be the case in any independent random matrix. Matrix E could be calculated from matrix P by powering or, more simply, from the overall composition of the section. For example, the number of sandstones divided by the total number of units is $43/180 \simeq 0.2388$. Matrix E^* can be derived from matrix E making use of [5.48]. According to [5.49], the factors $\alpha_1 ... \alpha_4$ are: 0.791, 0.578, 0.750 and 0.911 and these will give the transformation leading to E^*.

TABLE 5.4

Independent random matrix E and its transformation E^* for alluvial sediments (after Gingerich, 1969)

Random process					Transformation with zero diagonal			
	0.2388	0.4222	0.2500	0.0888		0.00	0.55	0.33	0.12
E =	0.2388	0.4222	0.2500	0.0888	E^* =	0.41	0.00	0.43	0.16
	0.2388	0.4222	0.2500	0.0888		0.31	0.57	0.00	0.12
	0.2388	0.4222	0.2500	0.0888		0.26	0.47	0.27	0.00

It is very difficult to think of any geological justification for carrying out this transformation except when one is dealing with incomplete data. It is quite possible that transitions between identical states are simply not recognized when measuring a section and, consequently, they are not recorded; in this case, one cannot estimate the probability structure of the sequence because of missing data.

That the thickness of the lithological steps is completely irrelevant in this type of structuring, needs some consideration. To obtain any benefit from such a model, one has to ensure that the lithological steps (and by implication the various episodes of sedimentation) are somehow genetically commensurate. This means that the steps should apply to events of roughly equal importance. Doveton (1971) achieved this by introducing a minimum thickness which a lithological unit must reach before it qualifies as a state in the Markov-chain model. In doing this, small-scale and presumably rapid alternations, such as silt—sand or shale—silt laminations, are ignored and the model concentrates on lithologies which are sufficiently established to represent important episodes in the history of the sediment. Just as the sampling interval in the previous method of structuring depends on sedimentation rates, so the minimum thickness also relies on geological judgement. Certain lithologies, such as rootlet beds in Doveton's study of coal measure sections, can be regarded as so important that they are recorded, regardless of their thickness. Structuring a section by equal lithologies can, therefore, hardly be regarded as an objective method. As in the equal-interval procedure, a good deal of preconceived geological ideas must enter the model. This is not necessarily undesirable, but it also means that such models will behave in many respects quite predictably. For example, Lumsden (1971) who analysed a "cyclic" carbonate sequence in terms of Markov chains, defines the states as a system of various facies which ultimately can be interpreted in terms of energy level of water depth. It is found that the episodes of varying water depth are related to each other and the model suggests the not unexpected result, that if a deep-water environment changes into shallow water, it has to pass through a state of intermediate water depth. Since the model only refers to the sequence of facies, very little else can be learned from it.

Finally, in assessing the three methods of structuring geological sections, it becomes clear that the Markov-chain model is best suited to those stratigraphical sequences which contain a natural step structure. Sequences which contain sedimentary bedding, have such a structure and, therefore, they can be examined directly in terms of Markov chains.

The ultimate aim of stratigraphic models would seem to be a sedimentation model which is related to time and for this reason the equal-interval method of structuring has been used. The method, however, makes assumptions about sedimentation processes and sedimentation rates which may be difficult to establish in geological terms.

Structuring, according to states with different lithologies, provides models with a considerably reduced information content. Like the equal-interval method, this way of structuring makes a-priori assumptions about sedimentation processes and by implication also about sedimentation rates.

5.11 THE MARKOV PROPERTY AND SEDIMENTATION

In order to appreciate the Markov-chain process as a stratigraphical model, it is useful to review some of the work which has been published on the subject. Particular emphasis will be given to the geological interpretation of observed sections in order to see whether the Markov property in stratigraphical sequences implies some special conditions of sedimentation.

The Markov-chain model was first introduced by Vistelius (1949) who investigated the problem of bed formation in flysch-type deposits. The series which he examined consisted essentially of three lithologies (sand, silt and shale) which were naturally subdivided by sedimentary bedding. Vistelius did not give any detailed sedimentological description or definition of the "beds" but they were apparently established during field work and must, therefore, correspond to sedimentation units or beds as normally recognized by mapping geologists. Each bed was classified into one of the three lithological states and the consecutively numbered beds represent a sequence for which a transition probability matrix can be estimated. Vistelius and Faas (1965) established that the observed series has the properties of a stationary ergodic Markov chain.

The geological evidence suggests that the sediments were deposited from turbidity currents and Vistelius tries to show that the Markov property is inherent with this mechanism of sedimentation. His arguments (A.B. Vistelius, personal communication, 1967) are as follows: assume a suspension cloud consists of a mixture of various grain sizes and that by some random process the quantity corresponding to the thickness of a bed is withdrawn from the mixture. It is assumed that the size fraction which is deposited at anyone time is fairly uniform and that its removal will thus change the composition of the remaining suspension. The deposition of a particular bed will, therefore, determine to a certain extent the composition of future beds; however, any beds which have been deposited earlier have no effect on the subsequent composition of the suspension. The process clearly has the Markov property. One may question whether the developed model is an adequate representation of turbidite sedimentation, but as far as the theory goes it shows that the mechanism incorporates the Markov property and, consequently, any rock sequence deposited by this process must follow the Markov scheme. Observation has shown this to be the case and the hypothesis of turbidite formation is, therefore, acceptable. Vistelius is very careful not to suggest that the presence of Markov properties proves an origin from turbidity currents and, indeed, he has analysed examples of terrestrial deposits (Vistelius and Faas, 1965) in which a different mechanism must have been responsible for the Markov property of the observed sections.

Subsequent papers making use of the Markov model as applied to stratigraphical problems, may be roughly subdivided into two groups. Papers by Carr et al. (1966), Potter and Blakely (1967) and Krumbein (1967) used the discrete Markov model in an attempt to simulate stratigraphical sections, sometimes making reference to observed sections. This early approach was based on the belief that, if one can produce an artificial section which "looks" like a real record, one may learn something about its origin. This proved to be rather optimistic assumption but, nevertheless, such simulation experiments led to the development of sedimentation models which were an improvement over either deterministic or independent random models. One particular study (Scherer, 1968) indicates that such simulation procedures may, indeed, have practical applications in predicting well log data.

The second group of papers, Gingerich (1969), Read (1969), Selley (1970), Doveton (1971) and Lumsden (1971), are specific applications of Markov-chain models to observed sections. The main aim of these investigations appears to be to demonstrate that so-called sedimentary cycles can be generated by random processes of the Markov-chain type. The Markov model is essentially used to describe sedimentation patterns in a vertical sequence, a problem which will be discussed more fully in a later chapter. At this stage, it is interesting that no stratigraphical section has been observed so far which could be attributed to an independent random process. Although only Doveton (1971) has tested his data for the Markov property, it is clear that the observations of the remaining workers could be reduced to first-order Markov chains if tests should prove this necessary. Most authors have commented on the geological reasons for the non-randomness of the observed sections.

Gingerich (1969) examined the sequence which has already been mentioned, of alluvial sediments (see Table 5.3) in which there are preferred transition probabilities from sandstone to mudstone, mudstone to lignite and lignite to sandstone. Essentially, this is attributed to an upwards fining cycle and, according to Gingerich, is most probably generated within the sedimentary system. Using a term introduced by Beerbower (1964) the cycles are called "autocyclic". This means that their formation does not require any change in the total energy or the material input of the sedimentation system, but only involves the redistribution of these elements within the system. For example, channel migrations and diversions or barrier migrations could produce such redistribution of material and energy. The changes of lithology are probably determined by the rules of sedimentary transport and, in this sense, Gingerich follows to a certain extent Vistelius (see earlier) when explaining the Markov property of the recorded sequence.

Both Read (1969) and Doveton (1971) examined Carboniferous coal-bearing strata, which, as is generally accepted, formed in a deltaic environment. Read observed in his data high transitions from mudstone to siltstone, and from siltstone to sandstone, which constituted a general upwards coarsening sequence which follows the coal seams. This is tentatively interpreted as the advance of possibly quite local delta lobes over mudstones which are rich in organic components. Once again, the mechanism of sedimentation is resolved into a transport problem and the Markov property of the sequence is a consequence of the lateral migration of subenvironmental facies belts. Doveton found, in addition, that following a seat-earth or rootlet bed, sedimentation gradually decreases in grain size and the following inset of renewed clastic sedimentation, which is Read's coarsening sequence, is gradual. Doveton suggests that the "programming" of such sedimentary sequences could be the result of the gradual establishment and disappearance of a vegetation cover. When vegetation is on the increase, the transport of clastic material decreases and is resumed when vegetation is again on the decrease. The Markov property of the sedimentary succession is caused simply by the inertia of environmental processes. This means that once a vegetative cover has been initiated, it will continue to grow and decline over a certain time, and thus influence all sedimentation processes which occur during this time.

It is obvious that there must be many environmental processes which have this general property of inertia and which will persist for certain time periods. Consequently, there will always be elements in an environmental system which progress in a "programmed" manner. For example, the development of a drainage system, a period of glaciation, or the rise and fall of a small ecological community, could lead to Markov structures in the sedimentary record. Indeed, it would be surprising if one did not find this structure in sediments if one searched for it.

To obtain some real benefits from the transition probability analysis, it will be necessary to develop much more specific models which are geared to special problems of sedimentation or stratigraphy, such as has been seen in the classical study of Vistelius.

Most other geologists have attempted to analyse very complex systems in terms of Markov chains. Because the Markov process is so commonly realized in nature, it is not surprising if what one may learn from very general Markov models is comparatively trivial, if not predictable.

The next chapter will discuss models which are based on the Markov process but which are slightly more complex and can, therefore, be adopted as more specific sedimentation models.

Chapter 6

Renewal processes and semi-Markov processes

6.1 INTRODUCTION

The independent random process was introduced in Chapter 4 as a model for the phrase "no pattern", and it was argued that this model is often applied when there is no information about the relationships between the various components in a geological system. Thus, in the basic stratigraphic problem of linking time with the accumulated sediment, both quantities are regarded as independent random variables. It was found on further examination (Chapter 5), that independent random processes are probably not very common in nature and the Markov process was recognized as the simplest model for a more realistic probability structure. At the same time, it became clear that the Markov property is so universally present in natural processes, that its mere existence is of relatively little geological significance and the simple Markov model must be very general. Improved models ought to explain more than just one aspect of an observed stratigraphic section and therefore more specific mathematical models must be developed.

The development of models by making them more specific, is well illustrated by the work of Krumbein and Dacey (1969) and Dacey and Krumbein (1970) who have drawn attention to the following. The Markov-chain model is primarily concerned with the sequence of states. Information about the thickness of the lithological units can only be obtained by structuring the section into equal intervals. This structure may be artificial but providing it has the Markov property one can calculate the thickness distribution of lithological units from such data. A lithological unit, L_i, is any part of the stratigraphic section which has a uniform lithological composition. For example, if i represents a sandstone, then the thickness distribution of the lithological sandstone unit is given by the probability that the system stays for $1,2...k$ steps in the sandstone state. The probability that a sandstone has thickness k is given by:

$$P(L_i = k) = p_{ii}^{(k-1)} \; p_{ij} \; (k = 1,2 \; ...) \hspace{2cm} [6.1]$$

This clearly implies that all lithological units which are generated by the Markov model must be geometrically distributed. It is, therefore, possible to compare the observed thickness distributions with the theoretically calculated ones and if they agree, one can have reasonable confidence in the model, but if they do not agree, then the model will have to be modified. The possibility of testing more than one aspect of the observed section, that is the sequence of rock types and their thickness distribution, makes the model automatically more specific.

Vistelius and Faas (1965) and Schwarzacher (1968, 1972) attempted to make models more specific by considering the actual mechanism of sedimentation. It is argued that the formation of a stratigraphic section may be regarded as a two-stage process. The environment determines the type of lithology which can be deposited at any time, potentially, but the actual process of sedimentation, such as various sedimentation rates (Schwarzacher, 1969) or erosion within the section (Vistelius and Feigelson, 1965), will modify the record which is preserved in the stratigraphic section. In practice, this means that two models are used. The first is intended to represent the environmental history and the second model contains any data which refer to the sedimentation process in the system. Once again, more data or assumptions can be introduced by this method and the model therefore becomes more specific.

The studies by Dacey and Krumbein (1970), Vistelius and Faas (1965) and Schwarzacher (1972) show, as will be seen, a very pleasing convergence towards similar models and this can only indicate that real progress in understanding the structure of stratigraphic sections is being made. Further improvements, it seems, will come from even more detailed sedimentation models and, therefore, it is useful to return to the probabilistic aspect of sedimentation processes.

A number of stochastic sedimentation models were discussed in Chapter 4 but the Poisson process was regarded as the most basic model. Random processes, which have properties similar to the Poisson process, are known as renewal processes and these will be discussed in the following section.

6.2 RENEWAL PROCESSES

Renewal theory originated from the study of problems connected with the failure and replacement of components (Cox, 1962). The following situation may be considered as an example. A new component, for example a light bulb, is installed at time zero. This component fails at time X_1 and is immediately replaced by a second component which has the failure time X_2 and so on. The failure times which represent the times between renewals, are random variables with a probability density distribution of $f(x)$. One is obviously dealing with the already familiar Poisson process if the failure times are negative exponentially distributed. The process is called a renewal process whenever the failure times are independently and identically distributed and $f(x)$ can therefore be any valid probability density function.

Associated with this ordinary renewal process, are a number of random variables which are of practical interest. These are in particular, the time up to the r-th renewal S_r and the number of renewals in a fixed time interval t, which is written as N_t.

The time up to the r-th renewal must be the sum of all the failure times $X_1 + X_2 + \ldots X_r$ and:

$$S_r = X_1 + X_2 + \ldots X_r \tag{6.2}$$

If $k_r(x)$ is the probability density of the random variable S_r with the cumulative distribution function $K_r(x)$, then it will have the transform:

$$k_r(s) = [f(s)]^r \tag{6.3}$$

This means that $k_r(x)$ is the r-fold convolution of $f(x)$, the distribution of the failure time. We have already seen that in the case of the Poisson process $k_r(x)$ will be the special Erlang distribution with r stages (see [4.41]). If the distribution of failure times $f(x)$ is of the special Erlang type with k stages, then $k_r(x)$ will be a special Erlang distribution with kr stages. The latter is an example of a renewal process which is not a Poisson process.

The number N_t of renewals in a fixed time can be found by using the connection between N_t and S_r. Clearly the number of renewals:

$$N_t < r \text{ if and only if } S_r > t$$

Therefore

$$P(N_t < r) = P(S_r > t) = 1 - K_r(t) \tag{6.4}$$

where $K_r(t)$ is the cumulative distribution function of S_r. It follows that:

$$P(N_t = r) = K_r(t) - K_{r+1}(t) \tag{6.5}$$

with $K_0(t) = 1$. Applying [6.5] to the already familiar Poisson process, where it is known that S_r has the special Erlang distribution with r stages, one finds:

$$K_r(t) = 1 - \sum_{i=0}^{r-1} \frac{(e^{-\rho t})^i}{i!} \tag{6.6}$$

and therefore:

$$(N_t = r) = e^{-\rho t} \frac{(\rho t)^r}{r!} \tag{6.7}$$

which is the Poisson distribution as it should be.

It is sometimes useful to introduce the probability generating function of N_t and if one writes:

$$G(t, \xi) = \sum_{r=0}^{\infty} \xi P(N_t = r) \qquad [6.8]$$

then according to [6.5]:

$$G(t, \xi) = 1 + \sum_{r=1}^{\infty} \xi^{r-1} (\xi - 1) K_r(t) \qquad [6.9]$$

The Laplace transform of $k_r(t)$ with respect to t may be written as $k_r^*(s)$ and consequently the transform of $K_r(t) = k_r^*(s)/s$. Applying the Laplace transformation to [6.9], one can write:

$$G^*(s, \xi) = \frac{1}{s} + \frac{1}{s} \sum_{r=1}^{\infty} \xi^{r-1}(\xi - 1) k_r^*(s) \qquad [6.10]$$

and since $k_r^*(s)$ equals $[f^*(s)]^r$ for the ordinary renewal process (see [6.3]) then one can write finally for the generating function of N_t:

$$G^*(s, \xi) = \frac{1 - f^*(s)}{s[1 - \xi f^*(s)]} \qquad [6.11]$$

To illustrate the use of the generating function, one may calculate the number of renewals in an interval which in itself is randomly distributed. One may think of a line, which could represent time, along which points are scattered at random (see Fig. 6.1). It may be assumed that the distance between the points, which represent renewals, are Poisson distributed with density ρ. The time interval $(0, T)$ over which the incidents are counted is also assumed to be a random variable with distribution $q(t)$. It will be assumed at first, that $q(t)$ is again exponential with the parameter λ. The generating function $G(\xi)$ for renewals $N = 1, 2 \ldots$ within the random time T is:

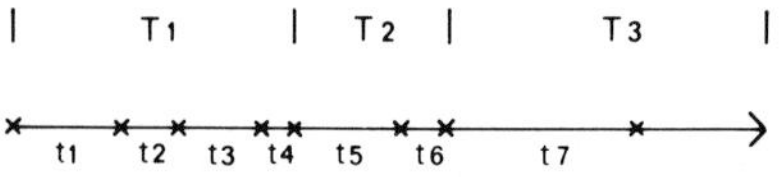

Fig. 6.1. Random points distributed along a line.

$$G(\xi) = \int_0^\infty G(t, \xi) \, q(t) \, dt = \int_0^\infty G(t, \xi) \, e^{-\lambda t} \, dt \qquad [6.12]$$

Comparing [6.12] with [6.8], it may be seen that:

$$G(\xi) = \lambda G^*(s, \xi) \; s = \lambda \qquad [6.13]$$

and by [6.11]:

$$G(\xi) = \frac{1 - f^*(\lambda)}{1 - \xi f^*(\lambda)} \qquad [6.14]$$

If $f^*(s) = \rho/(\rho + s)$ as is the case with Poisson distributed renewals, one can write [6.14] as:

$$G(\xi) = \frac{\lambda}{\lambda - \rho(1 - \xi)} \qquad [6.15]$$

which can be expanded to:

$$G(\xi) = \frac{\lambda}{\lambda + \rho} \sum_{r=0}^{\infty} \left(\frac{\rho}{\lambda + \rho} \right)^r \qquad [6.16]$$

and therefore:

$$P(N = r) = \frac{\lambda}{\lambda + \rho} \left(\frac{\rho}{\lambda + \rho} \right)^r \qquad [6.17]$$

The result of [6.17] could have been obtained in a much more elementary fashion, as will be seen later. However, [6.14] is interesting because it shows that the distribution of N will always be exponential, regardless of $f(t)$, as long as the sampling times T are exponentially distributed. It is interesting to examine next the distribution of N where the sampling time is not exponentially distributed and a particularly interesting case is given when $q(t)$ is of the special Erlangian form. If in this case $f^*(s)$ is again $\rho/(\rho + s)$ then:

$$G(\xi) = \frac{\lambda}{(\rho + \lambda - \xi\rho)^k} \qquad [6.18]$$

which is the generating function of the negative binomial distribution (see Cox, 1962, p. 44).

Two more random variables which will be used later, are the forward and

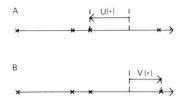

Fig. 6.2. The definition of backwards recurrence time (A) and forwards recurrence time (B).

backward recurrence times of a renewal process. These are defined as follows. Let t be a fixed time point, then the random variable U_t is the backward recurrence time which measures the time from point t to the last renewal which occurred in the process (see Fig. 6.2A). U_t could also be called the age of the component. In a similar way V_t is the forward recurrence time which measures the time from point t to the next renewal (see Fig. 6.2B). In this sense, V_t is the remaining life span of the component. It has been already commented upon that the Poisson process has no memory and, therefore, the components in such a process do not age. This is clearly expressed by the fact that the probability density function of V_t equals the distribution of distances between the renewal points and therefore:

$$f(V_t) = f_t = \rho e^{-\rho t} \tag{6.19}$$

Applications of such renewal processes and their associated random variables will be discussed in the following section.

6.3 SEDIMENTATION AS A RENEWAL PROCESS

The renewal processes which have been introduced in the previous section, are examples of point processes. This name is given to stochastic processes in which the main interest centres around the occurrence of certain events which differ from each other only by their position in time. Therefore, any sedimentation model which considers the process of deposition as a repetition of individual acts of sedimentation, can be simplified into a point process model. The simple random walk, which was introduced as a first stochastic model of sedimentation in Chapter 4, is a typical example. The steps of deposition in this model are the point events which are geometrically distributed in discrete time (see [4.22]). It is clear from the definition of a random walk, that the process is a renewal process because the times between the steps are independently distributed; one is led to the Poisson process if one makes the slightly more general assumption that the time between the depositional steps is a continuous variable. This model may be

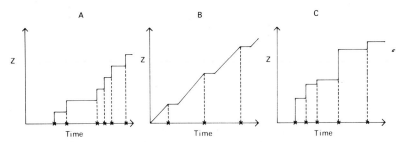

Fig. 6.3. Random-walk models of sedimentation. A. Equal steps of sedimentation. B. Interrupted sedimentation. C. Irregular steps at irregular time intervals.

interpreted in a variety of ways (see Fig. 6.3). For example, one may assume that sedimentation proceeds in discrete steps, each step occurring after a random time of non-deposition. Alternatively, sedimentation can be constant, but at random intervals it is interrupted for short periods, in which case the accumulated sediment again depends on the random variable N_t, that is the number of renewals in time over which the process is active. A further modification of the random walk is obtained by assuming that the depositional steps are variable in size and follow some probability distribution which is independent of the time distribution.

Before analysing some of these sedimentation models in detail, it seems appropriate to consider some geological situations which can lead to step-like sedimentation. A very much simplified example is given in Fig. 6.4A, where it is assumed that a delta builds forward, into a stagnant water body to such an extent that the delta slope becomes unstable. The slump which will develop removes the shaded area of sediment in the diagram and distributes it practically instantaneously as a layer of sediment. If conditions of sediment supply remain constant, then the process can be repeated over a long period and each slump leaves a layer in the stratigraphic record. Perhaps it is unrealistic in this example to assume that the times between slumps are independently random distributed but, whichever distribution is chosen,

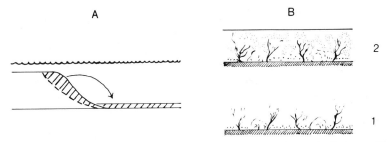

Fig. 6.4. Two sedimentation processes producing steps. A. Delta sedimentation. B. A limestone environment.

sedimentation will always proceed in steps. The mechanism leading to laminations in sandstones (see Chapter 2) may, in some respects, be similar to this model and in this case, one may well assume independent random fluctuations in the flow pattern of the transporting current as the primary cause for step-like sedimentation.

The second example (Fig. 6.4B) is based on a detailed petrographic analysis of a Carboniferous limestone sequence in the NW of Ireland (Schwarzacher, 1964) for which the following hypothetical mode of formation is proposed. Under normal conditions (stage 1 in Fig. 6.4B), there develops a rich benthonic fauna of corals, bryozoans and brachiopods. The surface profile of the sea floor probably does not change for a long time and remains essentially the same through several life cycles of the benthos. Possibly, the debris of dead organisms is removed by current activity towards the more central parts of the basin. Occasionally some catastrophic event kills the whole benthos population and this is achieved by extremely rapid sedimentation which preserves the fauna in their life position. It is possible that storms are responsible for this smothering of the benthos with sediment. The bottom fauna acts during such periods of rapid sedimentation (stage 2 in Fig. 6.4B) as a baffle and the thickness of the sediment which is deposited during this step-like event, is determined by the organic relief of the sea floor prior to the storm.

This interpretation is of necessity hypothetical and there are alternatives which one might take into consideration. For example, one might assume that after a certain time the benthonic community dies in an ecological sense of old age and that a step is produced when after a certain period of rest a new community becomes established. The rapid sedimentation during the step formation is against this hypothesis in this particular example, but it is interesting to speculate how such an assumption would affect the assumed time distribution of the step events. Clearly, if an ecological ageing of the community is assumed then the events of steps cannot be Poisson distributed and one would automatically choose a distribution of failure times which has a non-zero mode.

There are many more examples of step sedimentation which one could discuss but this small selection already illustrates that a careful sedimentological analysis is the prerequisite for constructing any theoretical sedimentation model. Because our knowledge of the sedimentation processes, is as a rule, incomplete, the sedimentary analysis will generally suggest more than one interpretation and one of the values of a theoretical investigation lies in the possibility of exploring such alternative interpretations by different models. Vice versa the theoretical models sometimes suggest new ways of looking at the sediment and in this way help to sharpen sedimentological investigations. Thus the concept of step-like sedimentation, in a very general way, is an outcome of Markov-chain and renewal-type processes. More specific examples will be given later.

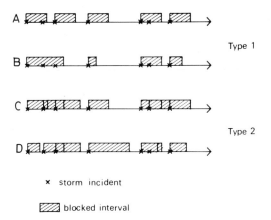

× storm incident

▨ blocked interval

Fig. 6.5. Electronic counter models.

If one accepts the proposed model of storm disturbed benthonic growth for the limestone example, which has been discussed, some further elaborations follow logically. Assuming that a step has just been formed by smothering the carpet of benthonic growth, then clearly, if a second "storm" follows this immediately, it will have no effect on sediment formation because the bottom fauna had no time to develop. This leads to a variety of different renewal models (see Fig. 6.5), the least complicated of which assumes that each active "storm" is followed by a recovery period of constant time (Fig. 6.5A). If any renewal or "storm" falls into this recovery period it becomes inactive and, therefore, has no effect on the sedimentation. A somewhat more flexible model is obtained by making the length of the recovery period a random variable (Fig. 6.5B) which is independent of the distribution of the intervals between the storm incidents. A different type of renewal process is obtained by assuming that each incident is followed by a recovery period and if a storm falls into the recovery period of a preceding incident, it will extend the inactive period by an amount which is measured from the last renewal. Once again the recovery periods can either be of constant length (Fig. 6.5C) or they may be random variables (Fig. 6.5D). The first two cases A and B are known as a "Type 1" mechanism and examples C and D belong to a "Type 2" mechanism. This terminology is borrowed from the study of electronic counters where precisely the same problem is met with. Counting devices such as Geiger counters which are used for counting α particles and cosmic rays, rely on the ionizing effect of such particles causing a discharge. After each count the charge has to be built up again and the counter needs a certain time before it can operate once more. This effect is known as the dead or blocked time of the counter. Because of its practical importance, the theory of counters is well known (cf. Feller, 1957; Bharucha-Reid, 1960) and many of the results can be applied directly to geological problems.

In the specific example of limestone formation either a type-1 or a type-2

model could apply. This will depend largely on how a once destroyed benthonic fauna will re-establish itself. If there is a waiting time with no organic
activity after the "storm" which is followed by a comparatively rapid growth
of organism, then a type-1 model is indicated. If the recovery is gradual, but
possibly starting immediately after the storm incident, then a renewed storm
will destroy any progress which has been made and the dead time will be
extended which suggests a type-2 model. Here again, two alternative hypotheses are suggested by the theoretical analysis and such possibilities can be
investigated by a more detailed petrographic examination.

6.4 THE THICKNESS DISTRIBUTIONS OF BEDS REPRESENTING EQUAL TIME
INTERVALS

In the following discussion, we will consider the thickness of sediment
which has been deposited between two marker horizons representing two
time events in the stratigraphic history. At first it will be assumed that the
events are separated by a constant time interval T. Such a sedimentation unit
will be called simply a bed. Keeping to these assumptions, one can calculate
the bed-thickness distributions which will result from the previously discussed sedimentation models. Unfortunately it will be found that some models lead to identical distributions and practically all the discussed sedimentation processes lead to very similar distributions of bed thicknesses.

Clearly all sedimentation processes which are based on discrete steps of
constant thickness (Fig. 6.3A) or processes in which deposition is interrupted by a constant gap (Fig. 6.3B), will lead to discrete bed-thickness
distributions. The bed thickness is strictly proportional to the random variable N_t, which if the storm incidents are exponentially distributed, will be
the Poisson distribution (see [6.7]). As an example for such a distribution,
the data of Table 4.1 are plotted in Fig. 6.6 where it is assumed that $\rho = 3.0$
and $T = 1.0$. The distribution remains discrete if each storm incident is
followed by an inactive recovery time. Choosing a type-1 mechanism with a
constant recovery time τ (see Fig. 6.4A) one can argue as follows. All open
intervals are foreward renewal times and if the storm incidents are exponentially distributed, then the open intervals are identically distributed (see
[6.19]). The sum S_r in time $T - r\tau$ is given by the r-fold convolution of the
exponential intervals and this is the special Erlang distribution which in its
cumulative form ([6.6]) may be written as:

$$K_{(r)} = 1 - \sum_{r=0}^{r-1} \frac{(t - r\tau)^r}{r!} \, \rho^r e^{-\rho(t - r\tau)} \qquad [6.20]$$

and after entering into [6.5] the thickness distribution is:

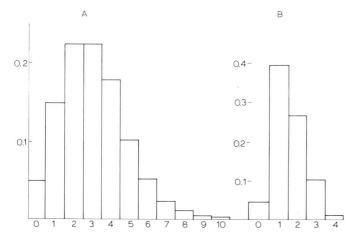

Fig. 6.6. Bed-thickness distributions. A. The Poisson distribution. B. A type-1 model.

$$P(Z = z, T = t) = \rho^z \frac{(t - \tau)^2}{z!} e^{-\rho(t-\tau z)}, z = 0, 1, 2 \ldots \qquad [6.21]$$

Using the same parameter ρ as before and assuming that $\tau = 0.2$, the bed-thickness distribution has been calculated and is shown in Fig. 6.6B. As one would expect, the maximum thickness of this distribution is four units because five recovery periods would completely take up the total time T. Any other sedimentation mechanism which is based on the electronic counter model would have a smaller dispersion when compared with the unrestricted step sedimentation. As long as the steps are of equal thickness, the step structure will be preserved in the bed-thickness distribution.

A continuous thickness distribution may be obtained by assuming that the sedimentation step is itself a continuous random variable, let us say with the distribution $f(z)$. If the intervals between steps follow a Poisson process of density ρ, then the general thickness distribution will be a compound Poisson distribution:

$$P(Z = z, T = t) = e^{-\rho t} \sum_{n=0}^{\infty} \frac{(\rho t)^n}{n!} [f^*(z)]^n \qquad [6.22]$$

where $[f^*(z)]^n$ stands for the n-fold convolution of the $f(z)$. The generating function of this distribution is:

$$G(s, t) = e^{-\rho t} + \rho t \, f(s) \qquad [6.23]$$

which has the property that:

$$G(s, t_1 + t_2) = G(s, t_1) \, G(s, t_2) \qquad [6.24]$$

One may interpret this in the following way (Feller, 1957, p. 270). Associated with each period of duration t, there is a random variable with the generating function $G(s, t)$ which is called the contribution of that period. The contributions of two non-overlapping periods are independent, which means that a partitioning of $t = t_1 + t_2$ decomposes the random process Z_t into two parts whose contributions are two independent increments. It follows that any process with independent variables is a Markov process and it can be proved (see Feller, 1957, p. 272) that amongst integral valued random variables, only the compound Poisson distribution has the property which is defined by [6.24]. The Poisson distribution itself is of course a special case of a compound distribution in which the steps are constant, and it is also easy to see from the generating function [6.18] that the negative binomial distribution must be a compound Poisson distribution. Therefore a very wide variety of sedimentation models and indeed all models which are based on the random-walk principle, lead to distributions of the type given in [6.24].

A few examples will illustrate the wide use of the compound Poisson distribution in different sedimentation models. The simplest case of a variable-step model may be obtained by assuming the step thickness to be exponentially distributed, that is:

$$f(z) = \lambda e^{-\lambda t} \tag{6.25}$$

This implies that consecutive steps are completely unrelated to each other and are generated by an independent random process. Applying [6.22] we find for the bed-thickness distributions:

$$P(Z = z, T = t) = \lambda e^{-(\rho t + \lambda z)} \sum_{n=0}^{\infty} \frac{(\lambda \rho z t)^n}{(n!)^2} \tag{6.26}$$

This expression can be written more conveniently as:

$$P(Z = z, T = t) = \lambda e^{-(\rho t + \lambda z)} J_0(2i \sqrt{\lambda \rho z t}) \tag{6.27}$$

in which J_0 is the Bessel function order zero (cf. Whittaker and Watson, 1969). A numerical example which assumes $T = 1$, $\rho = 2.0$ and $\lambda = 1.0$ is shown in Fig. 6.7 and this distribution is quite typical for many similar curves which can be generated by this model. The maximum will always be found at zero thickness when $\lambda > \rho$, but when $\lambda < \rho$ the maximum shifts towards positive thicknesses and there is a positive skewness.

The thickness distribution of [6.26] is the result of compounding the special Erlang distribution with the Poisson distribution, it is important to note that the same result can be obtained by situations which are geologically quite different. For example, consider an interrupted sedimentation

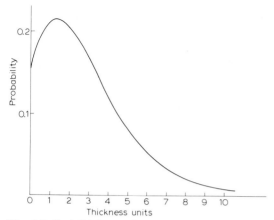

Fig. 6.7. Bed-thickness distribution based on a variable-step model.

model as shown in Fig. 6.3B. Each "storm" is followed by a period of non-deposition, which at the same time represents a blocked interval which means that if a "storm" falls into this period, it becomes inactive. The length of the recovery period will be again a random variable which is exponentially distributed with density λ. In this way, the interrupted sedimentation model becomes the problem of a type-1 electronic counter with a variable recovery period (see Fig. 6.5B). Using counter terminology, one can see that the interval between two registrations must be given by the convolution of the blocked and open times and since the latter are the forward renewal times between "storm" incidents which occur with density ρ, one can write the transform of [6.5] as:

$$\Gamma^*(n, s) = \frac{1}{s}\left[\left\{\frac{\rho\lambda}{(\rho + s)(\lambda + s)}\right\}^n - \left\{\frac{\rho\lambda}{(\rho + s)(\lambda + s)}\right\}^{n+1}\right] \qquad [6.28]$$

This Laplace transform is not easily inverted but backward transformation becomes simple in the special case when $\lambda = \rho$ and one can write the probability:

$$P(n, t) = \frac{(\rho t)^{2n}}{(2n)!}\, e^{-\rho t} \qquad [6.29]$$

which is again the Poisson distribution with the dummy variable replaced by $2n$. Remembering next, that the amount of deposited sediment after each recorded "storm" is proportional to the open time which is exponentially distributed, one finds, as in [6.26], that the bed-thickness distribution must be the compounded Poisson and the special Erlang distribution as before.

In the previous discussion of how step-like sedimentation may originate it became clear that such steps are not likely to be precisely the same thickness whenever a process of sedimentation is repeated. On the other hand if step formation is attributed to a repetition of essentially similar sedimentation processes, then it is equally unlikely that successive steps will be distributed as independent random variables. It is therefore important to investigate step models for which the distribution of step $f(z)$ is intermediate between the deterministic constant and the independent random distribution. A distribution which can portray a great variety of such situations is the gamma distribution ([4.41]). It has been shown previously that this distribution has two parameters, let us say k and λ and it was noted that if $k = 1.0$ then the gamma distribution becomes the negative exponential. Furthermore with increasing k and λ the gamma distribution approaches the normal distribution. If the mean of this distribution, which is k/λ, is kept constant and both parameters are increased, then the standard deviation, which is $\sqrt{k}/\lambda$, will decrease. The limiting form is a degenerated distribution with zero dispersion and all the probability concentrated around the mean value. This latter case can be used to describe a constant step thickness which leads to the discrete Poisson bed-thickness distribution. A good deal of this flexibility is maintained if the gamma distribution is replaced by the special Erlang distribution which differs from the former only by having an integral valued parameter k. By doing this, the theory of a variable-step process becomes much simpler and one can now write [6.22] as:

$$P(Z = z, T = t) = \lambda e^{-(\rho t + \lambda z)} \sum_{n=0}^{\infty} \frac{(\lambda^k z^k \rho^t)^n}{n!(kn)!} \qquad [6.30]$$

Calculated examples of such distributions with a fixed $\rho = 2.0$ and various values of k and λ are shown in Fig. 6.8 and one can clearly see that this compound distribution becomes polymodal when the dispersion of the step distribution decreases.

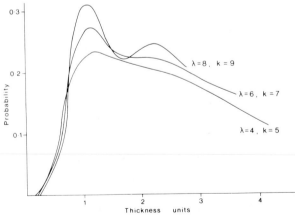

Fig. 6.8. Examples of a compound Poisson distribution; the first figure gives parameter λ, the second parameter k (see [6.30]).

6.5 THE THICKNESS DISTRIBUTION OF SEDIMENTARY BEDS

The distributions which were calculated in the previous section can be taken as the basis for more realistic models of bed formation by making only a few modifications. In the previous section it was assumed that the time period T over which the sedimentation was active had a constant length. However, it is much more likely that the time represented by a bed is itself a random variable and this variation must enter the model. The effect of a variable time interval on the counting distribution has already been investigated (see [6.17] and [6.18]) and it was found that exponentially distributed time intervals lead to exponential counts, whereas time intervals which had the special Erlang distribution led to counts which are negatively binomial distributed. It has been recognized by the criterion of [6.24] that the negative binomial distribution is a compound Poisson distribution. Therefore, any bed thickness arising from a sedimentation process which operates during time intervals, which have a modal length different from zero, again will be some type of compound Poisson distribution.

The sedimentologist might hope that a very detailed analysis of bed-thickness distributions would allow him to infer something about the physical processes which have led to the formation of sedimentary beds, but the chances of achieving this are not very good. One can think of a great variety of basically quite different sedimentation processes, which lead to similar compound Poisson distributions and it has been demonstrated, for example, that the model of variable-step sedimentation ([6.26]) leads to precisely the same distribution as the model of interrupted steady sedimentation, which physically is quite different. It is not difficult to elaborate such models but this usually involves introducing more parameters to describe more complex sedimentation processes. It has been found that most of the observed distributions of bed thicknesses are extremely well approximated by two-parameter distributions like the gamma distribution, and it is impossible to extract more than these two parameters from the data. Any further input into the models would have to be largely guess work.

The gamma distribution is undoubtedly the simplest one and it is still a very instructive distribution which can be used for examining bed thicknesses. Although the distribution is mentioned by Krumbein and Graybill (1965) as a possible distribution for bed-thickness data, it has not been used for this purpose until recently (Schwarzacher, 1972), largely because it is traditionally assumed that the bed thicknesses are lognormal distributed (cf. Pettijohn, 1957). Indeed the lognormal distribution provides a reasonably good fit for many data but it would be very difficult to test statistically, with the available material, which of the two distributions is the more appropriate one. Simple χ^2-tests for the examples which follow, give better fits for the gamma distribution. The discrepancies of the observations with the lognormal distribution, occur mainly in the tail and it should be noted that the gamma distribution has a very much shorter tail than the lognormal one.

To obtain an estimate of the two parameters (K, ρ) of a gamma distribution, one can use several methods. In the following examples a maximum-likelihood estimate has been obtained by a method which is discussed in Cox and Lewis (1966). The estimate for the parameter ρ is:

$$\hat{\rho} = \sum_{i=1}^{n} x_i/n \tag{6.31}$$

in which x's are the observations and n is the number of observed frequencies. The maximum-likelihood estimate for the index k is the solution of the equation:

$$n[\log \hat{k} - \psi(\hat{k})] = n \log \hat{\rho} - \sum_{i=0}^{n} \log x_i \tag{6.32}$$

in which $\psi(\hat{k})$, the digamma function, is defined as:

$$\psi(x) = \frac{d \log \Gamma(x)}{dx} \tag{6.33}$$

The solution of [6.32] can be found by an iterative procedure which is easily programmed for automatic computers; the digamma function has been tabulated (Davis, 1933) but it can also be evaluated in the computer programme.

The three examples of bed-thickness distribution, which are shown in Fig. 6.9 are all measurements of limestone beds whereby bedding was defined in the sense of Section 2.5 as sedimentation units which on the scale of outcrops, are enclosed by synchronous time planes. The Benbulbin Shale lime-

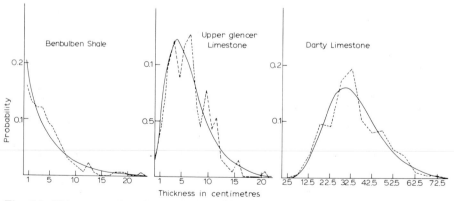

Fig. 6.9. Three examples of bed-thickness distributions approximated by gamma distributions (after Schwarzacher, 1972).

stones are represented by 137 measurements, giving a gamma distribution which is nearly exponential and with the parameters $k = 1.16$ and $\rho = 0.2381$. The Glencar Limestone ($n = 206$) gave $k = 2.53$ and $\rho = 0.4363$ and the Dartry Limestone ($n = 175$) gave $k = 6.56$ and $\rho = 0.1978$. These three examples belong to the previously mentioned (Section 6.3) Carboniferous series from NW Ireland and the model of benthos development which has already been discussed, was developed specifically for a part of the Glencar Limestone. The fact that the bed-thickness frequencies can be approximated by a gamma distribution, means that the sedimentation can be treated as if it occurred in stages. It does not mean that such stages had any physical reality, although it is often interesting to accept this as a working hypothesis. How such a hypothesis can be further developed, may be seen from the Glencar Limestone example. A part of this section was collected as hand specimens and a continuous 10 m long peel of suitably etched sectioned surfaces, was prepared. This provided a very detailed stratigraphic record, which could be scrutinized under the microscope. The smallest stratigraphic unit which was recognized and measured in this study, was any thickness of limestone which was enclosed between two horizons and which could be correlated across the width of the peel (10—15 cm) and which was judged to represent a time stage in the sediment. Naturally, the stratigraphic division under such conditions is much more detailed than the differentiation which is possible in field studies. This is clearly seen in the distribution of the thicknesses which were derived from such measurements (see Fig. 6.10). The frequency curve of microstratigraphic units is clearly split up into three maxima which one might regard as sedimentation steps and which with their average thickness of approximately 30 mm, represent roughly the height of benthonic growth within each stage. In retrospect it is seen that the k-value of 2.5 which was estimated from the gamma distribution and which indicates 2—3 steps per bed was a very meaningful parameter. It is further interesting that quite a number of the beds which were recognized during field work could not be subdivided, even when they were examined carefully under the microscope. This is largely due to severe bioturbation which occurred in these sediments and which has eliminated the more detailed stratigraphic record. Surprisingly, indications of step-like sedimentation can be obtained by a study of this bioturbation. Many of the beds contain vertical U-shaped borings of the type *Corophioides* and such borings invariably start from the surface of individual limestone beds. If the maximum depth penetration of such borings is measured and plotted as a frequency distribution (see Fig. 6.11), one obtains a curve which still shows noticeably the three maxima corresponding to the stages of Fig. 6.10. One can only assume that the burrowing organisms (Crustacea) preferred three levels of depth penetration because something checked their progress at discrete intervals.

In the examples which have been discussed here, a particular search was made for step-like sedimentation and it is quite likely that other mechanisms

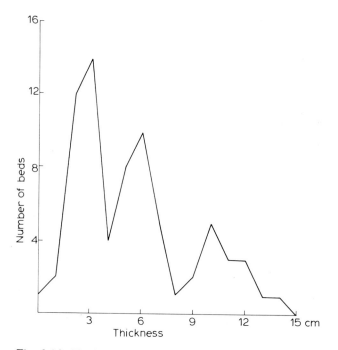

Fig. 6.10. Thickness distribution of the microstratigraphic units in the Glencar Limestone (after Schwarzacher, 1972).

have operated in a variety of environments. Indications of polymodal bed-thickness distributions can be found in a number of published data, including both clastic and carbonate sediments. A particularly clear example is the bimodal distribution of the Upper Chester Limestone beds from Illinois, U.S.A. (Potter and Siever, 1955). Such observations, however, must be followed up by detailed petrographic work before any conclusions can be

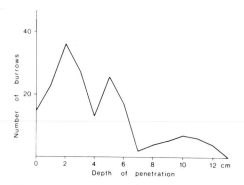

Fig. 6.11. The depth of penetration of U-shaped burrows.

reached about their origin. In this context, an observation by Fiege (1937) is particularly interesting. This author measured a large number of predominantly clastic sedimentary beds from the Kulm (Carboniferous) in NW Germany. The beds were classified according to petrographic types and the bed-thickness distributions clearly show that the types which are predominantly coarse clastic have a high positive skewness, and consequently a low k-value, whereas the fine-grained sediments are more nearly symmetrical and can therefore be interpreted as consisting of a relatively large number of stages. Again, further examination is needed before an interpretation of this interesting finding can be attempted. It is possible that the Kolmogorov mechanism of bed truncation (Section 4.5) played an important role in modifying the distributions as the chance of erosion is much higher in a high-energy environment (coarse clastic beds) than with the fine-grained beds.

The negative exponential distribution as a limiting case to the gamma distribution, deserves some special consideration. Out of the three examples which were analysed, only the limestones of the Benbulbin Shale (Fig. 6.9) come close to this distribution. Geologically, these limestones are different as they occur as isolated thin beds in a sequence of predominantly black shale. Very often a layer rich in coarse fragmented organic debris forms the base and this is followed by very argillaceous limestone which has been intensely bioturbated. Although no detailed work has been done on this formation, it seems possible that each limestone bed represents one single event which initiated the limestone formation and which then lasted for an exponential time interval. However, it is well known from practical statistics that it is very difficult to establish whether a frequency curve has its maximum at zero or somewhere near zero. If the class intervals of a highly asymmetric distribution are not sufficiently detailed it may well appear that the distribution is exponential when in fact it is not. This together with the general lack of reported exponential bed-thickness distributions, makes it likely that the exponential model is only applicable in special cases. It is suggested that the model is useful for relatively shortlived phenomena which are caused by a single event.

This last remark is of particular relevance in connection with the thickness distributions, which are predictable from the simple Markov model. It has been shown in [6.1] that, provided the step thickness is constant then the thicknesses of lithological units must be geometrically distributed. The unlikelihood of constant sedimentation steps, makes it clear why such geometrical distributions are not readily observed. The examples which Krumbein and Dacey (1969) use to illustrate geometrical or near geometrical distributions, are not very convincing. They are based on thickness class intervals of two feet (60 cm) which is rather coarse by any standards.

A method which maintains the Markov-chain model but allows for more realistic bed-thickness distributions will be discussed in the next section.

6.6 SEMI-MARKOV CHAINS

In [6.26] we have introduced a model that assumed continuously variable sedimentation steps which occurred at intervals that themselves have been exponentially spaced in time. This model will now be slightly modified by assuming a discrete time. Instead of postulating that "storms" are exponentially distributed it is assumed that time is discontinuous and that sedimentation steps can only occur at times $T = n$ $(n = 0, 1, 2...)$. This means that the times between events of sedimentation are geometrically distributed. The new model of "storm" distribution can be written in the form of a Markov matrix containing two states. State 1 is associated with deposition and state 2 with non-deposition. One may write, for example, the matrix:

$$\mathbf{A} = \begin{matrix} & \begin{matrix} 1 & \quad 2 \end{matrix} \\ \begin{matrix} 1 \\ 2 \end{matrix} & \begin{bmatrix} p & q \\ 1 & 0 \end{bmatrix} \end{matrix}$$

whereby it is accepted that an act of deposition only occurs if a transition into state 1 takes place. The times at which deposition may occur are a random variable and similar to [6.1], we have:

$$P \text{ (transition into } p \text{ occurring at time } t = n) = pq^{(n-1)}$$

To give a full description of the process, one has to specify some random variable with the distribution $f_1(z)$ which represents the step increment in this model and which, as in [6.25], could be the negative exponential distribution with density λ. The subscript 1 in $f_1(z)$ is added to indicate that this distribution only becomes operative when state 1 is entered. The model is illustrated in Fig. 6.12 where a short realization of such a process is shown. At discrete time 0, 1, 2, ... transitions occur in the system and between these times the system is in either state 1 or 2. The Markov chain A determines the states and every time state 1 is entered, sedimentation moves upwards in a step-like jump, the height of which is determined by $f_1(z)$. The stratigraphic record which is produced in this way consists of a series of point events (entering of state 1) which are separated from each other by distances which are random samples from the distribution $f_1(z)$.

The Markov matrix A which governs the occurrence of steps is known as the embedded matrix and the distribution $f(z)$ is usually called the waiting time, since it could be interpreted as a random time interval which elapses before a new state is entered. In the example under discussion, the waiting time is interpreted as step thickness. A process which operates in this two-fold way involving an embedded Markov matrix and some specified waiting

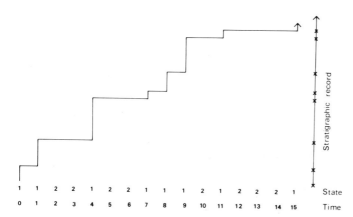

Fig. 6.12. The exponential step model as a semi-Markov process.

time functions, is known as a semi-Markov process. A generalization of the previous example is obtained by introducing more than one depositional state. For example, state 1 may stand for limestone sedimentation and state 2 for shale, giving an embedded matrix of:

$$A = \begin{bmatrix} \alpha_{11} & \alpha_{12} \\ \alpha_{21} & \alpha_{22} \end{bmatrix} \qquad\qquad [6.34]$$

Naturally, any number of states could be chosen but the two-state model will be treated in some detail. Associated with the embedded matrix are the waiting times for each type of transition and these are also conveniently written in matrix form:

$$F(z) = \begin{bmatrix} f_{11}(z) & f_{12}(z) \\ f_{21}(z) & f_{22}(z) \end{bmatrix} \qquad\qquad [6.35]$$

For the deposition of two different lithologies, one is justified in simplifying [6.35] by assuming only two types of waiting times, let us say $f_{11}(z) = f_{21}(z) = f_1(z)$ and $f_{12}(z) = f_{22}(z) = f_2(z)$. A realization of the process has been constructed in Fig. 6.13 using these assumptions. Note again that it is the state just to be entered which determines the choice of the step distribution and in the stratigraphical record, each event of type 1 is preceded by an interval of type 1 and events of type 2 are preceded by intervals of type 2. This is a conventional rule and although it is sometimes inconvenient, it will be adhered to throughout this discussion.

The probability structure of a stratigraphic record resulting from a semi-Markov process will now be investigated in detail. We wish to know the

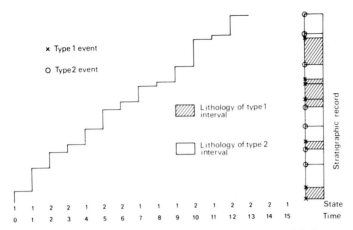

Fig. 6.13. Generating a stratigraphic section by a semi-Markov process.

probability of a transition between two states, i, j as a function of the waiting time z. Or, in more geological terms, suppose a transition from lithology i to lithology j has just been observed in a section, what are the probabilities of observing, let us say, a transition from j to k within the vertical thickness interval z? This problem can be solved by various methods and the following arguments ([6.36]—[6.46]) are taken from Cox and Miller (1965) who develop the theory of semi-Markov processes based on the methods of renewal processes.

We first introduce the renewal density which is defined as:

$$h_{ij}(z) = \lim_{\Delta z \to 0} \frac{P(\text{event of type } j \text{ in } (z, z + \Delta z)|\text{event of type } i \text{ at } 0)}{\Delta z} \quad [6.36]$$

The expected number of type-j events in the interval $(0, z)$ is:

$$\int_0^z h_{ij}(u)\, du \qquad [6.37]$$

One has to solve four integral equations to obtain the renewal densities ([6.36]) for the two-state case and the first one is:

$$h_{11}(z) = \alpha_{11} f_{11}(z) + \alpha_{11} \int_0^z h_{11}(z - u)\, f_{11}(u)\, du + \alpha_{21} \int_0^z h_{12}(z - u)\, f_{21}(u)\, du$$

$$[6.38]$$

This equation arises from the following considerations. In order that an event of type 1 happens in $(z, z + \Delta z)$ and assuming that an event of type 1 occurred at $t = 0$, then one of the three following must have occurred:

(a) A transition to state 1 took place and this has the probability α_{11} (see [6.34]). Associated with this is an interval of length nearly z.

(b) A sequence of events led to a type-1 event near $z - u$ and this was followed by a transition to a type-1 event in an interval of length nearly u.

(c) A sequence of events led to a type-2 event near $z - u$ and this was followed by a transition to a type-1 event in an interval of length nearly equal to u.

Each of these conditions accounts for one term in [3.38] which may be solved by applying the Laplace transformation and one can write:

$$h_{11}^*(s) [1 - \alpha_{11} f_{11}^*(s)] - h_{12}^*(s) \alpha_{21} f_{21}^*(s) = \alpha_{11} f_{11}^*(s) \qquad [6.39]$$

If the renewal densities are h_{12}, h_{21} and h_{22} then three further integral equations are obtained in a similar way and the Laplace transform of these can be written in matrix form as:

$$\mathbf{h}^*(s) [\mathbf{I} - \mathbf{g}^*(s)] = \mathbf{g}^*(s) \qquad [6.40]$$

where:

$$\mathbf{h}^*(s) = \begin{bmatrix} h_{11}^*(s) & h_{12}^*(s) \\ \\ h_{21}^*(s) & h_{22}^*(s) \end{bmatrix} \qquad \mathbf{g}^*(s) = \begin{bmatrix} \alpha_{11} f_{11}^*(s) & \alpha_{12} f_{12}^*(s) \\ \\ \alpha_{21} f_{21}^*(s) & \alpha_{22} f_{22}^*(s) \end{bmatrix} \qquad [6.41]$$

One can develop equations for the probability $P_{ij}(z)$ that the system is in state j at z in a similar way, provided an event of type i has occurred at $z = 0$. Use is made of the so-called survivor function in order to simplify the writing of these equations. The survivor function is defined as:

$$F_{ij}(z) = \int_z^\infty f_{ij}(u) \, du \qquad [6.42]$$

and it is simply the complement to the cumulative distribution of $f_{ij}(z)$. One can write for the first term $P_{11}(z)$, based on the same arguments as before:

$$P_{11}(t) = \alpha_{11} F_{11}(z) + \alpha_{11} \int_0^z h_{11}(z - u) \, F_{11}(u) \, du$$

$$+ \alpha_{21} \int_0^z h_{12}(z - u) \, F_{21}(u) \, du \qquad [6.43]$$

There are four similar equations which can be written in matrix notation as:

$$\mathbf{P}^*_{(s)} = \mathbf{G}^*_{(s)} + \mathbf{h}^*_s \, \mathbf{G}^*_{(s)} \qquad [6.44]$$

where:

$$\mathbf{G}^*_{(s)} = \begin{bmatrix} \alpha_{11}F^*_{11}(s) & \alpha_{12}F^*_{12}(s) \\ \alpha_{21}F^*_{21}(s) & \alpha_{22}F^*_{22}(s) \end{bmatrix} \qquad [6.45]$$

The Laplace transform of the survivor function $F_{ij}(s)$ is $F^*_{ij}(s) = [1 - f^*_{ij}(s)]/s$ and it follows that:

$$s\mathbf{G}^*(s) = \mathbf{A} - \mathbf{g}^*(s) \qquad [6.46]$$

in which $\mathbf{A}$ is, of course, the embedded matrix $\mathbf{A} = (\alpha_{ij})$. From [6.46], [6.45] and [6.40] we obtain:

$$s\mathbf{P}^*(s) = [\mathbf{I} - \mathbf{g}^*_{(s)}]^{-1} \, [\mathbf{A} - \mathbf{g}^*_{(s)}] \qquad [6.47]$$

The inverse of [6.47] is the required solution to the problem of finding the probability structure of the stratigraphical record. The solution comes in the form of a transition probability matrix $\mathbf{P}_{(z)}$ in which each element $p_{ij}(z)$ is a function of z.

A simple numerical example will show the use of [6.47]. Let $\mathbf{A}$ be given by:

$$\mathbf{A} = \begin{bmatrix} 0.5 & 0.5 \\ 1.0 & 0 \end{bmatrix} \qquad [6.48]$$

which is a Markov chain in which state 1 will follow as soon as state 2 is entered. It will be assumed, in order to simplify the calculations, that the distributions of the waiting time are identical for each type of transition which means that $f_{ij}(z) = f(z)$. Equation [6.47] can be written under these conditions as:

$$s\mathbf{P}^*(s) = \frac{1}{1 - 0.5f^*_{(s)} - 0.5[f^*_{(s)}]^2} \begin{bmatrix} 0.5[1 - \{f^*_{(s)}\}^2] & 0.5[1 - f^*_{(s)}] \\ 1 - f^*_{(s)} & 0.5f^*_{(s)}[1 - f^*_{(s)}] \end{bmatrix}$$

$$[6.49]$$

Next it is assumed that $f(z)$ is negative exponentially distributed with a transform $f^*(s) = \rho/(\rho + s)$. One obtains, by substituting into [6.49]:

$$
\mathbf{P}^*_{(s)} = \frac{1}{s(s + 1.5\rho)}
\begin{bmatrix}
0.5s + \rho & 0.5(s + \rho) \\
s + \rho & 0.5\rho
\end{bmatrix}
\tag{6.50}
$$

The backward transformation of this expression causes no difficulty and a very simple solution is obtained by setting $\rho = 1$, which gives:

$$
\mathbf{P}_{ij}(z) =
\begin{bmatrix}
0.666 - 0.166e^{-1.5z} & 0.166e^{-1.5z} + 0.333 \\
0.333e^{-1.5z} + 0.666 & 0.333 - 0.333e^{-1.5z}
\end{bmatrix}
\tag{6.51}
$$

The matrix $\mathbf{P}_{ij}(z)$ gives a very compact description of a stratigraphic record if the individual elements of the matrix are interpreted as follows. To consider an example, the element $P_{11}(z)$, should be read as: provided the base of a section, $z = 0$, is chosen where a layer of sediment type 1 has just been completed, $P_{11}(z)$ gives the probability that at a distance z, another interval of type 1, is completed. It may be seen from the example [6.51] that when z becomes zero, that is when the transitions are instantaneous, matrix [6.51] becomes identical with the embedded matrix $\mathbf{A}$ ([6.48]). This is in perfect agreement with the model because it is assumed that $\mathbf{A}$ represents time events which are not necessarily connected with any stratigraphic record. On the other hand, it may be seen that if z approaches infinity then the transitions into state 1 and state 2 take on a constant probability of 0.666 and 0.333, respectively. It is not difficult to verify that these values represent the stable probability vector of $\mathbf{A}$. From [5.13] we find:

$$
\Pi = \frac{\alpha_{21}}{\alpha_{12} + \alpha_{21}} , \frac{\alpha_{12}}{\alpha_{12} + \alpha_{21}} = 0.666, 0.333
\tag{6.52}
$$

Clearly if ρ had been chosen differently, then the process would have converged towards $\rho_i \Pi_i$. More generally still, if one assumes that all waiting-time distributions are different, let us say with means μ_{ij}, then the limiting distribution is found by the following arguments. Over a very large number of transitions there is a proportion $\Pi_1 \alpha_{11}$ which is from state 1 to state 1. Therefore, the proportion of thickness which is taken up by passing from state 1 to state 1 can be written as:

$$
\frac{\Pi_1 \alpha_{11} \mu_{11}}{\Pi_1 \alpha_{11} \mu_{11} + \Pi_1 \alpha_{12} \mu_{12} + \Pi_2 \alpha_{21} \mu_{21} + \Pi_2 \alpha_{22} \mu_{22}}
\tag{6.53}
$$

State 1 can also be reached from state 2 in the two-state system and the limiting probability for state 1 is therefore given by:

$$P_{11}(z) = P_{21}(z) = \frac{\Pi_1 \alpha_{11} \mu_{11} + \Pi_2 \alpha_{21} \mu_{21}}{\Pi_1 \alpha_{11} \mu_{11} + \Pi_1 \alpha_{12} \mu_{12} + \Pi_2 \alpha_{21} \mu_{21} + \Pi_2 \alpha_{22} \mu_{22}} \qquad [6.54]$$

This probability plays exactly the same role as the stable probability of Markov chains, meaning that it gives the proportion of thickness which is occupied by individual states, or more simply the overall lithological composition of the section.

It is evident now, that the matrix $P_{ij}(z)$ contains roughly the same information as the matrices P_{ij} and the higher powers $P^n{}_{ij}$ in the case of the Markov-chain model. Plots of $P_{ij}(z)$ as functions of z are directly comparable with the plots of the powers of P_{ij} and examples of this will be given shortly.

6.7 THE SEMI-MARKOV CHAIN AS A STRATIGRAPHIC MODEL

It is useful to discuss some geological implications of this model, which up to this point was developed strictly as a sedimentation model before giving further examples and applications of semi-Markov chains. We have introduced two separate matrices: the embedded matrix, which was used to describe the environmental history and the transition probability matrix, which was used to describe the rock sequence of the section. The two are connected by the distribution of waiting times $f(z)$ and it is now fairly clear that this distribution must be somehow determined by the mechanism of sedimentation. This somewhat sedimentologically biased approach to the semi-Markov process, which was developed by Schwarzacher (1972) is not the only one. Indeed, Krumbein and Dacey (1969) and Dacey and Krumbein (1970), who have introduced the semi-Markov chain as a stratigraphic model were not particularly concerned with genetical problems. Krumbein and Dacey turned to the semi-Markov process because it was realized that the previously used discrete Markov-chain model must, by definition, yield geometrical thickness distributions for lithological units. A more flexible model for describing stratigraphic sections was needed. Introducing the semi-Markov model, has led to a critical re-examination of the structure of stratigraphic sections and it has forced the geologist to associate the Markov property with either some process or some structural element of the section which can be defined clearly. The work of Dacey and Krumbein (1970) is a very good example for illustrating this clear and improved way of structuring stratigraphic sections.

It is recognized that a sedimentary series consists of a hierarchy of layering, each layer having homogeneity at its scale of observation. Dacey and

Krumbein restrict their theory to two units: the sedimentary bed and the lithological unit which consists of one or more beds which have an identical lithology. The model which is introduced is the following. A discrete Markov process decides the type of bed to be deposited and the thickness of this bed is an independent random variable which is the waiting time in the semi-Markov process. The main problem concerns the thickness distribution of the lithological units which, as will be seen shortly, is very similar to the problem of finding the thickness distribution of beds which are generated by a renewal process. Indeed, bed formation may be regarded as equivalent to the formation of lithological units simply by sharpening the level of observation. One may substitute for the lithological unit, the sedimentary bed and for the sedimentary bed, the sedimentation step as has been postulated earlier. Krumbein used electric well logs as his primary data and lithological units were easily recognized on the scale of his observations but the beds could not be established with equal precision. Similarly, Schwarzacher's (1972) study is based on the Carboniferous limestone—shale sequence, which has been discussed previously. Here the bed thicknesses have been measured in outcrop exposures and were clearly recognized but the sedimentation steps were at this stage only inferred from a variety of petrographic observations. It is possible that this sharpening of the observations could be increased further, but it has been argued (see Chapter 2) that there are limits to the stratigraphic resolution which ultimately justify a discrete time scale.

One finds only one striking difference in comparing the model of Krumbein and Dacey with the one which has been developed in the previous section. It was assumed earlier that the embedded Markov process operates at discrete and precisely equal time intervals. No such, apparently very restrictive condition, has been postulated by Dacey and Krumbein. In fact, the latter authors are not concerned with time implications. The traditional attitude of geologists shying away from any time problem cannot be maintained if a physical interpretation is to be given to the model and it is therefore important to give further discussion to the point. It is quite clear that the embedded Markov process must have operated somehow in time and one can consider two extreme alternative cases. Either the time intervals at which the embedded chain operates, are constant and discrete $T = 1, 2...n$: then the time spent in one particular state, let us say i, is determined by the geometrical distribution:

$$P(T = n) = p_{ii}^{n-1} q; q = (1 - p_{ii})$$ [6.55]

It can be assumed alternatively that the points at which the Markov process operates are random points along the time axis, in which case they are exponentially distributed, let us say, with density λ. In this case the length of time spent in state i is clearly given by:

$$P(T = t) = q\lambda e^{-\lambda t} \sum_{1}^{\infty} \frac{(\lambda \rho t)^{n-1}}{(n-1)!} = q\lambda e^{-\lambda t(1-\rho)}$$ [6.56]

and the only difference which is found between the constant time and the variable time model is that in the first instance the time spent in a particular state is geometrically and discretely distributed, and in the second case the times occupied by a particular state are continuously distributed. The latter is, of course, identical with the renewal process based on the Poisson distribution of "storm" incidents.

One may now well ask, what are the advantages of the semi-Markov model over some alternating renewal process? The answer lies in the intermediate position of the semi-Markov process. As a model it provides a bridge as it were between the essentially discontinuous Markov-chain approach and the model which was introduced at the beginning of this chapter. The semi-Markov model takes on a very special meaning when it is applied to bed formation. In this case one can identify the constant time intervals, at which the model operates with the periods of the ultimate time resolution which are determined by the quantum-like nature of the stratigraphic record. One may use the following deliberately simple analogue to illustrate this situation. Let it be assumed that a small basin receives only one type of sediment, for example, sand, and that this only happens during periods of rain. A bed of sandstone will be formed during each spell of rainy days and fine days will only be recorded by bedding planes. Records of rainfall are usually based on daily measurements and the period of one day is therefore the lower limit of time resolution. It is known that spells of rainy days are, in fact, often geometrically distributed (Gabriel and Neuman, 1957) and a Markov matrix gives a good description of the distribution of wet and dry days. The history of the environment is therefore presented as a discrete process which, keeping the available time units in mind, gives the best description of the environment at any one time. A semi-Markov model is therefore ideally suited to describing the sequence of sandstone beds in this hypothetical basin.

Perhaps attention should be drawn to the one basic restriction of the semi-Markov-chain model which is so self-evident that it might be overlooked. The embedded matrix is a Markov matrix and so it has the property that the system it describes remains in each state for an exponential time. It follows, for example, that the time it took to complete the lithological units in Dacey's and Krumbein's study must have been exponentially distributed, whether the embedded Markov chain operated at constant or random time intervals (see [6.55] and [6.56]). The model is therefore found to be quite flexible enough for describing the section but the genetical interpretation is still restricted to the exponential time distribution. One feels intuitively that such a distribution may be an apt description for relatively short random bursts of sedimentation such as sedimentary steps but it seems less likely that all the time intervals represented by lithological units can be of a similar nature. We have previously used the argument that the completion of similar geological processes will take an approximately similar time. This implies that the time distribution of geological processes centres around a mean which is different from zero and that it is not in general exponential. Renewal theory differentiates between components which age and which do not age and only the latter have an exponentially distributed life-time. If one is talking about the time needed for completing a geological process one usually deals with events which have a built-in mechanism for ageing. The examples which have been used to explain step-like sedimentation are quite typical. The rise and decline of an ecological community will take a finite time because such communities age as soon as they start developing. Similarly, slumps on a delta front will become more likely when a large amount of sediment has accumulated already and in this way, the age of the delta margin becomes important. It would be premature to suggest models for such complex processes except in a very general way and a model of failure which is used in renewal theory may be applicable. To account for the failure of a component it is sometimes assumed (Cox, 1962) that as soon as it is put into use it becomes subjected to a series of random blows. It is a property of the component that it can only take a limited number of such blows. If the blows are exponentially distributed, then the life-time of the component will be gamma distributed and this is the general justification for using this distribution for geological time intervals (see Section 6.4). The idea implies that the environment itself contains certain thresholds which have been discussed in Section 2.6.

6.8 THE ANALYSIS OF SECTIONS IN TERMS OF THE SEMI-MARKOV MODEL

To illustrate the methods which can be used for analysing sections in terms of semi-Markov processes, an artificial series of 1000 beds was generated using the embedded matrix A of [6.48] and identically negative exponentially distributed waiting times. A sample of this section which consists of only two rock types is shown in Fig. 6.14. It may be seen that the lithological units in this section can be composed of 1, 2, ... individual beds.

It is instructive to analyse this section by the methods which were introduced in the last chapter. We have discussed three ways of structuring previously and these involve sampling the lithologies bed by bed, sampling lithologies at equally spaced intervals and finally sampling by lithological units. Clearly, the first method of structuring would fully recover the embedded matrix but there would be no information about the thickness distribution of the beds. It is clear in a similar way that the last method of structuring by lithological units would yield no information at all since the transition matrix for this section is a 2 × 2 matrix with 1's in the leading diagonal. The second method of structuring takes account of thicknesses of lithological units but there is no provision for recognizing the bedding planes which subdivide these units. Spot samples are simply taken at equal intervals, as indicated in Fig. 6.14 and the higher transition probabilities can be estimated

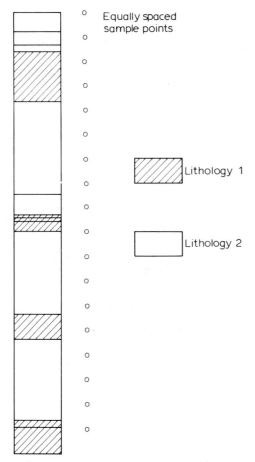

Fig. 6.14. A simulated two-state section showing the sampling points used in conventional stratigraphic analysis.

for such points. The procedure involves simply testing the lithology which is found at a multiple distance of the original sample interval. The results of such tests are shown in Fig. 6.15 and the higher transition probabilities are equivalent to the $P_{ij}(z)$ functions if proper attention to the definition of the semi-Markov process is given. We recall that a point event of type i marks the completion of a bed with type-i lithology. In the Markov chain P_{ij}, we record the transition into a bed with type-i lithology. To make the transition probabilities P_{ij} comparable with $P_{ij}(z)$, one has to exchange the rows of matrix P_{ij} and this has been done in Fig. 6.15.

The effect of not recognizing individual beds within the lithological unit can be analytically examined by substituting for matrix A ([6.48]) the matrix:

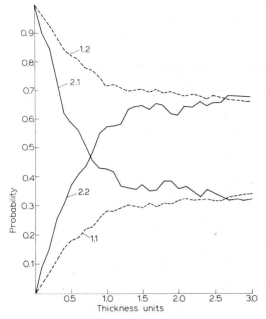

Fig. 6.15. Transition probabilities at 0.1 unit steps using a Markov-chain programme (after Schwarzacher, 1972).

$$\mathbf{A}' = \begin{bmatrix} 0 & 1 \\ 1 & 0 \end{bmatrix}$$

This induces the two states to alternate, but the densities with which state 1 and 2 occur are unaltered and they are given by the stable probability vector $\mathbf{\Pi}$ ([6.52]). One can modify the waiting-time distribution to make allowance for this and write:

$$f_1(z) = \rho_1 \Pi_1 \, e^{-\Pi^1 \rho^1 z} \text{ and } f_2(z) = \rho_2 \Pi_2 e^{-\rho_2 \Pi_2 z} \qquad [6.57]$$

Equation [6.47] now simplifies to:

$$s\mathbf{P}^*_{(s)} = \frac{1}{1 - f^*_{1(s)} \, f^*_{2(s)}} \begin{bmatrix} f^*_2(s) \, [1 - f^*_1(s)] & 1 - f^*_2(s) \\ 1 - f^*_1(s) & f^*_1(s) \, [1 - f^*_2(s)] \end{bmatrix} \qquad [6.58]$$

Setting $\rho_1 = \rho_2 = 1.0$ and using [6.57], one obtains the inverse:

$$\mathbf{P}'(z) = \begin{bmatrix} \Pi_2 - \Pi_2 e^{-z} & \Pi_1 + \Pi_2 e^{-z} \\ \Pi_2 + \Pi_1 e^{-z} & \Pi_1 - \Pi_1 e^{-z} \end{bmatrix} \qquad [6.59]$$

The elements of [6.58] give precisely the same transition probabilities as those which were estimated in Fig. 6.15. For example, $P_{11}(z)$ is zero when z is zero and approaches the stable probability which was calculated as 0.333 previously. Therefore, the method of equal-interval structuring can only recover the stable probability vector of the embedded matrix but not the matrix itself.

One has to modify the analysis in order to estimate the proper semi-Markov matrix $\mathbf{P}_{ij}(z)$. Samples are still taken at equal intervals but a transition is only counted if the interval which is being tested contains a bedding plane. This modified programme was tried on the artificial series and the resulting $P_{ij}(z)$ functions are shown in Fig. 6.16. Obviously the number of transitions which will be counted in this way is smaller than in the previous experiment and this accounts for the more irregular appearance of the curves. However, it can be clearly seen that the $P_{ij}(z)$ functions of [6.51] have been fully recovered and that the embedded matrix can be obtained

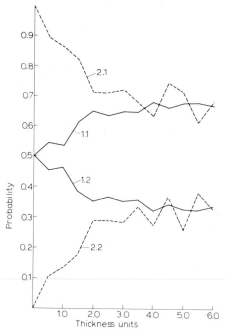

Fig. 6.16. Transition probabilities at 0.5 unit steps using a point event programme (after Schwarzacher, 1972).

from the diagram by letting the thickness be equal to zero. In this type of analysis both **A**, which is the embedded matrix, and $f(z)$, which is the waiting time distribution, can be reconstructed.

The embedded matrix could also be recovered from the thickness distribution of the lithological units if $f(z)$ is known, and vice versa: it is possible to calculate the thickness distribution of the lithological units if the waiting time and the embedded matrix is known. This problem has been dealt with by Dacey and Krumbein (1970) and it is very similar to the problem of bed thicknesses generated by renewal processes ([6.22]).

Considering that the thickness of a lithological unit L_i depends on the number n, of beds, which have the thickness W_1, W_2...W_n, one can write the conditional probability:

$$P(L_{i(n)} = z | N_i = n) = P(W_{1i} \ ... \ W_{nt} = z) \tag{6.60}$$

Writing $h_i(s)$ and $g_i(s)$ for the generating function of the lithological unit and the bed-thickness distribution gives:

$$h_i(s) = \sum_{n=1}^{\infty} P(N_i = n) \, [g_i(s)]^{*n} \tag{6.61}$$

The distribution of $P(N_i = n)$ is easily found since it is the number of times an identical state is revisited by a discrete Markov chain and, therefore:

$$P(N_i = n) = p_{ii}^{n-1} \, (1 - p_{ii}) \tag{6.62}$$

from which [6.60] becomes:

$$h_i(s) = \sum_{n=1}^{\infty} g_i^n(s) \, p_{ii}^{(n-1)}(1 - p_{ii}) = \frac{(1 - p_{ii}) \, g_i(s)}{1 - p_{ii} \, g_i(s)} \tag{6.63}$$

For example, if the bed-thickness distribution is the negative exponential with parameter λ, then the transform $\lambda(\lambda + s)$ can be substituted into [6.63] and this yields:

$$h_i(s) = \frac{(1 - p_{ii}) \, \lambda}{s + (\lambda - p_{ii}\lambda)} \tag{6.64}$$

which on inversion gives again a negative exponential distribution so that:

$$P(L_i = z) = \lambda(1 - p_{ii}) \exp\left[-\lambda z(1 - p_{ii})\right] \tag{6.65}$$

p_{ii} can be calculated if this distribution, λ and $f(z)$ are known.

Any calculation involving the negative exponential distribution is relatively easy but it has been recognized earlier that this distribution is only a special form of the more general gamma distribution. It has already been mentioned that sedimentary steps or bed thicknesses are more likely to be gamma distributed with maxima different from zero, than having distributions with maxima at zero. The type of result which is obtained by introducing the gamma distribution can be seen by considering the relatively simple case of an alternating renewal process of the type [6.58] and by choosing the parameters of the gamma distribution as $2,\rho_1$ and $2,\rho_2$. One can then write the transforms:

$$f_1(s) = \frac{f_1^2}{(\rho_1 + s)^2}, \quad f_2(s) = \frac{\rho_2^2}{(\rho_2 + s)^2} \qquad [6.66]$$

which on entering [6.58] gives:

$$sp_{11}(s) = \frac{\rho_2^2(\rho_1 + s)^2 - \rho_1^2 \rho_2^2}{(\rho_1 + s)^2 (\rho_2 + s)^2 - \rho_1^2 \rho_2^2}$$

which can be simplified to:

$$p_{11}(s) = \frac{\rho_2^2}{s} \frac{2\rho_1 + s}{(1 + s)(\alpha + s + s^2)} \qquad [6.67]$$

where $\alpha = 2\rho_1\rho_2$. This can be inverted to:

$$p_{11}(z) = A\rho_2^2\left[(1 - e^{-z}) - \frac{e^{-z/2} \sin Bz}{B}\right] \rho_2^2\left[\frac{1}{\alpha} - \frac{e^{-z/2}}{\sqrt{aB}}\right.$$

$$\left. \sin\left(Bz + \tan^{-1}\sqrt{4a - 1}\right)\right] \qquad [6.68]$$

where $A = (\rho_1 - \rho_2)/a$ and $B = \sqrt{4a - 1}/2$ when $4a > 1$. This expression is rather unwieldy but it does show that the transition probabilities are again exponential but with a superposed, strongly damped sine wave. The leading term $p_{ii}(z)$ is zero when z is zero and it approaches ρ_2 when z becomes large. The oscillating nature of the higher transition probabilities is not very pronounced when the parameter k of the gamma distribution is low but it increases with increasing k and in order to demonstrate this, a new series was generated with a gamma distribution and the parameters $\rho = 1$ and $k = 16$

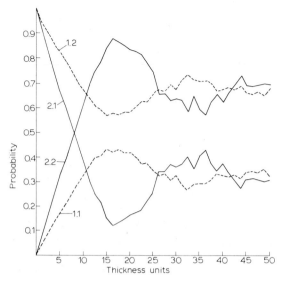

Fig. 6.17. Transition probabilities of a semi-Markov process with gamma-distributed waiting time (after Schwarzacher, 1972).

and matrix **A**. This series was analysed by the point-event method to give the $p_{ij}(z)$ functions which are shown in Fig. 6.17. It should be noted that the analytical treatment of this last example would lead to great difficulties in inverting the Laplace transforms and it is much more profitable to examine such models by simulation techniques. This is also true in the analysis of the thickness distribution of lithological units which in principle could be solved from [6.62], but which would require numerical methods for inversion. In

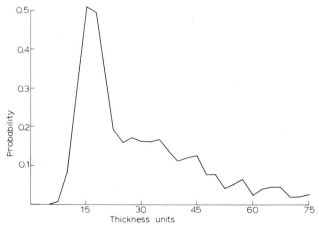

Fig. 6.18. Thickness distribution of beds generated by a **gamma-distributed semi-Markov** process (after Schwarzacher, 1972).

the case of the simulated example ($\rho = 1$, $k = 16$), one would expect a thickness distribution of lithological units which has maxima at $(k-1)$, $2(k-1)$, ... that is, at 15, 30, 45 ... Indeed, this is found to be so when analysing the artificial series. The lithological unit thickness distribution is shown in Fig. 6.18.

6.9 SEDIMENTATION—EROSION PROCESSES

A process which is related to the semi-Markov chain, has been proposed by Vistelius and Feigelson (1965) to account for the alternating stages of deposition and erosion. Each lithological state is split into two, one of which is associated with deposition, and a second one of which will cause the erosion of the previous bed. For example, the two-state system which consists of sandstone S and clay C, may be split into four states by introducing the following:

(1) Sd indicates that sandstone is available and will be deposited as a single layer.

(2) Se indicates that sandstone is available but instead of being deposited, the previous layer is eroded.

The states Cd and Ce are defined in a similar manner for clay. Vistelius proposes the following transition matrix:

$$
\mathbf{P} = \begin{array}{c} \\ \text{Sd} \\ \text{Se} \\ \text{Cd} \\ \text{Ce} \end{array}
\begin{array}{cccc}
\quad \text{Sd} & \quad \text{Sc} & \quad \text{Cd} & \quad \text{Ce} \\
\left[\begin{array}{cccccccc}
P_{ss}(1-q_s) & P_{ss} & q_s & P_{sc}(1-q_c) & P_{sc} & q_c \\
P_{ss} & & 0 & P_{sc} & & 0 \\
P_{cs}(1-q_s) & P_{cs} & q_s & P_{cc}(1-q_c) & P_{cc} & q \\
P_{cs} & & 0 & P_{cc} & & 0
\end{array}\right]
\end{array} \qquad [6.69]
$$

in which the transition probabilities P_{ij} are modified by a factor of q_s or q_c to give the probability of a bed being eroded prior to the non-deposition of either sandstone or clay. It may be seen from the structure of the submatrices, that in this model, only one bed can be removed at a time because the erosion positions are set equal to zero.

The connection of this model with the semi-Markov model may be seen by assuming that any state of type 'e' stands for non-deposition rather than active erosion. In this case, the model is a simple semi-Markov chain in which the depositional states have a geometrical waiting time of $(1-q)^n$. A problem which is of practical interest, is to find the transition frequencies between various lithologies which are preserved in a section that has been deposited by this mechanism. The combinations of a great number of transition probabilities have to be calculated in order to find these frequencies. For example, if the transition sand—clay is recorded in the section, then the Markov chain could have gone through the following states: Sd Cd or Sd Sd Se Cd or Sd Sd Se Sd Cd ... or a combination of Sd Se Cd Ce states. In practice, it is best to calculate such transition frequencies by simulation experiments which make various assumptions about the factors q. Vistelius and Feigelson made such an analysis for flysch-type sediments and found good agreement with the observed data by setting $q_{silt} = 0.11$ and $q_{clay} = 0.001$. The Vistelius model was expanded by Rivlina (1968) to allow for multiple stages of erosion but the combinatorial problems in such models become quite formidable and again, simulation techniques would provide simpler solutions.

6.10 CONCLUDING REMARKS

The models which have been discussed in this chapter could be classified from a geologist's point of view into two types: the sedimentation and the stratigraphic models. The difference is subtle. One can speak of a sedimentation which is controlled by the environment acting as an outside factor. For example, the various models of bed formation which are based on the renewal process, are sedimentation models because it was assumed that the duration of the process was dictated by either a constant time interval, or some other outside factor which interrupted sedimentation according to some probability pattern. In contrast to this, one can call a model a stratigraphic model if it incorporates the control of the sedimentation mechanism. Typical examples for such models are some applications of the semi-Markov process, where the models can be used to generate stratigraphical sections without any outside interference. The environmental history is modelled by the embedded matrix and the sedimentation processes are condensed into the waiting time functions. Of course, the difference between the two types of models is marginal and only one of degree. If a stratigraphical model operates over a relatively short time, it could be regarded as a sedimentation model and analogous to the hierarchy of layering in sedimentary rocks, one can think of a hierarchy of models which must be employed to generate stratigraphical sections. The simple stratigraphic models which have been discussed so far assume that the environment is controlled by a single Markov process and this is obviously an over-simplification when it comes to a realistic interpretation of environmental histories.

Chapter 7

Stationary stochastic processes and time-series analysis

7.1 INTRODUCTION

Many scientists deal with data which have been obtained by measuring some physical phenomenon at regular intervals. A series of observations, which has been collected in this way, is known as a time series. The daily temperature and rainfall, annual birthrate of a country, or the daily notation of share prices, are typical examples of time-series records. Time-series records are also created by dividing a stratigraphic section at equal intervals into lithological states and an embedded Markov chain operating at equal time intervals also leads to such a record.

In most of the previous examples we have considered discrete variables, or possibly only attributes, such as lithological states in a Markov chain. In this chapter, we will concentrate on continuous variables, and in particular, lithological parameters which can actually be measured. As before, we will regard time and stratigraphic thickness as equivalent unless it is otherwise stated; the term 'time series' is again used for convenience and the dimension time can be replaced by any dimension in space along which measurements can be carried out.

Sedimentologists and stratigraphers have developed a wide range of lithological parameters which can be expressed in numbers as continuous variables. For example, grain size, porosity, percentage of insoluble residue or electrical properties are all quantities which could be used for describing a stratigraphic section. However, there are two main reasons why relatively little progress has been made in quantifying the description of stratigraphic records by using these parameters. The first reason is simply the enormous amount of work which can be involved in this type of approach. To make a reasonable study using the methods of time-series analysis, one would need a large amount of data and possibly hundreds or even thousands of estimates of parameters would be required, depending on the type of problem. The second reason is perhaps even more discouraging. It is found that many lithological parameters mean very little, geologically speaking, if they are considered as isolated variables and they only become meaningful if they are taken together with other parameters. Well-logging procedures provide a well known example of this situation where a number of different logs are usually interpreted simultaneously. Such measurements would require multiple time-series analyses which are theoretically more complicated and which

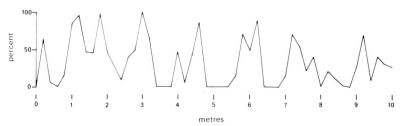

Fig. 7.1. Percent limestone measured at 20-cm intervals (Glencar limestone, NW Ireland).

would require even more data. Some of these difficulties may be overcome with the development of remote-sensing devices and so the methods which are to be discussed in this chapter will possibly become considerably more important.

One of the first steps which ought to be taken in any stratigraphic analysis is to display the lithological variation of the section in some graphic form. If measurements of a lithological parameter are available, then these can be shown as a function of the vertical position in the section. To illustrate this, Fig. 7.1 shows fifty limestone percentages which, in this case, were used as the descriptive parameter of a measured section. The percentages refer to the amount of limestone found in 20-cm intervals which at the same time is the spacing between the sampling points of the random variable X (limestone percentage) yielding the values $X_1 ... X_{50}$. The problem, which will be considered in this chapter, is to find some stochastic process (X_n) which could have generated the observed random variable X_n. We can assume that the values which were actually observed, represent only a small sample of what the random process could produce. It is useful to think of the random process as being defined for all real time, which means that it could have operated in the remote past as well as in the distant future and therefore n, in the process, can take on the values... $-1, 0, +1,...$ If the time variable is assumed to be continuous, then the process X_t is defined for $-\infty < t < +\infty$. Clearly any continuous time process can be treated as a discrete case if samples are taken at equal intervals. In most examples the random processes are assumed to be stationary. Stationarity will be defined more precisely later but, it means roughly that the probability structure of the process does not change throughout its range of definition and therefore, the properties of random variables will not be changed by an arbitrary translation in time. Before considering the specific random processes which can be used as sedimentation models, some of the statistical properties of random variables will be reviewed briefly.

7.2 THE DESCRIPTION OF RANDOM VARIABLES

A sequence of measurements U_n taken at points $n = 1... N$, represents a time series which can be described essentially by four statistical functions. These are the mean square value, the probability density function, the auto-correlation function and the spectral-density function.

Any sequence U_n will have a mean, let us say μ_U and the measurements can always be standardized in such a way that the mean becomes zero and this is achieved simply by the transformation:

$$X_n = U_n - \mu_U \qquad\qquad [7.1]$$

After this transformation, the mean square value becomes the variance of the process and this may be estimated in the normal way by calculating:

$$\text{var}\,(X) = \frac{1}{N} \sum_{n=1}^{N} X_n^2 \qquad\qquad [7.2]$$

The probability density function of the process may be any valid probability distribution but in the stratigraphic examples which will be discussed later, normal or approximately normal distributions, are very often found.

The autocorrelation and spectral-density functions are more typical features of time-series analysis and might not be as widely known as the other statistical functions. Both functions convey the same information because they are Fourier transforms of each other. However, it is found that in some problems one of the forms is more easily interpreted than the other. Most stratigraphical problems are best approached by first investigating the auto-covariance or autocorrelation of the sample.

The autocovariance is the covariance of two realizations of the same variable which is observed at two points, separated in time by the distance h. The covariance as a function of h, can then be written as:

$$C_x(h) = E(X_n, X_{n+h}) = \lim_{N \to \infty} \sum_{1}^{N} X_n X_{n+h} \qquad h = 0, 1, 2, \ldots \qquad [7.3]$$

or, if time is assumed to be a continuous variable:

$$C_x(\tau) = \lim_{T \to \infty} \int_{0}^{T} X_t X_{t+\tau}\, dt \qquad\qquad [7.4]$$

The quantity h in [7.3] or τ in [7.4] is often referred to as the lag which may be either positive or negative. The autocovariance function is symmetrical around zero and:

$$C_x(-h) = C_x(+h) \qquad\qquad [7.5]$$

If either h or τ become zero, [7.3] and [7.4] reduce to the mean square value [7.2] and one can write:

$$C_x(0) = \text{var}\,(X) \qquad [7.6]$$

The autocorrelation, that is the correlation coefficient between measurements of a given lag is given by dividing the covariance by var (x) and this function can therefore be written as:

$$R_x(h) = C_x(h)/C_x(0) \qquad [7.7]$$

If raw data are used for calculating the correlation coefficients of lag h, the following computational formula can be used:

$$r_h = \frac{(N-h) \sum_{1}^{N-h} x_i\, x_{i+h} - \sum_{1}^{N-h} x_i \sum_{1}^{N-h} x_{i+h}}{\sqrt{(N-h)\Sigma x_i^2 - (\Sigma x_i)^2} \cdot \sqrt{(N-h)\Sigma x_{i+h}^2 - (\Sigma x_{i+h})^2}} \qquad [7.8]$$

This type of calculation is normally carried out with automatic computers.

A special technique for obtaining autocorrelation functions from coded data has been used by Carrs and Neidell (1966). The authors made use of a result by Van Vleck and Middleton (1966) who calculated the correlation structure of time series which have been systematically distorted. Such problems arise in communication theory when an electrical signal is passed through a clipper or filter which is used essentially to reduce noise. The action of such a device is to change the amplitude of a signal x, to $f(x)$, a function of the amplitude. Writing x for the amplitude at time t, and y for the amplitude at time $t + h$, Van Vleck and Middleton find that the correlation coefficient R_h of the distorted signal is related to the correlation coefficient r_h before distortion by:

$$R_h = \frac{1}{2\pi \sqrt{1-r^2}} \int_{-\infty}^{+\infty} \int_{-\infty}^{+\infty} f(x)\, f(y) \exp\left[-(x^2 + y^2 - 2rxy)/2(1-r^2)\right]\, dx\, dy \qquad [7.9]$$

provided that the random element in the process is normally (Gaussian) distributed. For the case of extreme clipping which transforms the original signal into a square wave of amplitude ± 1, $f(x)$ has the form:

$$f(x) = +1,\ (x > 0);\ f(x) = -1,\ (x < 0) \qquad [7.10]$$

Introducing this into [7.9], one obtains the surprisingly simple result:

$$R_h = \frac{2}{\pi} \sin^{-1}(r) \qquad [7.11]$$

The relationship can be used in the following way. At first, a stratigraphic section is coded according to two states which are designated as +1 and −1. Next a correlogram is calculated which is based on the series of positive and negative ones and this provides the values R_h. Finally, r_h the original correlogram, is calculated using [7.11]. Carrs and Neidell (1966) applied this method to the analysis of a Carboniferous section of limestone, sands and coals (Archerbeck Borehole, Dumfrieshire, Great Britain) which was simply classified into marine and quasimarine (terrestrial?). Naturally, this method of estimating the autocorrelation function is considerably less accurate than a calculation based on formula [7.8], but on the other hand, it requires considerably less data.

The plot of correlation coefficients as a function of the lag h, is known as the correlogram of the series and it is obvious from [7.7] that $R_x(0) = 1.0$ and $-1.0 \leqslant R_x(h) \leqslant 1.0$. As a simple example, one can calculate the correlogram of a Markov-chain process. This may be done by associating a numerical value with each state. Let the first state in a two-state process equal zero and the second state equal one:

$$
\begin{array}{cc}
& \begin{array}{cc} 0 & \quad 1 \end{array} \\
\mathbf{P} = \begin{array}{c} 0 \\ 1 \end{array} & \begin{bmatrix} p_{00} & p_{01} \\ p_{10} & p_{11} \end{bmatrix}
\end{array}
$$

Then from [7.3]:

$$C_x(h) = E(X_n = X_{n+h} = 1) = P(X_n = 1 = X_{n+h}) = P(X_n = 1) P(X_{n+h} = 1 | X_n = 1)$$

and since:

$$P(X_n = 1) = E(X_n) = \pi_1$$

where π_0, π_1 is the stable probability vector of $\mathbf{P}$:

$$C_x(h) = \pi_1 p_{11}^h$$

and:

$$R_x(h) = p_{11}^h \ (h \geqslant 0) \tag{7.12}$$

The correlogram of this simple Markov chain corresponds precisely to the powers of the transition probability p_{11} and in general, for any Markov chain, the correlogram will be a simple function of $P^{(h)}$. Of course, the autocorrelation function is not a very suitable method for describing Markov chains since this function depends on the numerical values which have been attributed to the various states.

If the correlogram of an independent random process is calculated, in which $E(x_n) = 0$ then $R(h) = 0$ for $h = 1, 2...$, having only a spike-like maximum of 1.0 at $h = 0$. This, of course, is in agreement with the definition of an independent random process which assumes that there is no correlation between X_n and X_{n+h}.

For a second example, consider the entirely deterministic process:

$$X_t = A \sin (2\pi\omega_0 t + \theta k) \qquad [7.13]$$

which is a sine wave with an amplitude A and arbitrary phase shift θk. Formal integration (see [7.4]) gives:

$$C_x(\tau) = \frac{A^2}{2} \cos 2\pi\omega_0 \tau \qquad [7.14]$$

and:

$$R_x(\tau) = \cos 2\pi\omega_0 \tau \qquad [7.15]$$

The correlogram of a sine wave is, therefore, a cosine wave which when $\tau = 0$ equals 1.0. Typically, the information about the phase shift, which is time-dependent, has been lost and it will be noticed that all correlograms are independent of time and purely a function of the lag. This latter property makes the autocorrelation very suitable for defining the stationarity of a process more sharply.

A process is called stationary in the wide sense if:

$$E(X_n) = \mu \text{ and } E(X_n, X_{n+h}) = C_x(h) \qquad [7.16]$$

are constant for all values of n and h. A process is called stationary of order p if:

$$E(X_{n1}X_{n2} ... X_{np}) = E(X_{n1+h} X_{n2+h} ... X_{np+h}) \qquad [7.17]$$

for all $n_1 ... n_p, h$ and it is called stationary in the strict sense if all higher moments of the probability distribution are constant when translated in time. This latter condition would be very difficult to establish from observational data and it is usually sufficient to demonstrate that a process is stationary in the wide sense.

It is found in practice, that geological data are only very rarely, if ever, stationary in either the strict or the wide sense because most stratigraphic data incorporate long-term fluctuations which are commonly called trends. This trend component must be removed before the data can be treated as a stationary process. The problem of trend removal will be discussed later in some detail. If such an operation has been caried out, the stationarity of the residue must be tested. Tests of this type can only involve subdividing the observational series and calculating the statistics equation [7.16] for each subsection. Since it is very unlikely that a series changes its autocovariance without changing the variance, it is usually sufficient to investigate $E(X_n)$ and $E(X_n^2)$ of the subsections (Bendat and Piersol, 1966) which ought to be essentially the same as for the complete section. A non-parametric test such as the run test (cf. Bendat and Piersol, 1966) can be used to establish that the statistics derived from the subsections fluctuate only as might be expected from sampling variations. Any test for stationarity will clearly need a considerable amount of data.

7.3 SIMPLE STATIONARY TIME PROCESSES

This section will introduce a number of relatively simple stationary processes which have been used as models in time-series analysis. The processes to be discussed are stochastic processes because at any point X_n a stochastic element is incorporated and this will be denoted by Z_n. The random values Z_n are uncorrelated with each other and form a series with a zero mean and a constant variance S_z^2. If the Z_n's are normally distributed, then the process will be called a Gaussian random process. As before, $E(x_n) = 0$ which, as has been explained, can always be achieved by transformation.

A process which has an immediate appeal as a model for geological data is the scheme of finite moving averages. Let $a_0 ... a_r$ be a set of constants with the property $\sum_r a_i = 1$, then one may define a process by:

$$X_n = a_0 Z_n + a_1 Z_{n-1} + ... a_r Z_{n-r} \qquad [7.18]$$

Here, each individual realization of the random process is determined by the sum of r uncorrelated random variables which have been given the weights $a_0 ... a_r$.

As an example, consider a river which at regular intervals along its course, receives a tributary. Let it be assumed for the sake of simplicity, that the sediment which is carried by the river at any point is determined by the contributions of the last two tributaries and furthermore that the weight of these contributions remains constant, a_0, a_1. Thus if the sediment load is determined at three points, let us say at n, $n + 1$ and $n + 2$, one will find:

$$
\begin{aligned}
X_n &= a_0 Z_n + Z_{n-1} \\
X_{n+1} &= a_0 Z_{n+1} + a_1 Z_n \\
X_{n+2} &= a_0 Z_{n+2} + a_1 Z_{n+1}
\end{aligned}
\qquad [7.19]
$$

The equation illustrates that by the time position X_{n+2} is reached, the influx of sediments at position X_n has lost its effect on the sedimentation load. This situation is brought out by the autocovariance function. We have, by definition:

$$E(X_n) = 0 \text{ and } \text{var}(X_n) = (a_0^2 + a_1^2 + \dots a_r^2)\, S_z^2$$

and the covariance is:

$$C_x(h) = \left(\sum_{s=0}^{r-h} a_s a_{s+h}\right) S_z^2 \quad (h = 0, 1 \dots r)$$

$$= 0 \ (h = r + 1 \dots) \hspace{4cm} [7.20]$$

The correlogram of any finite moving average process will therefore decline from 1.0 to zero within the interval of r, which is the length of the moving average. The approach to zero can be oscillating whereby positive and negative correlation coefficients may alternate. A special case arises when $a_0 = a_1 = \dots a_r = 1/(r+1)$ in which case the random process X_n is a simple mean of $(r+1)$ uncorrelated random variables and it follows from [7.20] that the correlogram is given by:

$$R_x(h) = 1 - \frac{h}{r+1} \quad (h = 0, 1, \dots r)$$

$$= 0 \qquad (h = r + 1, \dots) \hspace{3cm} [7.21]$$

In this special case, the autocorrelation coefficients decline linearly from 1.0 at $h = 0$ to zero at $h = r + 1$.

Vistelius (1949) observed correlograms of this type when he investigated the correlation structure of bed thickness in the Redbed Formation of Cheleken (U.S.S.R.). He interprets this in the following way. For the formation of a single bed, a number of environmental conditions (u_1, $u_2 \dots u_k$) are necessary. Adding a new factor u_{k+1} leads to the formation of a new bed, but during its formation some of the old factors, let us say, $u_2 \dots u_k$ are still active and the thickness of the bed will, therefore, be determined by the sum of factors $u_2 \dots u_{k+1}$. This type of model can be developed in geological terms and Vistelius associated the formation of the Redbeds of Cheleken with a single source area in which random bursts of erosion occur. Any new sediment producing activity will inherit a number of conditions from the previous cycle of erosion and it is, therefore, related to the past history. In a very similar way, randomly occurring tectonic activity could lead to a sedimentation history and this can be modelled by a moving average scheme. Any upward movement which may create new source areas, will have a profound effect on the sedimentation, but previously existing configurations of basins, drainage systems and other similar factors will also contribute to the environment which determines sedimentation. The same applies to downward movements which may create new areas of deposition but the inherited geomorphology of the surrounding areas remains important.

A correlogram which is very similar to the one which has been reported by Vistelius (1949) was found by Schwarzacher (1964) when investigating the total shale content in successive "cycles" in the Glencar limestone (NW Ireland). In this instance, it was found that a moving average with equal weights and which extended over ten successive "cycles", accounted for the observed correlogram. The extent of this average scheme in terms of stratigraphical thickness is some 30—40 m and the "memory" which determines the clastic component in the sedimentary sequence, therefore, extends a considerable way back into the stratigraphical history.

7.4 AUTOREGRESSIVE PROCESSES

A second type of stochastic processes which have been used extensively in time-series analysis are the autoregressive processes which have the general form of:

$$X(n) = a_1 X_{n-1} + a_2 X_{n-2} + \ldots a_k X_{n-k} + Z_n \qquad [7.22]$$

As the name implies, the event X_n is determined by a linear regression on past events and a stochastic variable Z_n is incorporated at any time n. The simplest of these autoregressive processes is the first-order process:

$$X(n) = a_1 X_{n-1} + Z_n \qquad [7.23]$$

This is a Markov process in the sense that X_n depends, in this particular case, only on the previous value X_{n-1} and if in addition, Z_n is normally distributed, then it is also known as a Gaussian Markov process. By carrying out the successive substitutions:

$$X_n = a(a X_{n-1} + Z_{n-1}) = a^2 X_{n-2} + a Z_{n-1} + Z_n$$

$$= a^3 X_{n-3} + a^2 Z_{n-2} + Z_{n-1} + Z_n$$

$$\vdots$$

it may be seen that:

$$X_n = \sum_{i=0}^{\infty} a^i Z_{n-i}$$

[7.24]

Writing S_z^2 for the variance of Z_n, one can write the variance of the process X_n as:

$$\text{var}(X_n) = S_z^2 \sum_{i=0}^{\infty} a^{2i}$$

and if $|a| < 1$:

$$\text{var}(X_n) = \frac{S_z^2}{1 - a^2} = S_x^2$$

[7.25]

It may be noted that if $|a| \geqslant 1.0$, then the variance S_x^2 would go to infinity and this would contradict the definition of a stationary process.

The covariance function can be derived in a similar way, using [7.24] and one can write:

$$E(X_n X_{n+h}) = E\left[\left(\sum_{i=0}^{\infty} a^i Z_{n-i}\right)\left(\sum_{i=0}^{\infty} a^i Z_{n+h-i}\right)\right]$$

$$= S_z^2 \sum_{i=0}^{\infty} a^{h+i} a^i = a^h S_x^2$$

[7.26]

This leads to the autocorrelation function which by using [7.7] becomes:

$$R(h) = a^h$$

[7.27]

Neither the variance nor the covariance depend on n and this must be so if the process is stationary in time. Clearly the theoretical correlogram ([7.27]) of a first-order process is a simple geometrical curve (Fig. 7.2) which declines rapidly when the autocorrelation is weak and less rapidly when it is strong. Like the discrete Markov-chain model, the first-order autoregressive process can be fitted as a first approximation to a wide variety of time series, but most stratigraphic records seem to require autoregressive processes of higher orders for a full description.

The general autoregressive process of [7.22] can be written in a more

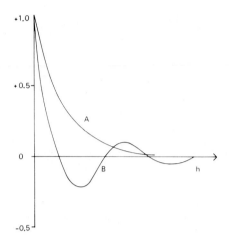

Fig. 7.2. Correlogram of a first (A) and second-order (B) autoregressive process.

compact form as:

$$Z_n = \sum_{i=0}^{k} a_i X_{n-1} \qquad\qquad [7.28]$$

where a_0 is taken as -1.0. Multiplying this by X_{n-h}, ($h = 1, 2, ...$) and taking the expectations, after dividing by S_x^2, one obtains:

$$R(h) = a_1 r_{(h-1)} + a_2 r_{(h-2)} + ...a_k r_{(h-k)} \qquad\qquad [7.29]$$

Setting $h = 1, 2,...$ and remembering that $r_i = r_{-i}$, one obtains a set of equations which are known as the Yule-Walker relations (c.f., Kendall and Stuart, 1968):

$$r_1 = a_1 + a_2 r_1 + a_3 r_2 + ...$$

$$r_2 = a_1 r_1 + a_2 + a_3 r_1 + ...$$

$$\vdots$$

$$[7.30]$$

The same equations can be written in matrix form as:

$$
\begin{bmatrix}
1 & r_1 & r_2 \dots r_{h-1} \\
r_1 & 1 & r_1 \dots r_{h-2} \\
r_2 & r_1 & 1 \dots \\
\cdot & \cdot & \\
\cdot & \cdot & \\
\cdot & \cdot & \\
r_{h-1} & r_{h-2} &
\end{bmatrix}
\begin{bmatrix}
a_1 \\
a_2 \\
\cdot \\
\cdot \\
\cdot \\
a_n
\end{bmatrix}
=
\begin{bmatrix}
r_1 \\
r_2 \\
\cdot \\
\cdot \\
\cdot \\
r_n
\end{bmatrix}
\qquad [7.31]
$$

The use of regression equations leads to some important general results regarding stationary processes which will be outlined without detailed derivation. Any value in a stochastic process can be regarded as consisting of two parts:

$$X_n = (\text{linear combination of } X_{n-1}, \dots X_{n-p}) + Z_{n,p} \qquad [7.32]$$

where $Z_{n,p}$ can be considered as the residual which is uncorrelated with the independent variables $X_{n-1}, \dots$ Equation [7.31] minimizes the residual in the least-square sense and if the correlation structure of the process is known, this will provide the best linear predictor for X_n. If it is next assumed that p goes towards infinity one can write:

$$X_n = R_n + Z_n \qquad [7.33]$$

where R_n is the regression term which is purely deterministic and Z_n the residual which has zero mean and which is uncorrelated with any past value of the series. The residual Z_n is also called the innovation at time n (Cox and Miller, 1965, p. 286). One can write:

$$E(Z_n) = 0 \text{ and } E(Z_n, X_{n-k}) = 0 \quad (k = 1, 2 \dots) \qquad [7.34]$$

Since X_{n-1} is incorporated into the process at time $n-1$ and at time n is already part of the linear system $X_{n-1}, X_{n-2} \dots$, one can also write that:

$$\text{cov}(Z_n, Z_{n-1}) = 0 \qquad [7.35]$$

This leads to a twofold decomposition of the process which is indicated by the two equations:

$$X_n = V_n + W_n \tag{7.36}$$

and:

$$W_n = b_0 Z_n + b_1 Z_{n-1} + \ldots \tag{7.37}$$

Any stationary process can be expressed as the sum of two processes which are uncorrelated with each other, the process V_n which is deterministic and the process W_n which is a moving average process of infinite extent containing all innovations. Equations [7.36] and [7.37] are known as Wold's decomposition theorem which can be applied to any stationary process, however, if the physical processes which generated the time series were non-linear the decomposition becomes rather artificial.

As a specific example we consider the second-order process:

$$X_n = a_1 X_{n-1} + a_2 X_{n-2} + Z_n \tag{7.38}$$

The Yule-Walker equations for this process are:

$$\begin{bmatrix} 1 & r_1 \\ r_1 & 1 \end{bmatrix} \begin{bmatrix} a_1 \\ a_2 \end{bmatrix} = \begin{bmatrix} r_1 \\ r_2 \end{bmatrix}$$

and:

$$a_1 = -\frac{r_1(1 - r_2)}{1 - r_1^2}, \quad a_2 = -\frac{r_2 - r_1^2}{1 - r_1^2} \tag{7.39}$$

which means that the two regression coefficients a_1 and a_2 can be calculated from the first two autocorrelation coefficients and vice versa:

$$r_1 = +\frac{a_1}{1 - a_2} \quad r_2 = a_2 - \frac{a_1^2}{1 - a_2} \tag{7.40}$$

This gives the correlation coefficients in terms of the regression coefficients. Equation [7.32] can be expanded into an expression:

$$X_n = W_0 Z_n + W_1 Z_{n-1} + W_2 Z_{n-2} + \ldots \tag{7.41}$$

giving the process as the weighted sum of all the previous random variables.

If [7.41] is compared with [7.38], it can be seen that:

$$W_0 = 1, \ W_1 = a_1 W_0, \ W_k = a_1 W_{k-1} + a_2 W_{k-2} \qquad [7.42]$$

If ξ_1 and ξ_2 are taken as the two roots of the quadratic equation:

$$\xi^2 - a_1 \xi - a_2 = 0 \qquad [7.43]$$

then:

$$W_k = c_1 \, \xi_1^k + c_2 \, \xi_2^k \qquad [7.44]$$

where c_1, c_2 are constants which can be determined in such a way that the initial conditions for [7.42] are fulfilled, that is:

$$W_0 = c_1 + c_2 = 1, \quad W_1 = c_1 \xi_1 + c_2 \xi_2 = a_1 \qquad [7.45]$$

If X_n in [7.41] is to be a random variable, its variance:

$$S_x^2 = S_z^2 \sum_{k=0}^{\infty} W_k^2$$

must be finite and this, by [7.41], will be the case when:

$$|\xi_i| < 1 (i = 1, 2) \text{ or by [7.38]} \ |a_1^2 + 4a_2| < 1 \qquad [7.46]$$

Under these conditions and using [7.44], one can write:

$$\sum_{k=0}^{\infty} W_k^2 = \frac{c_1^2}{1 - \xi_1^2} + \frac{2c_1 c_2}{1 - \xi_1 \xi_2} + \frac{c_2^2}{1 - \xi_2^2}$$

which, after taking account of the initial conditions, can be transformed into:

$$S_x^2 = \frac{(1 - a_2)}{(1 - a_2)\,[(1 - a_2)^2 - a_1^2]} \, S_z^2 \qquad [7.47]$$

This gives the variance of the second-order process in terms of S_z^2 and the two regression coefficients.

In order to obtain the autocorrelation coefficients $r(h)$, one can start from [7.29]. It can be seen that:

$$r_h = a_1 r_{h-1} + a_2 r_{h-2} \qquad [7.48]$$

This is a difference equation of precisely the same form as [7.41]. The initial conditions here are:

$$W_0 = r_0 = c_1 + c_2 = 1 \text{ and from } [7.40] \ W_1 = c_1\xi_1 + c_2\xi_2 = \frac{a_1}{1-a_2} \quad [7.49]$$

Subject to the condition for stationarity ([7.46]), we find that the roots ξ_1, ξ_2 are complex when $a_1^2 + 4a_2 < 0$. This can be conveniently expressed by:

$$\xi_1 = pe^{i\theta}, \ \xi_2 = pe^{-i\theta}$$

where $p = \sqrt{-a_2}$ and writing:

$$\tan \theta = \frac{a_1}{\sqrt{-(a_1^2 + 4a_2)}}$$

the autocorrelation function becomes:

$$R(h) = p^h \frac{\cos (h\theta + \psi)}{\cos \psi} \quad [7.50]$$

where $\tan \psi = [(1 + p^2)/(1 - p^2)]\tan \theta$ (see Kendall and Stuart, 1968). A typical correlogram is shown in Fig. 7.2 and it is evident from [7.50] that it consists of a cosine wave which oscillates with period $2\pi/\theta$ and which is exponentially damped, according to the damping coefficient p.

It is particularly instructive to compare the correlogram of the second-order autoregressive process with a stochastic process containing a deterministic harmonic component. For example, a series of observations on a sine wave can be disturbed by a random variable and this might be interpreted as an observational error. Such a process can be written as:

$$X_n = A \sin \omega_0 n + Z_n \quad [7.51]$$

Assuming as before that $E(x_n) = 0$, we obtain as in [7.16]:

$$E(X_n, X_{n+h}) = \frac{A^2}{2} \cos \omega_0 h \quad [7.52]$$

and the variance of the process:

$$E(X_n^2) = \frac{A^2}{2} + S_z^2 \quad [7.53]$$

and therefore the autocorrelation function:

$$R(h) = \frac{A^2}{A^2 + 2S_z^2} \cos \omega_0 h \quad (h > 0) \tag{7.54}$$

A slight modification of [7.45] makes the harmonic process somewhat more general and one can assume that the deterministic component consists not of one, but a number, let us say p, different sine waves which can be written as:

$$X_n = \sum_{i=1}^{p} A_i \sin (n\omega_{0i} + v_i) + Z_n \tag{7.55}$$

The different phases $v_1 \ldots v_p$ are independently and rectangularly distributed over $(0, 2\pi)$ and the A_i's and ω_{0i}'s are constants, Z_n is again a random variable and could represent an observational error. The autocorrelation function of [7.49] can be evaluated as:

$$R(h) = (1/\Sigma A_i^2 + 2S_z^2) \Sigma A_i^2 \cos h\omega_{0i} \ (h > 0) \tag{7.56}$$

This correlogram, like the one of [7.49], is a cosine wave with an amplitude which is smaller than 1.0, the reduction being due to the superimposed error. However, the amplitude of these correlograms remains constant and, in contrast to the autoregressive schemes, there is no damping of the oscillating correlogram. This difference between a partly deterministic and a completely stochastic model will be briefly discussed in the following section.

7.5 OSCILLATING AND CYCLICAL SEDIMENTATION MODELS

The second-order autoregressive model was introduced by Yule (1927) when he investigated the behaviour of a randomly disturbed pendulum. Yule visualized the following situation. A pendulum is suspended from the ceiling of a room which is full of little boys armed with pea shooters. The boys will soon begin to bombard the pendulum. Any of the random hits, $\epsilon(t)$, will cause the pendulum to get out of its position and its movement can be described by a stochastic differential equation:

$$\frac{d^2x}{dt^2} + a \frac{dx}{dt} + \omega^2 x = \epsilon(t) \tag{7.57}$$

The second-order difference equation of [7.38] provides an approximation to [7.57]. The movement of the pendulum will be irregular because the hits are irregular and both phase and amplitude changes will occur continuously. Yule originally used the second-order autoregressive process to describe an observed series of sunspot intensities. It has been known for a considerable time that the sun's activity shows a fairly regular pattern with the greatest number of sunspots occurring approximately every eleven years. A number of attempts had been made previously to analyse this phenomenon in terms of strictly cyclical movements and models of the type [7.49] had been fitted to the observations. Yule's interpretation of the sunspot activity as a stochastic process was very successful although Whittle (1954) showed, as will be discussed later, that the second-order autoregressive model can be improved. It is clear from the correlograms of stochastic and partially deterministic processes that the essential difference between the two models is the predictability of the processes. For example, if a harmonic model is adopted for the sunspot process and if an observation shows that there is a sunspot maximum at the present time, then the best prediction of when the maxima will occur again, for any times in the distant future, will be in multiples of eleven years. A similar prediction can be made for the immediate future if the autoregressive model is adopted but for the more distant future, only the mean intensity can be given as an estimate. The decrease in predictability can be seen clearly on the graphed correlograms which indicate that the maxima and minima soon disappear with damping. The physical cause of the difference between the two models is the different nature of the disturbance. In the harmonic model, the disturbances are superimposed and do not interfere with the actual process. For example, they may be observational errors which clearly have nothing to do with the physical phenomenon. In the autoregressive process the disturbances are an inherent part of the system and are therefore incorporated into the process. A simple example makes this situation clearer. Assume that a sinoidal a.c. voltage is recorded by some chart recorder. The voltage may be disturbed by some superimposed noise. This noise is caused by random voltages which can make the graph quite irregular. However, as long as there is still any evidence of the a.c. voltage, one can extract the originally sinoidal signal and the record will be that of a disturbed harmonic process. Next, the assumption is made that the disturbances or noise actually affects the chart drive of the recorder which may go at different speeds according to the random signal. In this case, if one considers the record strictly as a physical phenomenon, one is dealing with a stochastic process which incorporates the random disturbances.

Both the disturbed harmonic-process model and the second-order autoregressive-process model have something in common. They generate sequences which oscillate more or less regularly around the mean value. The models are therefore important in connection with the study of "cyclic" sedimentation which will be discussed more fully later, but attention can be drawn now to the different meaning of the word "cycle" when it is used by the geologist in contrast to the mathematicians' use of the word. The mathematician uses the word "cyclic" only in connection with strictly periodic processes. If:

$$X_n = X_n + r\omega \ (r = 1, 2, ...) \tag{7.58}$$

where ω is the period of the process, then the series is said to be cyclic in the mathematical sense. A series can be cyclic if it contains an added random element. If a series is generated by a number of harmonic terms, it does not necessarily follow that it is cyclic. After all, it is known from Fourier's theorem that any function within certain limits can be represented by a sum of trigonometric terms and this applies also for example to functions which are steadily increasing. For sequences which fluctuate more or less regularly, Kendall (1946) used the term oscillatory and other authors have used the term quasi-periodicity for such behaviour.

If the geologist uses the term "cyclic" he hardly ever means anything as accurately defined as a mathematical cycle. He usually refers to oscillations or fluctuations in quasi-periodic systems. The term "cyclic" sedimentation however, is very firmly embedded in the geological vocabulary and would be difficult to replace. To avoid confusion one can refer to "geological cyclicity" and "true or mathematical cyclicity" when the distinction becomes important.

It is clear from examining the second-order autoregressive process that the oscillating behaviour depends on the damping coefficients p in the correlogram ([7.50]). If p is quite large then the relatively regular oscillations will persist and the correlogram will damp very slowly. If p is small, then the oscillations are much more irregular. When p approaches zero, the second coefficient of regression a_2 also approaches zero. This means that the second-order process becomes a first-order regressive process which shows no oscillating behaviour.

The presence or absence of oscillating behaviour may give valuable information about the processes which were involved in generating the stratigraphic sections and if oscillations are observed then they must be capable of an explanation in geological terms. Environmental systems are usually more complex than the simple model of a randomly disturbed pendulum. Nevertheless, if the concept of oscillating behaviour is reduced to its basic essentials, it is easy to see that one is dealing with processes which often operate in physical systems. Consider the simple mechanical system which is shown in Fig. 7.3. A mass M is pulled by a spring towards the right and this force is resisted by the inertia of the mass. The spring force is assumed to be proportional to the mass and one can therefore write the well known equation of harmonic movement as:

$$M \frac{d^2x}{dt^2} = -ax \tag{7.59}$$

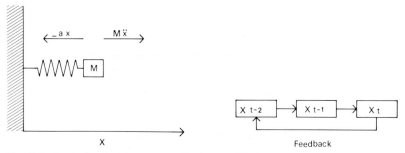

Fig. 7.3. A mechanical system producing oscillations.

Fig. 7.4. The general system for producing oscillating processes.

Equation [7.59] can also be written as a finite-difference equation:

$$x_{t+1} = cx_t - x_{t-1} \qquad\qquad [7.60]$$

where c is a constant depending on M and Δt. Equation [7.60] is also a second-order autoregressive process with the stochastic element missing. Fig. 7.4 shows diagrammatically the relationship between the stages if the oscillating system is considered as a time sequence. Stage t is seen to be linked, not only to stage $t-1$, but also to stage $t-2$. Because the latter stage has a controlling influence on the output at $x(t)$, one can visualize the linkage as a feed-back loop which is an important feature of many dynamic models in geology (see Harbaugh and Bonham-Carter, 1970). A number of examples will be discussed later where the self-regulating processes are actually part of the environmental system and it will be found that feed-back mechanisms are important in tectonic, biological and climatic control systems.

Once again, the autoregressive model is very general and cannot be used to identify any particular environmental factor which may have been responsible for oscillating sedimentation but it is a considerable help in understanding why geological cycles do occur.

7.6 THE ESTIMATION AND TESTING OF CORRELOGRAMS

The theoretical treatment of various stationary processes shows them to have characteristic correlograms and one might hope that it would be possible to identify a process, once an empirical correlogram has been calculated. Any empirical correlogram is an estimate of the true autocorrelation function and it can be shown that the variance of an estimated autocorrelation coefficient $\hat{r}_h$ depends on all the autocorrelations in the function (Bartlett, 1946), which means that it cannot be estimated from a finite series. Approximations can be made for large samples and one can calculate var $(\hat{r}_h)$ only if the original data were part of an independent random series. In this case, one can write for the variance of the covariance function:

$$E\left(\frac{1}{n-h}\sum_{i=1}^{n-h} x_i x_{i+h}\right)^2 = \frac{1}{(n-h)^2}\ E\left(\Sigma x_i^2\ x_{i+h}^2 + 2\Sigma_i\ x_{i+h}\ x_m\ x_{m+h}\right); i \neq m$$

$$= \frac{1}{(n-h)^2}\ E(\Sigma x_i^2\ x_{i+h}^2) = \frac{1}{n-h}\mathrm{var}^2\ (x) \qquad [7.61]$$

and for large samples one can therefore write:

$$\mathrm{var}\ (\hat{r}_h) = \frac{1}{n-h}\frac{\mathrm{var}^2\ (x)}{\mathrm{var}^2\ (x)} = \frac{1}{n-h} \qquad\qquad [7.62]$$

One is however, usually not interested in determining whether a single auto-correlation coefficient is significant. The main aim in interpreting an observed correlogram is to deduce something about the structure of the random process which generated the particular correlogram. Work with artificially constructed series has shown that the correlograms which were calculated from a relatively small number of observations can give a most misleading picture (Kendall, 1946). Any abnormally high value in a short series will be propagated throughout the correlogram, simply because the series is not long enough to compensate for an accidental disturbance like this. This may cause a second-order process to show no damping in an observed correlogram although theoretically, damping ought to occur. Damping is caused by the random element in the amplitudes and the phases of the oscillations. If only a few oscillations, perhaps in the order of five, are used to calculate the correlograms, then the disturbances are not large enough to cancel out subsequent oscillations and the correlogram will therefore be non-damping like that of a disturbed harmonic process. Clearly, some statistical tests are needed to help in the interpretation of empirical correlograms.

A stratigraphic section which was measured in the Dartry Limestone (Carboniferous, NW Ireland) is introduced as an example in order to illustrate the methods which can be used for interpreting observed correlograms. The data consist of bed-thickness indices which were determined at 20-cm intervals throughout the section and this thickness index happens to be a very good descriptive parameter in this particular case (Schwarzacher, 1964). The section is 60.8 m long and is represented by 304 observations. The data show the index to be oscillating and there are approximately twenty fairly regular oscillations. Estimates of the autocorrelation coefficients were calculated by the formula which was given in [7.8] and the first twenty of these estimates

TABLE 7.1

Autocorrelation of bed-thickness data from the Dartry Limestone

h	r	h	r	h	r
1	0.8463	11	−0.3536	21	−0.0046
2	0.6613	12	−0.1951	22	−0.0243
3	0.3631	13	−0.0426	23	−0.0194
4	0.0851	14	0.0734	24	0.0053
5	−0.1766	15	0.1439	25	0.0580
6	−0.3885	16	0.1659	26	0.1184
7	−0.5479	17	0.1564	27	0.1754
8	−0.6104	18	0.1208	28	0.2109
9	−0.6008	19	0.0744	29	0.2168
10	−0.7983	20	0.0276	30	0.1898

are given in Table 7.1. The correlogram (Fig. 7.5) is obviously that of some oscillating process and so an attempt is made to fit a second-order auto-regressive equation to the data. Solving the Yule-Walker equation [7.31] gives the values $a_1 = 1.0101$ and $a_2 = -0.1935$. A slightly better estimate can be made by making use of the first four correlation coefficients and if one solves:

$$\begin{bmatrix} r_2 & r_1 \\ r_3 & r_2 \end{bmatrix} \times \begin{bmatrix} a_1 \\ a_2 \end{bmatrix} = \begin{bmatrix} r_3 \\ r_4 \end{bmatrix} \qquad [7.63]$$

one obtains $a_1 = 1.2942$ and $a_2 = -0.5821$. These two values can now be used to calculate the theoretical correlogram of a second-order process from

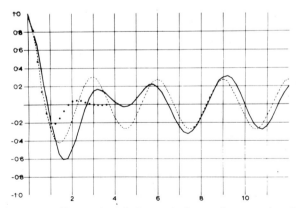

Fig. 7.5. Observed and theoretical correlograms for the Dartry Limestone, NW Ireland (after Schwarzacher, 1964).

[7.50] and this is also shown in graphic form in Fig. 7.5. Casual comparison of the two shows that the second-order process does not provide a very good fit for the empirical correlogram. Not only is the damping quite different for the two correlograms, but the theoretical correlogram seems to indicate a shorter wavelength of oscillations; 11.25 lag units, compared with approximately 15 lag units in the observed correlogram. The latter appears to show damping up to about lag 24 but the next positive maximum and subsequent maxima and minima (which are not shown on the diagram) keep to an almost constant amplitude. There is some evidence in the original data that well developed and less well developed oscillations alternate with each other. This suggests that in addition to the "geological cycle" which is indicated on the correlogram and which has a thickness of approximately 3 m (fifteen lag units), one may be dealing with another "cycle" twice this thickness. There would be only about ten of these long oscillations in the stratigraphic record and if they are disturbed by a few random fluctuations they would have to be regarded as a short observational series which would yield a correlogram which shows no signs of damping. Under such circumstances, statistical tests are needed to establish how well an autoregressive process of a given order fits a particular observed correlogram.

All the practical tests of correlograms are based on larger samples and the time series is in fact treated as if it were an ordinary regression problem. Accepting this limitation, the correlation coefficients can be tested and the order of the autoregressive system is best estimated by examining the partial correlation coefficients. In its normal notation, a partial coefficient $r_{1k,2,3\ldots k-1}$ denotes the correlation which exists between variable 1 and k, after the correlation between variable 1 and all other variables which are not k have been eliminated. For example, $r_{13,2}$ would indicate the observed correlation between x_n and x_{n-2} when the effect of x_{n-1} has been removed. Clearly, if one is dealing with a first-order (Markov) process $r_{13,2}$ should be zero and similarly $r_{14,23}$ should be zero for a second-order autoregressive process. The partial correlation coefficients are usually calculated by:

$$
r_{1k,23\ldots} = \frac{\begin{vmatrix} 1 & r_1 & r_2 \ldots r_{k-2} & r_1 \\ r_1 & 1 & r_1 \ldots r_{k-3} & r_2 \\ \cdot & \cdot & \cdot & \cdot \\ \cdot & \cdot & \cdot & \cdot \\ \cdot & \cdot & \cdot & \cdot \\ r_{k-1} & r_{k-2} & \ldots \cdot & r_1 & r_k \end{vmatrix}}{\begin{vmatrix} 1 & r_1 & r_2 \ldots r_{k-2} & r_{k-1} \\ r_1 & 1 & r_1 & r_{k-3} & r_{k-2} \\ \cdot & \cdot & \cdot & \cdot & \cdot \\ \cdot & \cdot & \cdot & \cdot & \cdot \\ r_{k-1} & r_{k-2} & \ldots & r_1 & 1 \end{vmatrix}}
$$

[7.64]

This is equivalent to fitting a $(k-1)$ coefficient to a process by means of the Yule-Walker equations. If one progressively determines:

$$x_n = a_1 + x_{n-1} + \epsilon_n$$

$$x_n = a_1 + x_{n-1} + a_2\, x_{n-2} + \epsilon_n$$

$$x_n = a_1 + x_{n-1} + \ldots\ldots\ldots\ldots a_{k-1}\, x_{n-k+1}$$

then:

$$r_{1k}, \ldots = a_{k-1} \qquad\qquad\qquad\qquad\qquad\qquad [7.65]$$

This calculation was carried out for the Dartry limestone data and the first ten partial autocorrelation coefficients are shown in Table 7.2. Quenouille (1947, 1949), who has developed the tests which are based on partial-correlation coefficients in time-series analysis, has shown that correlations can be tested for significant deviations from zero in the ordinary manner, provided that three extra degrees of freedom are allowed. The 5% significance which was calculated for the data in Table 7.2 assumes that the deviations are normal. Following [7.62] and the Quenouille correction, an approximate 5% significance level for the correlation coefficients is given by $1.96\,(n + 3 + p)^{-\frac{1}{2}}$ where $n = 304$ and $p = 1\ldots9$. Clearly, $r_{1\,4}$ ... is highly significant and

TABLE 7.2

Partial autocorrelation coefficients of the Dartry limestone data

Coefficient	Value	5%-level
r_1	0.8463	0.1118
$r_{13,2}$	−0.1935	0.1120
$r_{14,23}$	−0.5224	0.1122
$r_{15,23}$ ·	−0.1262	0.1124
$r_{16,23}$ ···	−0.0522	0.1126
$r_{17,23}$···	−0.1669	0.1128
$r_{18,23}$···	−0.1912	0.1130
$r_{19,23}$···	−0.3012	0.1132
$r_{1\,10,23}$···	−0.0181	0.1133

this indicates that a second-order autoregressive scheme is quite insufficient for representing the observations. There are also significant deviations from zero near the partial coefficient which corresponds to a lag of 7—8 and this is near the minimum of the correlogram. The tests therefore suggest some feed-back link in the generating mechanism of the series which extends over approximately half a "cycle" length; as yet there is no geological explanation for this.

A somewhat different testing procedure was proposed by Whittle (1952) and very successfully applied in an investigation of the sunspot intensity observation series (Whittle, 1954). Whittle studied the effect of fitting different orders of autoregressive schemes by comparing the residual variances after such models had been fitted. Consider for example, that α coefficients have been fitted by making use of the correlation coefficients, $r_p, r_q...r_r...r_u$, then the residual variance of the fitted scheme is proportional to:

$$
\mathbf{v} =
\begin{bmatrix}
1 & r_p & r_q & \cdots r_u \\
r_p & 1 & r_{q-p} & \cdots r_{u-p} \\
r_q & r_{q-p} & 1 & \cdots r_{u-q} \\
\cdot & \cdot & \cdot & \cdot \\
\cdot & \cdot & \cdot & \cdot \\
r_u & r_{u-p} & r_{u-q} & \cdots 1
\end{bmatrix}
\cdot
\begin{bmatrix}
1 & r_{q-p} & \cdots r_{u-p} \\
r_{q-p} & 1 & \cdots r_{u-p} \\
\cdot & \cdot & \cdot \\
\cdot & \cdot & \cdot \\
r_{u-p} & r_{u-q} & \cdots 1
\end{bmatrix}
\qquad [7.66]
$$

If, by comparison, a scheme with $\alpha-\beta$ coefficients is tested, then the residual variance, let us say v', is found by repeating the calculation of [7.66] and omitting the β-coefficients. If these coefficients are in fact, zero, then:

$$
\psi^2 = (n-\alpha)\ln\left(\frac{v'}{v}\right) \qquad [7.67]
$$

is approximately χ^2-distributed with β degrees of freedom. For example, in testing the Dartry limestone data one may be interested to compare a second-order with a third-order autoregressive scheme and one obtains for the residual variance of the second-order scheme:

$$
\mathbf{v} =
\begin{bmatrix}
1 & r_1 & r_2 \\
r_1 & 1 & r_1 \\
r_2 & r_1 & 1
\end{bmatrix}
\cdot
\begin{bmatrix}
1 & r_1 \\
r_1 & 1
\end{bmatrix}
= 0.2731
$$

In a similar way, the residual variance of the third-order scheme is found as $v' = 0.1987$. Using the statistic of [7.67] we find:

$$\psi^2 = (304 - 1) \ln \left(\frac{0.1987}{0.2731} \right) = 95.72$$

and from the χ^2-distribution with one degree of freedom $p \ll 0.001$. Clearly, the third-order contribution is highly significant and if one tests a scheme employing $r_1\ r_2$ against one containing $r_1\ r_2\ r_8$ (which incorporates a lag of one half "cycle" length) one obtains $\psi^2 = 29.61$ with $p \ll 0.001$. This is in good agreement with the partial-correlation coefficient tests. One may finally compare $r_1\ r_2$ with $r_1\ r_2\ r_{15}$ which represents a full "cycle" lag and find $\psi^2 = 4.45, p \approx 0.25$, which is of doubtful significance.

7.7 THE INTERPRETATION OF EMPIRICAL CORRELOGRAMS

Attention has already been drawn to the geological importance of recognizing the oscillating nature of stratigraphic records and this will be discussed in more detail in the chapter on cyclic sedimentation. As far as tests for this property are concerned, one can clearly use the methods which have been discussed in the previous section to establish whether second or higher-order autoregressive processes provide a significantly better fit than a first-order process, but more direct tests will be introduced later. The choice of a model which is fitted to a particular set of data however, depends largely on geological reasoning. The statistical tests can only be used to support that choice. The situation is rather similar to the one in standard regression analysis where observational data can be approximated by more and more complicated terms, without gaining anything towards the physical interpretation of the observed phenomenon. If there are good geological reasons for accepting a certain model, it could be adopted, even if other models provide a better fit for the observed data. This principle was followed by Schwarzacher (1964) when a randomly disturbed harmonic process was fitted to the Dartry limestone data which were used, in the previous examples. It was demonstrated for the data that a simple second-order autoregressive process is an insufficient model. On the other hand, a remarkably good fit is provided by the harmonic model. The analysis of partial-correlation coefficient and variance ratios, as illustrated in the previous section, shows that a third of fourth-order process incorporating lags of half a "cycle" length might also provide good or perhaps even better models. However, there is no concrete geological explanation for such behaviour and the harmonic model, although it is nowadays not considered as fashionable, must remain a simple alternative. The analysis of a time series may provide guide lines for constructing a model but it does not provide a ready-made explanation of a stratigraphic section. It can be very profitable to use time-series analysis as a method of testing specific models and also to decide between specific alternative models. This method of working is well illustrated by Whittle's (1954) analysis of sunspot-intensity observations.

It has already been mentioned that sunspot counts show a pronounced quasi periodicity of approximately eleven years between their maxima and Yule obtained a good result by fitting a second-order process to the data. Whittle used Alfvén's theory of sunspot formation as an alternative model to be tested against the disturbed pendulum hypothesis of Yule. The Alfvén theory claims that near the centre of the sun there are two regions of activity which can amplify any magnetic disturbance which passes through them. Once a disturbance has gone through this zone it will migrate towards the surface where it creates a sunspot. The amplification of the original signal produces a disturbance which travels to the opposite active region where it is amplified to produce eventually, a sunspot on the opposite hemisphere. The transit time between the two centres is assumed to be roughly eleven years and it is this time lag which is essential for the oscillation. If the total sunspot intensity is considered, meaning spots on both the southern and northern hemispheres, then the intensity at anytime ought to be strongly determined by what happened 22 years ago and tests must be made as to which of the two models:

$$x_t = a_1 \, x_{t-1} + a_2 \, x_{t-2} \text{ and } x_t = b_1 \, x_{t-1} + b_2 \, x_{t-22}$$

provides a better fit. The test of the residual variance ratio shows the Alfvén model to be very much better than the Yule model and it can therefore be said that the observational data support Alfvén's hypothesis.

Whittle's study is not only interesting for its clear definition of the problem and method of analysis but also because of the mechanism which could produce such a relatively long time lag. There is some evidence that some very long time lags are important in generating stratigraphic sections and the migration of subcrustal disturbances is one possible mechanism for explaining such long time lags. For example, in a preliminary investigation of the Upper Pennsylvanian cyclothems of Kansas (U.S.A.), it was found that if the section is represented in terms of lithological percentages at 5-feet intervals (150 cm), then an expression:

$$x_{(t)} = a_1 \, x_{t-1} + a_2 \, x_{t-2} + a_3 \, x_{t-30}$$

provides a fairly good model for the stratigraphic record (Schwarzacher, 1967). The lag term x_{t-30} implies that sedimentation at any particular point in the section is partly determined by what occurred 45 m previously in the section and when this is translated into time it must represent a considerable period. Any hypothesis trying to explain this type of sedimentation must incorporate a mechanism which provides for this "memory" effect and this must be one of the most important factors in çyclothem sedimentation, simply because it is one of the essential parts of the model.

7.8 MULTIPLE TIME-SERIES ANALYSIS

It has been stated at the beginning of this chapter that a single sedimentological parameter hardly ever provides a satisfactory description of a stratigraphic section. Most quantitative logs contain a number of variables which have been measured simultaneously and the appropriate analysis for such data is a multiple time-series analysis. No attempt as yet, has been made to examine a stratigraphic problem by these methods but Agterberg (1966) applied multivariate Markov schemes to the analysis of geochemical data. Quenouille (1957) has written a monograph on the analysis of multiple time series and the very short account which will be given here is based on his book and the paper by Agterberg. The subject is introduced not only because it is a potential method for analysing stratigraphic data, but also because it has a direct bearing on what has been said about the interpretation of single time series.

If a number of p variables are measured simultaneously at n successive times, then the data can be arranged in a $p \times n$ matrix $\mathbf{X}$ in which each row represents the measurements taken on a single variable and each column the state of all the variables at a given time t. This column vector is written as x_t. It is further convenient to introduce a shift operator D with the property $Dx_{i,t} = x_{i,t-1}$ and $D^h x_t = x_{t-h}$. Using this notation, one can write the covariance matrix $\mathbf{C}(h)$ as:

$$\mathbf{C}(h) = \mathbf{X}\, D^h\, \mathbf{X}' \qquad\qquad [7.68]$$

where $\mathbf{X}'$ stands for the transpose of $\mathbf{X}$. Corresponding to the univariate case [7.28] one can write the autoregressive scheme as:

$$z_{i,t} = \sum_{s=0}^{k} \sum_{j=1}^{p} a_{(ij)s}\, x_{j,t-s} \qquad\qquad [7.69]$$

or by making use of the operator D:

$$z_t = \sum_{s=0}^{k} \mathbf{A}_s\, D^s x_t = \mathbf{A}(D)\, x_t \qquad\qquad [7.70]$$

where:

$$A(D) = \sum_{s=0}^{k} A_s D^s \qquad [7.71]$$

The solution of [7.63] can be written as:

$$x_t = A^{-1} Dz_t \qquad [7.72]$$

In a similar way, one can find the multivariate case of a moving average scheme and can find:

$$x_t = B(D) z_t \qquad [7.73]$$

where: $$B(D) = \sum_{s=0}^{l} B_s D^s \qquad [7.74]$$

The terms A and B are polynomials in D. For example, in the Markov scheme where an observation at time t only depends on the previous observation x_{t-1}, one has for A:

$$A(D) = A_0 + A_1 D = A_0(I - U_1 D) \qquad [7.75]$$

where: $$U_1 = -A_0^{-1} A_1 \qquad [7.76]$$

Each element of A represents the predictable fraction in terms of a probability which a single variable contributes in a transition from t to $t + 1$. If the scheme by which this transition occurs is unknown, one depends on observations of the process as a whole and the multivariate Markov process can be written as:

$$X_{t+1} = Ux_t + Z_{t+1} \qquad [7.77]$$

This equation is equivalent to [7.23] and it stands for the p equations:

$$x_{1,t+1} = u_{11}x_{1t} + u_{12}x_{2t} + \dots z_{1,t+1}$$

$$\cdot$$
$$\cdot$$
$$\cdot$$

$$x_{p,t+1} = u_{p1}x_{pt} + \dots\dots\dots\dots z_{p,t+1}$$

Following [7.68] one has for the covariance matrix:

$$C_{(1)} = E(x_{t+1}\, x_t') = \frac{1}{n-1} \sum_{}^{n} (x_{t+1} x_t')$$

but from [7.76] it is seen that:

$$\sum_{}^{n} x_{t+1}\, x_t' = U\, \Sigma x_t x_t'$$

and as the independent random term disappears on summation, one can write:

$$C_{(1)} = UC_0 \text{ or } U = C_1 C_0^{-1} \tag{7.78}$$

The last equation enables one to obtain an estimate of matrix U from the calculated covariance matrices of lag zero and one. The matrix U is generally asymmetrical and its eigenvalues can be either real or complex. In the latter case, the Markov scheme will be oscillating and this means that an oscillating sequence can be generated by a multivariate first-order process. Considering the importance which has been attributed to oscillating systems earlier in this chapter, one has to keep in mind that a multivariate Markov scheme may be an alternative interpretation of more complex schemes which arose when stratigraphic sections were analysed as univariate problems. The following short analysis (Quenouille, 1957) will show what happens if an observed process is under-represented which means that a variable is incorrectly excluded. According to [7.69], an autoregressive process has the form:

$$\begin{vmatrix} K & \vdots & L \\ \cdots & \vdots & \cdots \\ M & \vdots & N \end{vmatrix} \begin{vmatrix} X_1 \\ \cdots \\ X_2 \end{vmatrix} = \begin{vmatrix} Z_1 \\ Z_2 \end{vmatrix} \tag{7.79}$$

where K, L, M, N are polynomials in D. Eliminating the variable X_2 gives the scheme:

$$(K - LN^{-1} M)\, X = Z_1 - LN^{-1} Z_2 \tag{7.80}$$

For example, consider the bivariate Markov process for which:

$$A = \begin{bmatrix} a_0 + a_1 D & b_0 + b_1 D \\ c_0 + c_1 D & e_0 + e_1 D \end{bmatrix}$$

with two variables x_1 x_2. If this bivariate process is represented as a single time series then by [7.80] the scheme becomes:

$$a_0 e_0 - b_0 c_0 + (a_1 e_0 - e_1 a_1 - c_1 b_1 - b_0 c_1) D + (a_1 e_1 - b_1 c_1) D^2 =$$

$$= z_1 e_0 + z_2 b_0 + (e_1 z_1 - b_1 z_2) D$$

which is a second-order scheme with correlated residuals. Quite generally, the omission of v variables from a Markov scheme changes it to a scheme of order $v + 1$ with correlated residuals.

The problem of including the right number of variables is however only one of the difficulties which arise in the analysis of multiple series. We have only concentrated here on autoregressive schemes. If moving average schemes and mixed moving average autoregressive schemes are also considered, then one runs into great difficulties in identifying the probability structure of the process. Even if such an identification is possible, it is still difficult to show that a chosen model is unique and once again the methods of multiple time-series analysis can only be developed when there is enough geological beckground knowledge to allow the development of reasonable models.

Chapter 8

The spectral analysis of geological data

8.1 INTRODUCTION

The previous chapter introduced some of the methods which can be used for analysing a time series and the autocorrelation function was chosen as the basic tool because of its immediately obvious applications to stratigraphy. This function connects the state of a random variable, such as a lithological parameter, with previous states, so it is ideally suited to the geologists' historical approach. One often refers to this type of analysis as an analysis in the time domain.

A great deal of geological thinking must be in historical terms. In electrical engineering or theoretical optics, where spectral-analysis methods were first developed, one is accustomed to thinking in terms of frequency rather than in terms of a sequence of events. The power-spectral function is the appropriate tool for this approach in the frequency domain.

An analogy is useful for giving a rough idea of what is meant by a power spectrum and this particular one is discussed by Granger and Hatanaka (1964). Their book was written primarily for economists but it gives a very good general introduction to the subject. It is assumed that one is interested in measuring the power that can be transmitted through a wireless receiver. In order to do this, one could connect a measuring instrument across the loudspeaker terminals and take readings while one tunes the receiver to different frequencies. A graph of such measurements is called a power spectrum because it represents the power output as a function of the frequencies which produced it. (See Fig. 8.1.) Three types of result might be obtained if the experiment were to be carried out in practice. If a portion of the waveband is chosen where no broadcasting station is active at the time of measurement, then one will record a steady output for all the frequencies. This is illustrated in Fig. 8.1A and is the general noise which originates partly in the receiver and is due also to atmospherics. Since this noise comprises all frequencies, it is also known as "white noise" and an independent random sequence, in which events follow each other at any frequency, will have a power spectrum of this type. The second possible result would come from an idealized situation. If it were possible to construct a transmitter which broadcasts only a single wavelength and if, in addition, it were possible to build a receiver with infinite selectivity, then the power spectrum would consist of a series of spike-like peaks. (See Fig. 8.1B.) This type of spectrum

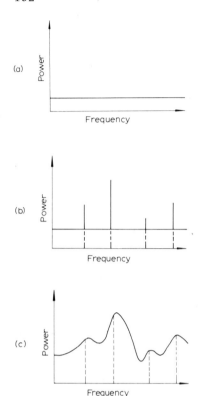

Fig. 8.1. The power spectrum.

is called a line spectrum because it is similar in its power distribution to the line spectrum which is obtained in emission spectroscopy. Such spectra are produced by entirely deterministic time series which consist of one or more pure harmonic terms. The third possible result comes about if random elements enter. The picture will then become blurred and the type of power spectrum which might be observed in reality is shown in Fig. 8.1C, where the power spreads around the maximum and this maximum is the specific wavelength of a particular broadcasting station. The spreading is due partly to random disturbances, but it also occurs because transmitters do not produce single frequencies. The spectrum is typical of the harmonic processes which were discussed in Section 7.4 (equation [7.55]) and it indicates that the process is partly deterministic and partly random.

The analogy which exists between radio communications and a power spectrum is introduced to give a very general idea about the meaning of spectral analysis. It would be wrong to conclude from this example that spectral analysis can only be applied to a time series which contains harmonic elements. Indeed, it will be seen that spectral analysis can be applied to all stochastic processes because, mathematically, the spectral function contains the same information as the autocorrelation function which was treated in the previous chapter.

It has been stated previously that the correlogram analysis is, the more logical approach to stratigraphic problems. However, there are exceptions. In cyclic sedimentation problems and particularly in the analysis of varved sequences, one may be interested primarily in the hidden periodicities or quasiperiodicities which such data could contain. In this case, spectral analysis is the proper tool. Furthermore, it turns out that the statistical significance of power spectra is assessed more easily than that of the equivalent correlograms; for this reason power spectra have also been used to test geological models which did not necessarily refer to cyclic sedimentation.

The theory of spectral analysis can be treated with considerable mathematical sophistication but that would not be in keeping with the purposes of this book. It is hoped that the relatively simple approach which is adopted in this chapter will at least provide some ideas about the potentialities of the method.

8.2 THE PERIODOGRAM

In order to introduce the principles of spectral analysis, one should first consider the entirely deterministic process:

$$X_t = A + B \cos (\omega t + \varphi) \qquad [8.1]$$

This is a cosine wave in which ω stands for the angular frequency and ϕ represents a phase constant. A and B are arbitrary constants. Interpreting the expression as an electrical signal, for example, would mean that A represents a d.c. voltage which is combined with an a.c. voltage of amplitude B. The power which such a signal can develop at any time across a resistance of one, is precisely X_t^2. To obtain the average power which is produced by a complete cycle one has to integrate the square of [8.1] to give:

$$\text{average } (X_t^2) = A^2 + \tfrac{1}{2}B^2 \text{ or average } (X_t - A)^2 = \tfrac{1}{2}B^2 \qquad [8.2]$$

The expression [8.2] is obviously the variance of the process X_t and therefore $\text{var}(X_t)$ can be expressed in terms of the frequency of the alternating component.

One can next assume that the signal consists of a large number of alternating currents which may have any kind of frequency. In this case the total variance of the process can still be expressed as the sum of power contributions from the individual frequencies and one can write:

$$\text{average } (X_t - A)^2 = \tfrac{1}{2} \sum_{i=0}^{\infty} B_i^2 \qquad [8.3]$$

The frequency components of the process can be found by harmonic analysis, and the series X_t may therefore be approximated by the Fourier expansion:

$$X_t = A_0 + \sum_{i=1}^{n-1} (A_i \cos \omega it + B_i \sin \omega it) + A_n \cos \pi t \qquad [8.4]$$

Assuming that the time series consists of N observations, then $n = N/2$ coefficients may be fitted to the series. The constant $A_0 = E(X_t)$ represents the mean of the process and the least-square estimates of A_i and B_i are given by the following (cf. Whittaker and Robinson, 1960):

$$A_i = \frac{2}{N} \sum_{t=1}^{N} X_t \cos \omega it; \quad B_i = \frac{2}{N} \sum_{t=1}^{N} X_t \sin \omega it \qquad [8.5]$$

Normally, less than n coefficients are fitted and the remainder is regarded as error. If one defines:

$$R_i^2 = A_i^2 + B_i^2 \qquad [8.6]$$

then one can write the total sample variance as:

$$\sum_{t=1}^{N} (X_t - \overline{X})^2 = \frac{N}{2} \sum_{i=1}^{n-1} R_i^2 + NA_n^2 \qquad [8.7]$$

The contribution of the i-th harmonic towards the variance is therefore $\frac{1}{2}R_i^2$ and from [8.4] to [8.7] one finds:

$$\tfrac{1}{2}R_i = \frac{2}{N^2} \left[\sum_{u=1}^{N} \sum_{t=1}^{N} X_t X_u (\cos \omega_i t \cos \omega_u u + \sin \omega_i t + \sin \omega_i u) \right] \qquad [8.8]$$

The trigonometric term in this, can be reduced to the cosine of the difference $(u - t) = k$ and therefore:

$$\tfrac{1}{2}R_i = \frac{2}{N^2} \sum_{k+t}^{N} \sum_{t=1}^{N} X_t X_{t+k} \cos \omega_i k \qquad [8.9]$$

The function:

$$C_k = \frac{1}{N} \sum_{t=1}^{N-k} X_t X_{t+k} \qquad [8.10]$$

is familiar from the previous chapter where it has been defined as the auto-covariance function of lag k. Putting as before $C_k = C_{-k}$, one may write [8.9] as:

$$\tfrac{1}{2}R_i^2 = \frac{2}{N} \sum_{k=N+1}^{N-1} C_k \cos \omega_i k \qquad [8.11]$$

The quantity $\tfrac{1}{2}R_i^2$ when expressed as a function of ω_i is called the periodo-gram and one can write the symbol $I_N(\omega)$ for it. The periodogram gives the power which each discrete frequency contributes to the total variance of the process. Since only a number of discrete frequencies are concerned in gener-ating the process, a line spectrum will result. This is similar to the type which was illustrated in Fig. 8.1B. If one considers the line spectrum to be the limiting case of the more general group of processes which produce smooth spectra, then one may argue in the following way. Subdividing the frequency range from $-\pi$ to $+\pi$ into N bands will reduce the height of the spike-like maxima from $\tfrac{1}{2}R_i^2$ to $NR_i^2/4\pi$. This will give a histogram-like power spec-trum which, after changing the limits of summation in [8.11], can be writ-ten as:

$$I_N(\omega) = \tfrac{1}{2}\pi \left(C_0 + 2 \sum_{k=1}^{N-1} C_k + \cos \omega k \right) \qquad [8.12]$$

$I_N(\omega)$ can approximate the continuous function $g(\omega)$, which is known as the power-spectral density curve, if one allows the frequency bands in this histo-gram to become quite narrow. Integrating $g(\omega)$ gives the power distribution function $G(\omega)$ and this clearly has the following properties:

$$G(-\pi) = 0 \quad \text{and} \quad G(+\pi) = \text{var} (X_t) \qquad [8.13]$$

It is often useful to standardize the power distribution function. This is done by dividing it by S_x^2, which gives the function:

$$F(\omega) = G(\omega)/S_x^2 \qquad [8.14]$$

From [8.14], one may obtain the power density distribution $f(\omega)$ by differ-entiation and this is called simply the spectrum. The relationship of $g(\omega)$ to $f(\omega)$ is exactly the same as the relationship between the autocovariance and the autocorrelation functions.

8.3 THE SPECTRAL ANALYSIS OF RANDOM PROCESSES

The periodogram analysis of deterministic processes, which was developed in the previous section, has been successfully applied to studying problems of tidal movements, astronomical measurements, and similar essentially harmonic phenomena. In order to gain a better understanding of the power spectra of general stochastic processes, it is convenient to introduce a time series X_n, in which the individual realizations are represented by complex values. This means that the state of the time series at time n, is given by a number pair:

$$X_n = X_1 + iX_2 \qquad\qquad [8.15]$$

and this gives, respectively, the real and imaginary part of the value X_n. In the notation which will be used, $\overline{X}$ stands for the conjugate of X, which is defined by:

$$\overline{X} = X_1 - iX_2 \qquad\qquad [8.16]$$

As before, it will be assumed that the mean of the process $E(X_n) = 0$. The variance and covariance of the process are only meaningful if they are real and positive. To ensure this, one finds from the multiplication rules for number pairs that:

$$E(X_n\overline{X}_n) = \text{var } (X_n) \text{ and } E(X_n\overline{X}_{n-k}) = C_k \qquad\qquad [8.17]$$

If one considers two processes, X and Y, then:

$$\text{var } (X + Y) = \text{var } (X) + \text{cov } (XY) - \text{cov } (YX) + \text{var } (Y) \qquad\qquad [8.18]$$

If series X is independent of series Y, then $\text{cov}(XY) = 0$ and it is said that X is orthogonal to Y. It follows that:

$$\text{var } (X + Y) = \text{var } (X) + \text{var } (Y) \qquad\qquad [8.19]$$

Equations [7.36] and [7.37] can be taken as examples from the previous chapter. A stochastic process was decomposed into a deterministic regression part and a set of orthogonal random variables Z_n, which were called innovations. If these innovations are regarded as complex random variables with zero mean, then the following condition must hold:

$$E(Z_{n+k} \, \overline{Z}_n) = 0 \ \ (h \neq 0) \qquad\qquad [8.20]$$

Spectral analysis makes use of a decomposition which is similar to the Wold decomposition, discussed in Chapter 7. The stochastic process is represented by a series of harmonic terms, together with a series of orthogonal random variables.

Consider the stochastic process:

$$X_n = R \, e^{i\omega n}, \quad (n = \ldots -1, 0, +1 \ldots) \tag{8.21}$$

R is a complex valued random variable and $e^{i\omega n}$ is a cosine series in which ω is a fixed angular frequency that may be any number $-\pi \leqslant \omega \leqslant +\pi$. If this process is to be stationary in the wide sense, its autocovariance must be independent of n. This is indeed the case, since:

$$E(R \, e^{i\omega n} \, \bar{R} \, e^{i\omega n-k}) = e^{i\omega k} \, E(R\bar{R}) \tag{8.22}$$

The autocovariance exists independently of n, as long as $E(R\bar{R}) = S^2 < \infty$.

One may next form a process which combines two terms of the type which was seen in [8.21] but with different angular frequencies, ω_1 and ω_2:

$$X_n = R_1 \exp(i\omega_1 n) + R_2 \exp(i\omega_2 n) \tag{8.23}$$

Once again, R_1 and R_2 are random variables with zero mean. The covariance with lag k is:

$$C_k = E[R_1 \exp(i\omega_1 n) \, \bar{R}_1 \exp\{i\omega_1(n-k)\}] + E[R_1 \exp(i\omega_1 n)$$

$$\times \bar{R}_2 \exp\{i\omega_2(n-k)\}] + E[R_2 \exp(i\omega_2 n) R_1 \exp\{i\omega_1(n-k)\}]$$

$$+ E[R_2 \exp(i\omega_2 n) R_2 \exp\{i\omega(n-k)\}] \tag{8.24}$$

The terms $E(R_1\bar{R}_2)$ and $E(R_2 R_1)$ must be zero for this to be independent of n and this means that R_1 and R_2 must be orthogonal. Thus the autocovariance for a stationary process can be written as:

$$C_k(R_1\omega_1, R_2\omega_2) = \exp(i\omega_1 k) S_1^2 + \exp(i\omega_2 k) S_2^2 \tag{8.25}$$

S_1^2 and S_2^2 are the variances of the two random variables R_1 and R_2. In general, a process composed of h harmonics and random variables R_i:

$$X_n = R_1 \exp(i\omega_1 n) + \ldots R_h \exp(i\omega_h n) \tag{8.26}$$

is only stationary if $E(R_i) = 0$ and $E(R_j R_i) = 0$ for $j \neq i$. If this is the case, then the autocovariance may be written as:

$$C_k = \sum_{j=1}^{h} \exp i\omega_j k \, E(R_j\bar{R}_j) \tag{8.27}$$

and the variance as:

$$\text{var}\,(X_n) = \sum_{j=1}^{h} E(R_j \overline{R}_j) \tag{8.28}$$

Thus the total variance is composed of the variances of the individual harmonic components. This is in agreement with the analysis which was given in the previous section. As before, the variance spectrum will be a discrete line spectrum if there is a finite number of frequencies. If a general stochastic process is to be represented, then h in [8.25] has to become infinity and the whole interval $-\pi$ to $+\pi$, may be covered densely with frequency points. Any stochastic process can be represented in the form:

$$X_n = \int_{-\pi}^{+\pi} e^{i\omega n}\, dS(\omega),\quad (n = \ldots -1, 0, 1\ldots) \tag{8.29}$$

or by a similar integral with limits from $-\infty$ to $+\infty$, if time is assumed to be continuous. This result is due to Cramér, Kolmogorov and Wiener (see Granger and Hatanaka, 1964, p. 28) and [8.29] is known as the spectral representation of the process X_n. The function $S(\omega)$ in [8.29] is a stochastic process which has the following properties because of stationarity conditions:

$$E[dS(\omega)] = 0,\quad G[dS(\omega_1)\,\overline{dS}(\omega_2)] = 0,\quad (\omega_1 \neq \omega_2)$$

$$\text{and } E[|dS(\omega)|^2] = dG_\omega \tag{8.30}$$

G_ω is again the power distribution function. It is possible to calculate the following because of the properties of [8.30]:

$$\text{var}\,(X_n) = E(X_n \overline{X}_n) = E \int_{-\pi}^{\pi} \exp\,(i\omega_1 n)\, dS(\omega_1)$$

$$\int_{-\pi}^{\pi} \exp\,(i\omega_2 n)\, dS(\omega_2) = \int_{-\pi}^{\pi} dG_\omega \tag{8.31}$$

It should be noted that in this equation, pairs $\omega_1 \neq \omega_2$ do not contribute to the total variance. The autocovariance function is found in a similar way:

$$C_k = E(X_n X_{n-k}) = E \int_{-\pi}^{\pi} \exp\,(i\omega_1 n)\, dS(\omega_1)$$

$$\int_{-\pi}^{\pi} \exp\,(i\omega_2 n - k)\, dS(\omega_2) = \int_{-\pi}^{\pi} e^{i\omega k}\, dG_\omega \tag{8.32}$$

The autocorrelation function is therefore given by:

$$R_k = \int_{-\pi}^{\pi} e^{i\omega k}\, dF\omega \qquad\qquad [8.33]$$

In order to calculate the spectral density, one may again assume first that F_ω is discrete and that there is a set of variables ω_1, ω_2 ... at which F_ω has positive increments f_1, f_2 ... with $\sum_j f_j = 1$. The process of [8.29] may then be written as:

$$X_n = \sum_j \exp(i\omega_j n)\, Z_j \qquad\qquad [8.34]$$

where the Z_j's have zero variance and are orthogonal. From [8.33] one can write:

$$R_k = \sum_j \exp(i\omega_j k)\, f_j$$

and for a fixed ω_j one finds:

$$f_j = r_1 \exp(-i\omega_j 1) = r_2 \exp(-i\omega_j 2) = r_a \exp(-i\omega_j a)$$

and therefore:

$$f_j = \lim_{a \to \infty} \frac{1}{2a} \sum_{k=-a}^{a} e^{-i\omega k}\, r_k \qquad\qquad [8.35]$$

One now considers a process in which F_ω is absolutely continuous and so [8.33] becomes:

$$R_k = \int_{-\pi}^{\pi} e^{i\omega k}\, f_\omega\, d\omega \qquad\qquad [8.36]$$

and this shows that the autocorrelation coefficients r_k, $k = \ldots -1,0,+1 \ldots$ are proportional to the Fourier coefficients of f_ω. Therefore:

$$f_\omega = \frac{1}{2\pi} \sum_{k=-\infty}^{+\infty} e^{-i\omega k}\, R_k \qquad\qquad [8.37]$$

The corresponding result for continuous time is:

$$f_\omega = \frac{1}{2\pi} \int_{-\infty}^{\infty} e^{-i\omega k} R_k \, dk \qquad\qquad [8.38]$$

Clearly [8.36] shows the autocorrelation function in terms of the spectral components and [8.37] shows the power spectrum in terms of the auto-correlation function. Mathematically, the correlogram and power spectrum functions are equivalent because they are Fourier transforms of each other.

The stochastic processes which concern us here are not complex but real and X_n will be real valued when the terms corresponding to ω and $-\omega$ in [8.29] are complex conjugates. This means that:

$$dS(\omega) = d\overline{S}(-\omega)$$

One can then write the autocovariance for real processes as:

$$C_k = \int_{0}^{\pi} \cos k \, \omega g'(\omega) \, d\omega \qquad\qquad [8.39]$$

where $g'(\omega) = 2g(\omega)$, $0 < \omega < \pi$ and $g'(0) = g(0)$, $g'(\pi) = g(\pi)$.

When making any analysis in practice, one is naturally concerned with the part of the spectrum that can be realised physically. This is found by simply doubling the positive part of $f_{+\omega}$.

8.4 EXAMPLES OF SPECTRAL DISTRIBUTIONS AND LINEAR OPERATIONS ON PROCESSES

It was seen in the previous chapter that a process Z_n is an independent random process when its autocovariance is zero for all lags with the exception of C_0, when it takes on the value of the variance and $C_0 = S_z^2$. Equation [8.37] for an independent random process, gives therefore:

$$f_\omega = \frac{S_z^2}{2\pi} \qquad\qquad [8.40]$$

which is a constant power spectrum showing an equal participation of all frequencies. (See Fig. 8.1A.)

The power spectra of autoregressive processes can be derived in a similar way. Consider the first-order process:

$$X_n = aX_{n-1} + Z_n$$

which has the autocorrelation function $R_k = a^{|k|}$ with $|a| < 1$, according to [7.27]. Equation [8.37] can therefore be written as:

$$2\pi f_\omega = \sum_{-\infty}^{\infty} \varphi^k a^{|k|} = -1 + \frac{1}{1 - a\varphi} + \frac{1}{1 - a\varphi^{-1}}$$

$$= -1 + \frac{1}{1 - ae^{i\omega}} + \frac{1}{1 - ae^{-i\omega}} = \frac{1 - a^2}{1 - 2a\cos\omega + a^2} \qquad [8.41]$$

In a similar way, the power spectrum of a second-order autoregressive process:

$$X_n = a_1 X_{n-1} + a_2 X_{n-2} + Z_n$$

can be derived, yielding the formula:

$$2\pi f_\omega = \frac{(1 - a_2)(1 - a_1^2 + a_2^2 + 2a_2)}{(1 + a_2)[1 + a_1^2 + a_2^2 - 2a_2 + 2a_1(1 + a_2)\cos\omega + 4a_2\cos^2\omega]}$$

$$[8.42]$$

Examples of calculated power spectra for the first and second-order processes are given in Fig. 8.2 (*A* and *B*). These examples correspond to the theoretical correlograms given in Fig. 7.2 (*A* and *B*). It is seen that the first-order Markov series, which has an exponential correlogram, has its maximum at zero frequency. In contrast to this, the oscillating second-order process has a distinct, although rather wide, maximum which corresponds to its oscillation frequency $2\pi/\theta$. This oscillation frequency may be obtained from [7.44].

The power spectrum can also be obtained directly from the spectral representation which was given in [8.29]. For example, the independent random process may be written as:

$$Z_n = \int_{-\pi}^{\pi} e^{in\omega} \, dS(\omega) \qquad [8.43]$$

and according to [8.30]:

$$E[dS(\omega)\,\overline{dS}(\omega)] = 0 \; (\omega_1 \neq \omega_2) = \frac{1}{2\pi} S_z^2 \; (\omega_1 = \omega_2)$$

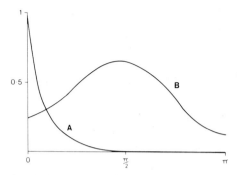

Fig. 8.2. Power spectra of a first-order (A) and second-order (B) autocorrelated process.

which is the same result as [8.40]. If one considers the moving average process:

$$X_n = \sum_{j=0}^{m} a_j Z_{n-j} \qquad [8.44]$$

then one can write the spectral representation as:

$$X_n = \int_{-\pi}^{\pi} \sum_{j=0}^{m} a_j e^{i(n-j)\omega} \, dS(\omega)$$

$$= \int_{-\pi}^{\pi} e^{in\omega} \left(\sum_{j=0}^{m} a_j e^{-ij\omega} \right) dS(\omega)$$

from which one can obtain the autocovariance function of [8.32] as:

$$C_k = \frac{S_z^2}{2\pi} \int_{-\pi}^{\pi} e^{ik\omega} \, l(\omega) \, \bar{l}(\omega) \, d\omega$$

where:

$$l(\omega) = \sum_{j=0}^{m} a_j e^{ij\omega}$$

The spectrum of the process X_n is therefore:

$$f_\omega = \frac{1}{2\pi} l(\omega) \, \bar{l}(\omega) \, d(\omega) \qquad [8.45]$$

The result in this case is achieved by transforming the time series Z_n into the time series X_n in [8.44]. The transformation is a typical example of a linear operation on a stationary process. It can be proved quite generally that if $Y(t)$ is a series which is derived from $X(t)$ by a linear operator L:

$$Y(t) = L[X(t)] \qquad\qquad [8.46]$$

then the effect of this operation on the power spectrum is surprisingly simple. It is found that:

$$f_y(\omega) = \frac{S_x^2}{S_y^2}\, f_x(\omega)\, A(\omega) \qquad\qquad [8.47]$$

where the subscripts to f indicate that the power spectra belong to the two processes X and Y and where $A(\omega)$ is a function of ω. Writing $X(t)$ in its spectral representation:

$$X(t) = \int_{-\infty}^{\infty} e^{i\omega t}\, dS_x(\omega) \qquad\qquad [8.48]$$

one can write from [8.46]:

$$Y(t) = L\left[\int_{-\infty}^{\infty} e^{i\omega t}\, dS_x(\omega)\right] = \int_{-\infty}^{\infty} [L(e^{i\omega t})]\, dS_x(\omega)$$

and setting:

$$L(e^{i\omega t}) = a(\omega)\, e^{i\omega t} \qquad\qquad [8.49]$$

$a(\omega)$ being complex, the process $Y(t)$ can be written as:

$$Y(t) = \int_{-\infty}^{\infty} e^{i\omega t}\, a(\omega)\, dS_x(\omega) = \int_{-\infty}^{\infty} e^{i\omega t}\, dS_y(\omega)$$

Therefore:

$$dS_y(\omega) = a(\omega)\, S_x(\omega)$$

and by definition (see [8.30]):

$$dG_y(\omega) = E[dS_y(\omega)\, d\overline{S}_y(\omega)]$$

$$= E[a(\omega)\, \overline{a}(\omega)\, dS_x(\omega)\, d\overline{S}_x(\omega)]$$

$$= |a(\omega)|^2\, dG_x(\omega)$$

Writing $g(\omega)$ for the increments of the power distribution functions, one has:

$$g_y(\omega) = |a(\omega)|^2\, g_x(\omega)$$

$$= A(\omega)\, g_x(\omega) \qquad\qquad\qquad [8.50]$$

and [8.47] follows directly from [8.50] if the power distributions are stan-dardized by dividing with S_x^2 and S_y^2, respectively (Cox and Miller, 1965, p. 321). The complex function $a(\omega)$ is known as the Transfer function and [8.49] shows that the linear operator will, in general, change the amplitude and phase of any harmonic on which it acts, but not its frequency. The quantity $A(\omega)$ is called the filter factor and it determines the amplitude changes which are caused by a linear operation. Practical applications of the filter theory will be discussed later.

8.5 THE ESTIMATION OF THE POWER SPECTRUM

The covariance of a real valued stochastic process was derived in [8.39] as:

$$C_k = 2 \int_0^\pi \cos \omega k g(\omega)\, d\omega$$

which by inversion gives:

$$g(\omega) = \frac{1}{2\pi}\left(C_0 + 2\sum_{j=1}^\infty C_j \cos j\omega\right) \qquad\qquad [8.51]$$

Therefore, it would seem reasonable to use a formula of the type:

$$\hat{g}(\omega) = \frac{1}{2\pi}\left(\hat{C}_0 + 2\sum_{j=N}^{N-1} \hat{C}_j \cos j\omega\right) \qquad\qquad [8.52]$$

where an estimate of the power spectrum is obtained by using all the available estimates of the autocovariance function which a limited record of N observations can yield. Comparison of [8.52] with [8.12] shows that this estimation formula is in fact the periodogram $I_n(\omega)$ of the process. Calculating periodograms from simulated random processes soon shows that the function $I_n(\omega)$ provides only very bad estimates of the power spectrum. From a theoretical point of view, one would expect to obtain a smooth curve but in practice, the graphs are usually very ragged and full of sharp peaks even when a large number of observations has been used. It can be shown, by examining the statistical properties of the periodogram, that the variance of the function does not become zero when n becomes large. Thus the variance of the function does not provide a consistent estimate of $g(\omega)$. However, a consistent estimate might be derived if the periodogram were to be split into individual frequency bands and if these frequency bands could be integrated. Unfortunately, the integration of individual bands can be only approximated. This is achieved by calculating a weighted or filtered periodogram which leads to estimates of the form:

$$\hat{g}(\omega) = \frac{1}{2\pi} \left[C_0 \lambda_0(\omega) + 2 \sum_{k=1}^{m-1} \lambda_k(\omega) C_k \cos k\omega j \right] \qquad [8.53]$$

The filter which is applied to the periodogram is defined by the filter factors $\lambda_k(\omega)$. A considerable amount of effort has been directed towards finding weighting functions with a filter factor close to 1.0 and this problem is discussed by Hannan (1960) while a summary of the various methods is given by Jenkins (1961). A very widely used filter is the Tukey-Hanning estimate in which the weights are defined as:

$$\lambda_k \ \tfrac{1}{2} \left(1 + \cos \frac{\pi k}{m} \right) \qquad [8.54]$$

where m is an arbitrary integer which is chosen by the investigator. The actual formula for calculating the power spectrum may then be written as:

$$L_j = \frac{1}{2\pi} \left(\hat{C}_0 + 2 \sum_{k=1}^{m-1} \hat{C}_k \cos \frac{\pi k j}{m} + \hat{C}_m \cos \pi j \right) \qquad [8.55]$$

The C_k's are estimates of the autocovariances which are obtained by calculating:

$$C_k = \frac{1}{N-k} \left(\sum_{n=1}^{N-k} X_n X_{n+k} - \frac{1}{N-k} \sum_{n=1+k}^{N} X_n \sum_{n=1}^{N-k} X_n \right) \qquad [8.56]$$

where N is the number of available observations. The estimate L_j of [8.55]
still yields a fairly irregular curve and a simple smoothing formula is employed
for the final estimate of the spectrum:

$$\hat{g}(\omega) = 0.25L_{j-1} + 0.5L_j + 0.25L_{j+1} \qquad [8.57]$$

The two end values $g(0)$ and $g(m)$ are given by $L_0/2 + L_1/2$ and $L_{m-1}/2 +$
$L_m/2$, respectively. The filters which are used in spectral analysis are known
as "windows" because, figuratively speaking, one is looking through them to
single out frequency bands. Unavoidably, these windows are not the perfect
rectangular slots which one would like them to be for estimating the power
spectrum in the form of a histogram. In fact, they distort slightly the
amount of power which passes through them. This effect is seen best by
analysing a pure harmonic wave to which the filter has been applied. Such a
power spectrum is shown in Fig. 8.3. A perfect window would respond to
this input of a sine wave by producing a square wave of width π/m, provided
that the power is determined at $m + 1$ points along the spectrum. It can be
seen from the calculated example of Fig. 8.3 that this is not the case with
the Tukey-Hanning window. A certain amount of power has leaked beyond
the limits $\omega_k \pm \pi/2m$ and this produces side peaks that, fortunately, are very
low. The side peaks are always less than 2% of the main peak in this particular form of estimate. Another type of window is due to Parzen and it is
slightly wider than the Tukey-Hanning window and yet it produces even less
power leakage. Estimation formulae which use the Parzen window are given
in Granger and Hatanaka (1964, p. 6). The estimates at $g(\omega_k)$ and $g(\omega_{k+2})$
are practically uncorrelated if the Tukey-Hanning window is used, although
there may be a slight correlation between adjacent estimates. It follows that
there is little point in estimating the power in between the equidistant frequency points because such estimates convey no real information. Clearly,

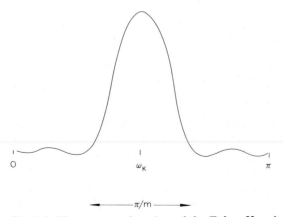

Fig. 8.3. The response function of the Tukey-Hanning window.

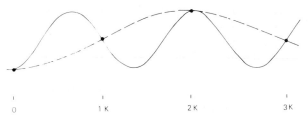

Fig. 8.4. The aliasing of power spectra.

the choice of the integer m in [8.54] and [8.55] will determine how detailed a particular power spectrum will be. If m is large, the frequency bands are narrow and the spectrum will be detailed but comparatively ragged. If a small value of m is chosen, the spectrum will be relatively smooth but less detailed. There must be limits to the resolving power of the spectral analysis if the number of observations N, is finite. It follows that if one made $m = N - 1$, then the estimate of the covariance C_k would be based on a single observation only and so the resulting estimate of $g(\omega)$ becomes the periodogram function which is statistically unstable. Granger and Hatanaka (1964, p. 48) avoided this by recommending that $m < N/3$ as a maximum although $m < N/5$ might be a more realistic choice. Statistically, both of these represent very small samples from which the autocovariances are to be estimated. Given a set of finite data, this minimum sample size determines the maximum resolving power of a spectral analysis.

A certain amount of care has to be taken in making sure that the highest frequency π/m can still be recognized from the record. In other words, the sampling points must be sufficiently closely spaced to allow for observations on this particular spectral component. Any time series which is obtained by measuring N at discrete time intervals of length k contains a maximum frequency of $\omega_0 = \pi/k$ and this can just be recognized from the record. This frequency is known as the Nyquist frequency. Any frequency above the Nyquist frequency that is sampled at regular intervals will contribute some power which belongs apparently to a much longer wave length. This can be seen when one studies Fig. 8.4. Such an unavoidable process is known as aliasing and, if there are powerful frequencies above the Nyquist frequency itself, they may produce spurious peaks in the power spectrum. A great deal of care must be taken in interpreting any estimated power spectra.

8.6 THE GEOLOGICAL APPLICATIONS OF SPECTRAL ANALYSIS

To date, most of the geological investigations which have used spectral or harmonic analysis have been concerned with the study of "cyclic" sediments. The study of varved sediments in particular has employed spectral methods for some time. The first extensive study of rocks which are older than the glacial deposits is probably that by Korn (1938) who used harmonic analysis to investigate laminated silt—shale deposits of Late Devonian and Early Carboniferous age from Thuringia in West Germany. Korn, who concluded that the laminae in the sediments which he examined were seasonal, found by traditional harmonic analysis, that the thicknesses of successive varves showed an 11-year and a 22-year periodicity which he attributed to the effect of the sunspot cycle. This finding confirms, to some extent, the annual nature of the laminae and it strongly suggests that the solar cycle, which was discussed in the previous chapter, was active during Devonian times. Korn's analysis was made under the assumption that solar activity can be represented by a harmonic and essentially deterministic process. This is not a very good hypothesis, as has already been shown in the previous chapter. However, the basic principle of Korn's work, which recognizes annual deposits by the spectral composition of the thickness measurements, has been used repeatedly. In particular, Anderson and Koopmans (1963) analysed a large number of glacial and ancient varves in order to examine whether any common pattern exists in varve deposition. The authors divided varved sediments into four groups: glacial varves, clastic lake varves, clastic marine varves, and evaporite varves. It is clear that the mode of formation must have been different in such diverse environments but all of them must have been influenced by the climatic conditions because the deposits are seasonal. The temperature-sensitive varves, which are the glacial and evaporite types, appear to show a considerable power increase in the lower frequencies. In particular, there seems to be a period of around eighty to ninety years which is found in the power spectrum of sunspot numbers as well as in the power spectrum of the growth rings of trees. Marine varves may be caused by a wide range of environmental factors and they too show a high proportion of low frequencies; this is less pronounced in the clastic lake varves, where it is thought that their formation is controlled mainly by run-off conditions. All the varve deposits show periodicities of 2—20 years in the higher frequencies but these are weak and not always persistent. No direct evidence about the existence of the 11-year solar period was found but the 22-year period occurred strongly in one sequence and weakly in six others. The 5-year period is present fairly persistently in the glacial varves and it could be the first harmonic to the 11-year period. This can be seen in Fig. 8.5.

It is not uncommon to find that the power spectra which contain a preferred frequency, for example ω, also show maxima representing the harmonics of this frequency at intervals of $\omega,...\omega/3, \omega/2, \omega, 2\omega, 3\omega$ ... and so on. This is because very few periodic or oscillating time series will have the precise shape of a sine wave and it follows that they must contain a certain number of frequencies that are themselves harmonics to the fundamental frequency. One can make use of this phenomenon to find evidence of a fundamental periodicity which might be hidden by noise or by a stochastic

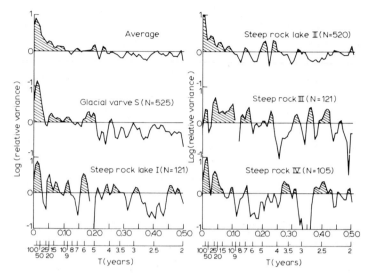

Fig. 8.5. Power spectra of glacial varves (after Anderson and Koopmans, 1963).

trend (Agterberg and Banerjee, 1969). One can use the occurrence of such multiple peaks to assess, at least qualitatively, the reality of the preferred maxima (Granger and Hatanaka, 1964, p. 63). The latter investigation rests on the argument that white-noise spectra may have spurious peaks but they are not likely to be distributed according to a definite pattern.

It is often asked, in stratigraphical analysis, whether a sequence represents cyclic sedimentation. The answer to this question is determined by the definition of "geological cyclicity" to a great extent and this will be discussed fully in Chapter 10. However, spectral analysis can answer the question of whether a defined lithological parameter occurs with a preferred periodicity within the series of measured data. Investigations of this type involve the statistical significance of an observed peak.

The statistical properties of the estimated power spectrum $\hat{g}(\omega)$ depend strongly on the type of window which is being used. For the Tukey-Hanning method of filtering, the variance of the estimate is given by:

$$\text{var} \left[\hat{g}(\omega) \right] = \frac{3m}{4n} \, \hat{g}(\omega) \approx \frac{m}{n} \, g(\omega), \, (\omega \neq 0) \qquad [8.58]$$

The sample distribution of the estimate can be approximated by the χ^2-distribution with k equivalent degrees of freedom, where:

$$k = \frac{8n}{3m} \approx \frac{2n}{m} \qquad [8.59]$$

These results are derived by assuming a normally distributed time series and the rounding off in [8.58] and [8.59], partly compensates for non-normality (Jenkins, 1961). In order to find the $(100-\alpha)$ percentage confidence band, each value of $\hat{g}(\omega_i)$ must be multiplied by a constant $T_\alpha(m,n)$ and $T_\alpha^i(m,n)$ to obtain the lower and upper boundaries. The two constants are defined by:

$$T_\alpha(m, n) = \frac{\chi^2 100 - \alpha}{k}, (k) \, df$$

$$T_\alpha^i(m, n) = \frac{\chi^2 \alpha}{k}, (k) \, df \qquad\qquad [8.60]$$

Consider the power spectra in Fig. 8.6 as an example. It represents the lithological sequence of the Pennsylvanian in Kansas (U.S.A.). This section is

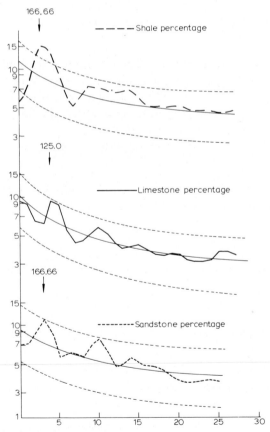

Fig. 8.6. Power spectra of Kansas cyclothems (after Schwarzacher, 1967).

2,000 feet (656 m) long and was compiled from records (Moore et al., 1951) in one of the classical areas of cyclothem studies. The section was subdivided into equal intervals in order to obtain a numerical record of the lithological composition. A percentage value of the three lithologies — sandstone, shale and limestone — was calculated for each interval. The following points have to be considered when designing the spectral analysis and deciding the spacing of the sample intervals. From a geological point of view, the section ought to be examined for periodicities ranging from a wavelength of 500 to 10 feet, the thickness of the Pennsylvanian cyclothems being of the order of 150 feet (45.7 m). Writing h for the sampling interval and ω_0 for the Nyquist frequency:

$$h = \frac{1}{2\omega_0} = \frac{1}{2 \times 0.1} = 5$$

The shortest wavelength which is required is 10 and $\omega_0 = 1/10$, so a sample interval of 5 feet is obtained. In order to find the number of lags m which is needed to calculate $m + 1$ estimates of the power spectrum, one simply divides the longest period by the shortest and so $m = 500/10 = 50$. Having chosen 5 feet as the sampling interval and knowing that the section is 2,000 feet long, one obtains for the number of sample points $N = 400$. One can calculate the standardized variance of the estimate $g(\omega_i)$ from [8.58], as $m/N = 50/400 = 0.124$ which is in fact better than the recommended error of 0.2 when $m \approx N/5$. To calculate the 10%-confidence limits, one determines k, the number of degrees of freedom from [8.59]; $k = 2N/m = 16$ degrees of freedom. Secondly, one finds the value of 23.54 for the upper limit and 9.31 for the lower limit from the χ^2-tables. Therefore, $T_\alpha(50,400) = 0.582$ and $T_\alpha^i(50,400) = 1.17$.

One can now multiply each estimate of the power spectrum by T_α and T_α^i and so obtain the confidence limits. It is often more convenient to plot the power spectrum on a logarithmic scale, making it simply a case of adding the logarithms of T_α and T_α^i to the power estimates. The confidence limits then form a band with a constant width on both sides of the estimates. A very useful graphical presentation of the confidence bands may be obtained by fitting a smooth line to the power spectrum and indicating the confidence limits by drawing the lower and upper boundaries to this line. This method was used for Fig. 8.6 and it shows a significant periodicity at a wavelength of 166 feet.

In the actual geological interpretation of power spectra, it is important to remember that, if the power is estimated at $m + 1$ discrete frequencies, absolutely nothing is gained by any interpolation between these frequencies. The wavelength determinations in the lower-frequency range become very coarse because of the very nature of the frequency scale. In order to show this clearly, the obvious relationship:

$$\text{wavelength} = \frac{m}{\omega\omega_0}$$

is given in Table 8.1. It can be seen from this table that, if the power
spectrum shows the Kansas cyclothem to be 166 feet thick, then this must
be taken together with the fact that the next estimated points are at 250 and
125 feet, which allows for a very wide range of thicknesses. One of the great
advantages of spectral analysis is that it brings out the limitations of inadequate data. In order to obtain a more accurate estimate of the wavelength of
the Kansas cycle, the spectral peak has to be brought into the higher-fre-
quency range. This could be done by using a much longer section, which of
course is not available.

A problem that will have to be discussed in connection with cyclic sedimentation, is the question of how often a lithological pattern must be repeated before it can be called "cyclic". This may be answered fairly precisely
in the context of the spectral estimation. Let it be assumed that the maximum standardized variance which is acceptable is 0.2. Taking the formula
$m/N = 0.2$, where $m = 50$, then $N = 250$ and, using the same sample interval
of 5 feet as before, the section must have a minimum length of 1250 feet.
The longest recognizable cycle of 500 feet is repeated 2.5 times and so any
lithological group must be repeated two to three times before it can be
recognized even vaguely as cyclic within the limits of the definitions which
are used in this context. It is clear that the length of a "cycle" cannot be
determined very accurately if it is repeated only two to three times and, in
econometric analysis, it has been recommended that a "cycle" should be
repeated at least seven times before it can be recognized as a so-called busi-
ness cycle (Granger and Hatanaka, 1964, p. 17).

TABLE 8.1

Frequency—wavelength conversion $\omega_0 = 0.1$, $m = 50$

Frequency	Wavelength	Frequency	Wavelength
1	500	11	45.5
2	250	12	41.7
3	166	13	38.5
4	125	14	35.7
5	100	15	33.3
6	83.3	16	31.3
7	71.4	17	29.4
8	62.5	18	27.8
9	55.5	19	26.3
10	50.0	20	25.0

8.7 FITTING MODELS TO OBSERVED POWER SPECTRA

The analysis of time series developed historically from a relatively small number of deterministic and stochastic models and this was brought out in the previous chapter. The deterministic models included both the disturbed harmonic process and the linear cyclic model which attempted to analyse an observed sequence in terms of a limited number of sine waves. The stochastic models, on the other hand, included the running average model and the autoregressive processes but there is a very wide range of alternative models which can be obtained either by combining different stochastic models, or stochastic and deterministic models. A definite identification of the various processes is hardly ever possible and it is very unlikely that a stratigraphical record can be fitted satisfactorily. Thus most of the observed data sequences will be time series which were generated by stochastic, rather than deterministic, processes with superimposed random variables. In spite of this, the difference is still of considerable importance from a theoretical point of view. Whereas in meteorology, econometrics and other subjects which employ time-series analysis there has been a tendency towards moving away from the concept of harmonic cycles, this has not been so in geology. It has been suggested by a number of workers that geological "cycles" can be ultimately explained by variations in the orbital elements of the earth. These astronomical cycles are probably best described by relatively few periodic functions in the strict sense. The proper tool for analysing such processes is the periodogram which was, indeed, successfully applied to such astronomical problems as the variation in the brightness of stars (Whittaker and Robinson, 1960). The geological reasons why the periodogram is not applicable to most stratigraphic problems will be discussed at some length later on. These same reasons make it unnecessary to discuss in detail tests which have been designed to differentiate between deterministic and purely random processes. Deterministic processes should, of course, give a line spectrum and, although this line spectrum may be blurred by superimposed noise or by the method of calculating the power spectrum, the purely random processes should contain a discontinuity at the frequency corresponding to the harmonic involved. Tests for a jump in the periodogram have been given by Hannan (1961). The problem of testing mixed and harmonic processes has been discussed by Whittle (1954).

The fitting of a stochastic model to an observed sequence of stratigraphical data is best illustrated by an example which was analysed in some detail by Agterberg and Banerjee (1969). The original data for this study were obtained by measuring glacial varves in Canada in the northern Ontario area. As usual, the time series was derived by plotting the thickness of the varve against the number of that varve in the series and the varve number may be taken as denoting successive years. In this particular study, the silt or summer layer and the clay or winter layer were measured independently. It was

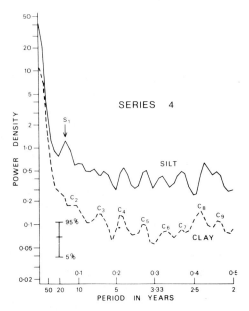

Fig. 8.7. Power spectrum of glacial varves, series 4 (after Agterberg and Banerjee, 1969).

found that the thickness of identically numbered varves decreased exponentially with the distance from the ice front which lay towards the north of the glacial lake in which they formed. The distance between the ice front and the locality at which the measurements were taken also increased with time because the ice front retreated generally towards the north during the formation of the sedimentary sequence. This introduces an exponential trend into the time series. The logarithms of the thicknesses were used in the power spectral analysis in order to counteract this. This logarithmic transformation reduced effectively the variance of the measured sections. Spectra were estimated from the transformed thickness data using the Tukey-Hanning window. Agterberg and Banerjee's results for series 4 which is the longest section ($N = 537$), are shown in Fig. 8.7. Only the silt series has a fairly pronounced maximum at 14 years but both the silt and the clay series have equally spaced maxima which are multiples of the frequency 0.05. Eight of these peaks are labelled in the diagrams as C_2 to C_9 and they were interpreted as being the harmonics of a fundamental periodicity of 19 years. The fundamental frequency is hardly visible in the power spectrum because it is hidden by the very high peaks of the low-frequency oscillations.

The behaviour of the time series in the low-frequency range may be clarified by applying filtering techniques. The hypothesis is made that the observed sequence X_n consists of two components the signal S_n and a superimposed noise Z_n which is assumed to be an independent and normally distributed random variable. Each value of the signal can be represented by the weighted sum:

$$S_n = \sum_{j=-m}^{m} a_j\, X_{n+j} \qquad\qquad\qquad [8.61]$$

If the correlogram of the signal $R_s(k)$ is known and if the values of Z_n are uncorrelated, then the weight factors a_j can be determined (Agterberg, 1967). For the varve model, it was assumed that the signal had an exponential autocorrelation function:

$$R_s(k) = c\, e^{-a|k|} \qquad\qquad\qquad [8.62]$$

where the two constants c and a can be determined from the observed correlogram $R_x(k)$. Under these assumptions, weight factors can be calculated by a formula which is derived by Agterberg (1967). It is found that:

$$a_j = q e^{-P|j|} \text{ with } P = \sqrt{a^2 + 2ac(1-c)} \text{ and } q = ac/(1-c)p \qquad [8.63]$$

Applying this symmetrical exponential filter to the data X_n resulted in an oscillating series which showed the 19-year period relatively clearly. It also showed a phase shift between the silt and clay curves, the latter leading over the former by approximately six years. Graphing the filtered data made it also obvious that the already logarithmically transformed data X_n contained a pronounced linear trend and it was decided to remove this and repeat the spectral analysis on the trend-free data. The power spectrum of series 4 after the removal of a linear exponential trend is shown in Fig. 8.8. It indicates

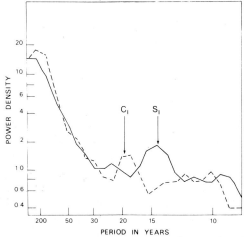

Fig. 8.8. Power spectrum of glacial varves after the removal of a linear, exponential trend (after Agterberg and Banerjee, 1969).

peaks at approximately 200, 19 and 10 years. The silt layers seem to have an additional cycle of approximately 14 years. Agterberg and Banerjee think it possible that their measured varve sections are incomplete; that is, that some varves have been missed during measurements. Applying a 10% correction factor to the shorter fluctuations results in an 11-year and a 20-year periodicity which could be an approximation to the sunspot cycle. The 15-year cycle may be caused by the topographical relief of the bedrock over which the glaciers moved on their retreat but there are as yet insufficient data to establish this hypothesis.

The separation of cycles into a low- and a high-frequency group, which is indicated by the power spectra, is seen also in the correlograms of the trend-free components. In particular, the autocorrelation function of the clay looks like a damped cosine wave on which the shorter fluctuations are superimposed. It is, therefore, reasonable to fit an autoregressive process in an attempt to explain the low-frequency fluctuations of the correlogram. Agterberg and Banerjee decided on a second-order model of the type which was seen in [7.38]:

$$X_n = a_1 X_{n-1} + a_2 X_{n-2} + Z_n$$

This has the correlation structure of a damped cosine wave (see [7.50]) which can be written alternatively as:

$$R_k = c \exp \left(-\alpha |k|^k\right) \left[\cos \omega_0 k + \alpha / \omega_0 \sin \omega_0 |k|\right] \qquad [8.64]$$

in which c, α, ω_0 are constants which were obtained by fitting various sine waves directly to the observed correlogram. A good fit is provided for the clay data by $\alpha = 0.01$ and $\omega_0 = 0.05$, $C = 1$ being kept constant, and from the fundamental frequency ω_0, the periodicity $2/\pi/\omega = 126$ years can be calculated. Alternatively, the two regression coefficients a_1 and a_2 can be estimated by making use of the Yule-Walker equations (see [7.31]). The estimate for the clay series leads to the process:

$$X_n = 1.978 X_{n-1} - 0.980 X_{n-2} + Z_n$$

The basic period which can be calculated from this equation is approximately 143 years. The correlogram of the process permits one to calculate the theoretical power spectrum and this has been done for the silt and clay layers and illustrated in Fig. 8.9. In comparing the observed power spectrum with the theoretical one, it must be remembered that the transfer function of the spectral window, which is used in estimating the power spectrum, will change the theoretical values somewhat. This effect can be calculated and the relevant points for the Tukey-Hanning window are shown as crosses in the diagram. The corrected theoretical spectrum can be directly compared

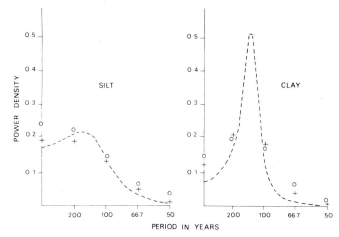

Fig. 8.9. Theoretical power spectrum for varve sequence (after Agterberg and Banerjee, 1969).

with the observed values which are taken from Fig. 8.8 and which are shown as open circles in Fig. 8.9. As can be seen from the diagrams, the autoregressive model provides a reasonably good fit for the low-frequency fluctuations of the series. Geologically this "cyclicity" may be caused by climatic variations which control the advance and the retreat of the glacier.

8.8 THE SPECTRAL ANALYSIS OF MULTIPLE TIME SERIES

Just as the correlogram analysis can be expanded from the univariate to the multivariate case, so also can power spectral analysis be applied to observations on two or more different time series. Once again, it is assumed that observations are available on, let us say, two variables X_t and Y_t. The full correlation structure of the two processes is given by the two autocovariances $C_{xx}(k)$ and $C_{yy}(k)$ together with the so-called cross-covariance $C_{yx}(-k) = C_{xy}(k)$ which may be estimated from the formula:

$$C_{xy}(k) = \frac{1}{N-k}\left(\sum_{t=1}^{N-k} Y_t\, X_{t+k} - \frac{1}{N-k}\sum_{t=k+1}^{N} X_t \sum_{t=1}^{N-k} Y_t\right) \qquad [8.65]$$

In accordance with [8.38] one may write:

$$C_{xx}(k) = 2\int_{0}^{\pi} \cos k\omega g_x(\omega)\, d\omega$$

$$C_{yy}(k) = 2 \int_0^\pi \cos k\omega g_y(\omega) \, d\omega$$

$$C_{xy}(k) = \int_{-\pi}^\pi e^{ik\omega} C_v(\omega) \, d\omega \qquad [8.66]$$

In the latter equation, $C_v(\omega)$ is the cross-spectrum of process X and Y and, since this is a complex quantity, it may also be written as:

$$C_v(\omega) = c(\omega) + iq(\omega) \qquad [8.67]$$

Now, since it is assumed that the cross-covariance function is real, it can be written as:

$$C_{xy}(k) = 2 \int_0^\pi \cos k\omega c(\omega) \, d\omega - 2 \int_0^\pi \sin k\omega q(\omega) d\omega \qquad [8.68]$$

Inverting [8.68] gives:

$$c(\omega) = \frac{1}{2\pi} C_{xy}(0) + \frac{1}{\pi} \sum_{k=1}^\infty [C_{xy}(k) + C_{yx}(k)] \cos k\omega$$

$$q(\omega) = \frac{1}{\pi} \sum_{k=1}^\infty [C_{xy}(k) - C_{yx}(k)] \sin k\omega \qquad [8.69]$$

The function $c(\omega)$ is known as the co-spectrum and the function $q(\omega)$ as the quadrature spectrum. They represent, respectively, the real and imaginary part of the cross-spectrum.

The estimation of the cross-spectrum is obtained by methods which can be compared directly with the estimation procedure of the single power spectrum. The "window" is the same for all the estimates: $g_x(\omega) g_y(\omega)$ and $C_v(\omega)$. If the Tukey-Hanning weights are used, one first calculates the raw estimates:

$$L_c(j) = \frac{1}{2\pi} \left[\hat{C}_{xy}(0) + \sum_{k=1}^{m-1} \{\hat{C}_{xy}(k) - \hat{C}_{yx}(k)\} \cos \frac{\pi jk}{m} + \frac{1}{2} \{\hat{C}_{xy}(m) \right.$$

$$\left. + \hat{C}_{yx}(m)\} \cos \pi j \right]$$

and:

$$L_q(j) = \frac{1}{2\pi} \left[\sum_{k=1}^{m-1} \{(\hat{C}_{xy}(k) - \hat{C}_{xy}(k)\} \sin \frac{\pi jk}{m} + \frac{1}{2} \{\hat{C}_{xy}(m) - \hat{C}_{xy}(m)\} \sin \pi j \right]$$

By further smoothing one obtains the estimates:

$$\hat{c}(\omega_j) = 0.25\, L_c(j-1) + 0.5\, L_c(j) + 0.25\, L_c(j+1)$$

$$\hat{q}(\omega_j) = 0.25\, L_q(j-1) + 0.5\, L_q(j) + 0.25\, L_q(j+1)$$

with:

$$L_c(-1) = L_c(+1), \quad L_c(m+1) = L_c(m-1)$$

$$L_q(-1) = L_q(+1), \quad L_q(m+1) = L_q(m-1)$$

It is difficult to interpret the cross-spectrum directly, but if one writes [8.67] as:

$$C_v(\omega) = |C_v(\omega)|(\cos \varphi + i \sin \varphi)$$

then one has, for the absolute value of $C_r(\omega)$:

$$|C_v(\omega)| = \sqrt{c^2(\omega) + q^2(\omega)} \text{ and } \varphi(\omega) = \arctan \left[\frac{q(\omega)}{c(\omega)} \right] \qquad [8.70]$$

The absolute value is usually given in a standardized form and is called the coherence $\gamma^2(\omega)$ which is defined as:

$$\gamma_{xy}^2(\omega) = \frac{c^2(\omega) + q^2(\omega)}{g_x(\omega)\, g_y(\omega)} \qquad [8.71]$$

where $g_x(\omega)$ and $g_y(\omega)$ are the covariance spectra of series X and Y. It is well known that the overall correlation between the two time series X and Y could be expressed by the correlation coefficient r, which in our notation would be written as:

$$r^2 = \frac{C_{xy}^2}{C_{xx}(0)\, C_{yy}(0)}$$

The square of the correlation coefficient is that part of the variance of Y

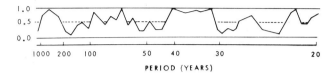

Fig. 8.10. Coherence between clay and calcium carbonate contents in an evaporite series (after Koopmans, 1967).

which can be attributed to the linear dependence or regression of Y on X. The coherence $\gamma_{xy}{}^2(\omega)$ is the proportion of the power of the Y series at frequency ω which is attributable to the linear dependence of the Y series on the X series (Koopmans, 1967). Alternatively, if one remembers that spectral analysis decomposes the random process into various harmonics, all with different amplitudes, then the coherence may be understood as the squared correlation coefficient between two amplitudes belonging to harmonics which have equal frequencies although they are taken from two different processes.

The phase angle $\varphi(\omega)$ of [8.70] is the lead of the harmonic component of the X series over the Y series at the given frequency ω.

Both coherence and phase-angle diagrams can be very useful tools in the examination of stratigraphic sections. The coherence is particularly suitable for the examination of the inter-relationship of two or more variables, which may be taken from one or more different sections. For example, Koopmans (1967) compared the analysis of calcium carbonate and clay content which were obtained from a varved sequence of the Rita Blanka Lake, Texas (U.S.A.). It was to be expected that these two quantities should show a strong negative correlation but the calculated correlation coefficient for the two series was only -0.12. The coherence, however, between the clay and carbonate series shows high values in all areas in which the power spectra of the individual series have peaks (see Fig. 8.10). Therefore, the coherence gives a very much better description of the relationship between the two variables than the conventional correlation coefficient. The reason why the latter is so small can be seen from the phase diagram (see Fig. 8.11), which shows that for the longer periods (70—1000 years) the phase angle is almost consistently $-180°$, whereas for the shorter periods the phase angle is near $0°$ or minus $-360°$, and this cancels out the overall correlation. Koopmans did not give any reason for this phase relationship, but it may be seen from the work of Agterberg and Banerjee (1969), which has already been mentioned, that a closer study of the phase relationships may eventually contribute, in an important way, to the understanding of the sedimentation mechanism of layered rock sequences. In the Canadian varve study, it was found that the thickness variations of the clay layers invariably came earlier than the equivalent variation in the silt layers. Agterberg and Banerjee suggest two possible geological explanations for this. Either the clay fraction responds much more

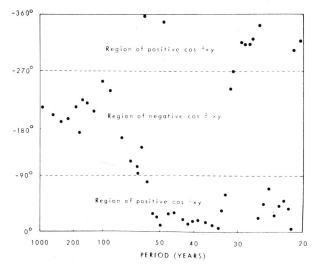

Fig. 8.11. Phase angle for the clay, calcium carbonate data (after Koopmans, 1967).

quickly to a suddenly increased amount of melt water being produced by a more rapidly retreating glacier, or alternatively, one may assume that increased melt water produces denser turbidity currents. This would result in almost all the coarse fraction being deposited near the ice margin and the increased clay content could spread over the basin. Later, when the topographical profile has become steeper, silt is spread in increasing quantities. Of course these are as yet speculations but they indicate that the results from co-spectral analyses can be much more elaborated and that the method can lead to interesting sedimentological and stratigraphical problems.

The analysis of two time series can be generalized to a multivariate analysis of n time series. In this case, the power spectrum is represented by an $n \times n$ matrix $\mathbf{G}(\omega)$ which goes together with similar coherence and phase matrices. As yet, no geological problem has been treated by such a multiple spectral analysis.

Chapter 9

The analysis of stratigraphical trends and the smoothing of data

9.1 INTRODUCTION

Most field or laboratory measurements of geological data are highly irregular and fluctuate so that they are often unsuitable for direct analysis. Sometimes such irregularities or noise may be removed by so-called "smoothing procedures". If stratigraphical data contain slow changes which occur systematically throughout the whole section, then the variation is referred to as a "trend".

It is easy to understand how these characteristic features develop. The sedimentation processes which make up the stratigraphic record can be complex random functions which operate on very different time scales and therefore, produce oscillations of every conceivable frequency. Short and rapid fluctuations very often appear as irregular noise, while slow, low-frequency oscillations naturally appear to be more systematic.

The gradual changes in the section reflect changes in the general sedimentation conditions and so the statistical properties of the record will not be constant. This means that the time series is not likely to be stationary. Since stationarity had to be assumed in all the previously discussed methods of analysis, attempts must often be made to make sequences stationary and this can be achieved by trend removal. Smoothing procedures, which deal with high-frequency fluctuations, will be taken to be different from trend removal but the distinction is relative and the mathematical methods which are used in both operations are often the same.

A trend involves a smoothly changing variation which either increases or decreases steadily. Clearly, such a definition must take into account the absolute length of the observed series. For example, if air-temperature measurements are taken at 15-min. intervals throughout the day, then starting in the morning, one will obtain a fluctuating record which increases fairly regularly until noon. (See Fig. 9.1.) If the measurements are discontinued before midday, the record will show a trend which is steadily rising. Of course, if the measurements are continued over several days, this "trend" will be recognized as the daily variation which is oscillating, but at the same time a new "trend" representing the seasonal variation may be discovered. Once again, examination over a longer interval of time will show this to be an oscillating component and new trends may turn up, depending on how long the series is. Therefore, it may appear that the separation of a trend

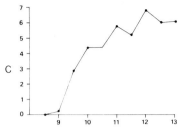

Fig. 9.1. Air temperatures at half-hourly intervals in Belfast on 23rd January, 1974 (measurements by W. Schwarzacher Jr.).

component is always a somewhat artificial procedure and this is true, unless a definite physical process is recognized which can be made responsible for the slow change. If the physical nature of the trend is not identified, arbitrary assumptions have to be made. One may accept, for example, that the trend is a simple polynomial or a low-order harmonic series but the type of function to be fitted must be decided on beforehand. It is difficult to lay down any firm rules about how complex a trend function should be, but one may be guided by the following considerations. If a record consists of n observations, then any oscillation which has a wavelength in excess of $2n$ will lead to a monotonic increase or decrease in the values which are determined by it. Such low frequencies can be taken as the trend (see Granger and Hatanaka, 1964). In cases where it is necessary to have the cycle repeated several times, as is desirable in spectral analysis, one may choose the frequency which arbitrarily divides the trend from the oscillations somewhat higher.

9.2 TESTS FOR STATIONARITY AND TREND

It is always useful to test a series of observations for the presence of a trend or for indications of non-stationarity before one embarks on a more detailed analysis. However, such tests invariably need a large amount of data. In stratigraphical analysis, one is usually dealing with a single record and stationarity can only be established by subdividing the original series into subsections and comparing the statistical parameters of the shortened sections. Indeed, this method was used to establish the stationarity of Markov chains (see Chapter 5), where transition matrices were calculated from subsections and were compared with each other by χ^2.

If one is dealing with a series of numerical values like grain size or shale content, it is usually sufficient to test whether the mean $\bar{x}_t$ and the variance $\bar{x}_t^2$ remain constant throughout the measured series. It is important that the covariance function C_k should be independent of time as well, but it is unlikely that C_k will change throughout the sequence unless there is a change in C_0. The latter is identical with the variance. Provided that N

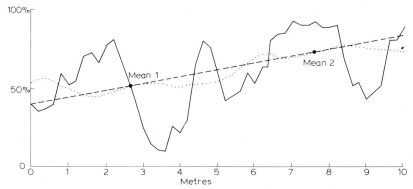

Fig. 9.2. Limestone percentages in the Glencar Limestone (solid line). Linear trend based on two averages (broken line). Running average of twenty five points (dotted line).

subsections are available, one can calculate $\bar{x}_1$, $\bar{x}_2...\bar{x}_N$ together with $\overline{x_1^2}$, $\overline{x_2^2}...\overline{x_N^2}$ and test whether the fluctuation in these two series exceeds what would be expected from random sampling. For this test, it is best to use non-parametric methods such as the "run test" or the "trend test" (Bendat and Piersol, 1966). Further methods for detecting trend from the observed data directly are given by Kendall and Stuart (1968).

In many geological problems, trend will be obvious from an inspection of the graphed data and this can be examined in a crude fashion by calculating a few mean values for the subsections. For example, Fig. 9.2 shows a graph of the percentage of limestone in 20-cm intervals of a Carboniferous sequence (Middle Glencar Limestone, Co Sligo, NW Ireland) and it may be seen that the series is strongly oscillating. If the total series of fifty measurements is subdivided into two series with 25 observations each, then the two mean values for the lower and upper sections are 50% and 73%, respectively, and a straight line drawn through these points gives a fair approximation to a linear trend. A change in the variance or covariance of a series is less easily detected in the graphed data. Nevertheless, a graphical comparison of the correlograms, which can be calculated from two or more subsections, will bring out any departure from stationarity (Kheiskanen, 1964; Schwarzacher, 1968).

9.3 THE FITTING OF A TREND BY REGRESSION METHODS

An essential feature of the trend concept is that it represents a slow and continuous variation which affects the whole of the observed series and, so, it is reasonable to approximate the trend by a polynomial in time. If the trend only concerns the mean of the investigated variable, then the time series can be generated by:

$$Y_t = m(t) + X_t \qquad [9.1]$$

where X_t is a stationary process and $m(t)$ is a steady function of t called the "trend" which can be approximated by the polynomial:

$$m(t) \approx P_t = a_0 + a_1 t + a_2 t^2 + \ldots a_p t \qquad [9.2]$$

Clearly, if X_t in [9.1] is an independent random variable, then the procedure of estimating the coefficients of the polynomial [9.2] is a straightforward regression problem. It is likely that nothing is known about X_t and then one has no alternative but to assume that it is an independent random process and to use the observed values of Y_t for the estimation of the coefficients $a_0 \ldots a_p$. This means that the trend which has been determined in this way, must incorporate some of the variation that is caused by the random process. The effects of this will be discussed later. However, it can be shown that the least-square estimation of the trend function is efficient and is the best unbiased linear procedure.

Instead of using the polynomial P_t of [9.2] to approximate the trend function $m(t)$, it is sometimes convenient to find a harmonic regression, that is, a trigonometric polynomial which approximates the trend in the form of:

$$m(t) \approx T_t = \sum_{j=0}^{n} \left[a_j \cos \left(\frac{\pi j t}{n} \right) + b_j \sin \left(\frac{\pi j t}{n} \right) \right] \qquad [9.3]$$

The Fourier coefficients a_j and b_j in this equation are again estimated by the least-square method. The computational formulae for calculating the coefficients from n equally spaced data y_t are given by (Whittaker and Robinson, 1960):

$$a_j = \frac{2}{n} \sum_{k=0}^{n-1} y_k \cos \frac{2k\pi j}{n}$$

$$b_j = \frac{2}{n} \sum_{k=0}^{n-1} y_k \sin \frac{2k\pi j}{n} \qquad [9.4]$$

As it was shown before, if the raw data y_t are used rather than the unknown trend $m(t)$, then some of the random variation which is contained in the stationary process will be incorporated during the calculation of the coefficients. In both the polynomial and the harmonic-regression analysis of trends, one is interested mainly in expressions of relatively low orders; one may consider linear, quadratic or possibly cubic polynomials or trigonometric expansions up to the orders of, let us say, five or six. The calculation of higher orders may be considered in smoothing operations where much more complicated functions are needed than for the trend analysis. A disadvantage of the polynomial fitting, which uses all the observational points, lies in the fact that the whole quite extensive computational work has to be repeated if the series is extended or shortened. Apart from this, polynomial and trigono-metrical regressions can be used very effectively to detect a trend. When the trend is subtracted from the original data, this will lead to residuals which are stationary, providing that the series is of the type which was defined in [9.1].

9.4 TREND FITTING BY FILTERS

One of the methods which has been used extensively in data analysis is the fitting of local polynomials by the process of moving averages. The theory is easily understood if one considers the simplest of all cases, namely the fitting of a straight line to three points (see Fig. 9.3). A series of measurements x_t are taken at constant time intervals. One particular point is singled out, let us say at t_0, which is central to the two points t_{-1} and t_{+1}, whereby:

$$\sum_{j=-1}^{+1} t_j = 0 \tag{9.5}$$

We are searching for the straight line $u_t = a_0 + a_1 t$, which shows the minimum departure from the three points in the least-square sense. Therefore:

$$\sum_{-1}^{+1} (x_t - a_0 - a_1 t)^2 = \text{minimum} \tag{9.6}$$

Differentiating for a_0 and setting the equation to zero gives:

$$\Sigma x_t - k a_0 - a_1 \Sigma t = 0 \tag{9.7}$$

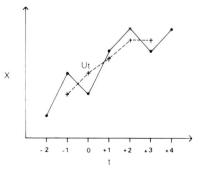

Fig. 9.3. Smoothing by averaging three points.

where $k = 3$ in this particular example. Because of [9.5], one finds:

$$a_0 = \frac{1}{k} \Sigma x_t \qquad [9.8]$$

Since a_0 is the origin of the straight line, it will pass through it at $t = 0$. The original position $x(t_0)$ may, therefore, be replaced by $u(t_0) = a_0$ which, as can be seen from [9.8], is simply the average of the three points. The point $u(t_0)$ is shown by a cross in Fig. 9.3. The same procedure is repeated for the points x_{t+1}, x_{t+2}, t..., meaning that a moving average is fitted to the series.

Higher-order curves can be fitted in a very similar manner. If a parabola is to be fitted, for example, one needs a minimum of five points. Again, only the origin of the curve a_0 is of interest, and calculation shows that:

$$a_0 = \frac{1}{35} (-3x_{t-2} + 12x_{t-1} + 17x_t + 12x_{t+1} - 3x_{t+2}) \qquad [9.9]$$

which is again an average but this time with unequal weights. The method can be elaborated in two directions. One either calculates higher-order polynomials or one uses an increasing number of points to calculate the low-order polynomials. In each case, the transformation from x_t into u_t involves the formation of an average and one can write the general formula:

$$u_t = \sum_{j=-m}^{m} a_j x_{t-j} \qquad [9.10]$$

where the a_j's are the weights of the moving average. The great advantages of this method are that the weights remain constant for the whole series and that the computational work is very simple. It is a disadvantage, however, that one loses m points each on the beginning and the end of the series and this can become important when m is relatively large compared with the length of the series.

To illustrate the effectiveness of a moving average calculation a first-order trend was fitted to the data in Fig. 9.2. Each point was replaced by a simple average of 25 points whereby it was necessary to know the twelve limestone percentage values preceding and following the graphed part of the section, and it may be noted that without this information the trend could only have been established for the interval of 2.5—7.5 m. It can be seen from Fig. 9.2 that the trend which was calculated in this manner approaches a straight line.

The smoothing formulae for higher-order polynomials are given in various text books (Kendall and Stuart, 1968; Whittaker and Robinson, 1960) and computer programmes for geologists are available (Fox, 1964; Davis and Sampson, 1967). A formula which is used particularly often in geological and econometric investigations is the Spencer twenty-one-point formula (Vistelius, 1967). This formula consists of a running average using the weights:

$$1/350 \ (-1, -3, -5, -5, -2, 6, 18, 33, 47, 57, 60, 57, 47, ...)$$

which are symmetrically arranged around the central value of weight 60. The formula provides an accurate fit for a cubic function but it is also a good approximation for higher orders. The Spencer formula is particularly useful for removing short fluctuations such as random disturbances. It is less successful in the determination of trends unless the series is relatively short.

The general formula ([9.10]) shows that the trend is obtained through the application of a linear operation on the original series (see [8.46]). Such linear operators L have been called "filters" in Chapter 8 and if the original series is given by [9.1], then one may write the trend-fitting procedure which uses weighted averages or filters as:

$$L(Y_t) = L[m(t)] + L(X_t) \tag{9.11}$$

9.5 TREND REMOVAL

Let it be assumed that a chosen polynomial represents a trend $m(t)$ precisely. This component will be removed completely when [9.11] is subtracted from [9.1]. The random process X_t will be:

$$Y_t - L(Y_t) = X_t - L(X_t) \tag{9.12}$$

The effect is illustrated most clearly by assuming that X_t consists of a series of independent random numbers. Clearly, $L(x_t)$ and $L(x_{t+1})$ are not independent because they have $2m - 1$ values in common. The trend removal transforms the independent random process into a stochastic process of moving averages (see Section 7.3). This has been called the Slutzky-Yule effect (Kendall and Stuart, 1968) and for a time it was taken as a reason for avoiding the method of moving averages, particularly in econometric work. Fortunately, this unavoidable effect can be controlled very well. It was seen in section 8.4 that one can evaluate the effect of filtering operations on power spectra by the transfer function $\zeta(\omega)$. Calling the original process X and the filtered process X', then the power spectrum of the filtered process $g_{X'}$ is, according to [8.50]:

$$g_{X'}(\omega) = \zeta(\omega) \, g_X(\omega) \tag{9.13}$$

where $g_X(\omega)$ is the power spectrum of the original process and $\zeta(\omega)$ the transfer function of the filter which was used. Clearly, the Slutzky-Yule effect can be compensated if the filtered spectrum is divided by the transfer function.

The filters which are employed in trend elimination are usually symmetrical which means that any weight $a_j = a_{-j}$. The transfer function is therefore real and consists of cosine terms only. If a moving average of $2m + 1$ terms is used (see [9.10]), then its transfer function will be:

$$\zeta(\omega) = a_0 + 2 \sum_{j=1}^{m} a_j \cos j\omega \tag{9.14}$$

In the special case of the simple moving average, where all the weights are equal, [9.14] can be written as:

$$(2m + 1)\, \zeta(\omega) = 1 + 2 \sum_{k=1}^{m} \cos \omega k = \frac{\sin (m + \frac{1}{2})\omega}{\sin \omega/2} \tag{9.15}$$

The function has been calculated for a simple moving average of 5 in which $m = 2$ and $a_{-2} = a_{-1} = a_0 = ... = 0.2$ and plotted in Fig. 9.4. The transfer function may be regarded as the power spectrum which results when an independent random process with unit variance is subjected to the above filtering procedure. It may be seen that the smoothing has produced high-power contributions in the low-frequency range and that there are also considerable side peaks, which in this case create an oscillation of four units at the frequency $\omega = \pi$.

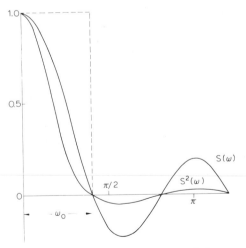

Fig. 9.4. Transfer function after applying one and two smoothing operations.

In trend removal, the aim is to eliminate the low frequencies and to interfere as little as possible with the higher frequencies. One would like to have a filter for this purpose that has a transfer function like the one which is shown in Fig. 9.4 by a dashed line. This would eliminate any frequency below a chosen value ω_0 and would not affect any higher frequency. A finite moving average can only approximate the ideal transfer function but some very sophisticated filters can be constructed for special situations. In most stratigraphical work, it is sufficient to use simple moving average filters as defined by [9.15]. The properties of such filters may be adjusted so that they can deal with most of the trends which are close to linear functions. It can be seen from [9.15] that the first zero is at:

$$\omega = \frac{2\pi}{2m + 1} \qquad\qquad\qquad\qquad\qquad\qquad [9.16]$$

and that by increasing m, the main peak can be made as narrow as necessary. The ratio of the main peak to the first side peak is given by:

$$(2m + 1) \sin \left[\frac{3\pi}{2(2m + 1)} \right] \bigg/ 1 \qquad\qquad\qquad\qquad [9.17]$$

and this ratio decreases with increasing m, but very soon approaches a limit. If $m = 2$ the ratio is 4/1. If $m \to \infty$ the ratio is 4.7/1 (see Granger and Hatanaka, 1964, p. 143) and very little is gained in this respect by increasing the length of a moving average. However, the effect of the side peaks can be suppressed considerably by a repeated application of the filtering procedure. For example, if a moving average of 5 is applied to an observational series, then the transfer function $\zeta(\omega)$ is the same as the one which is shown in Fig. 9.4. If the smoothed series is subjected once more to filtering with the same average, the new transfer function will be $\zeta(\omega) \zeta(\omega) = S^2(\omega)$ which is also shown on Fig. 9.4. Granger and Hatanaka (1964) recommend, from their practical experience in working with economical time series, that a trend should be estimated by using two or more moving averages. If the series consists of n observations, the length of the average should be between 40 and $n/10$ and this trend can be given a final smoothing by using an average of length 3 or 5. The effect of the trend removal on the residuals can be shown by examining the transfer function of [9.12]. The residual $Y_t - L(Y_t)$ consists essentially of a summation process which is followed by a differencing process. In the case of the simple moving average of 5, the residual R_t is given by:

$$R_t = -0.2 \, (X_{t-1} + X_{t-2}) + 0.8 \, X_t - 0.2 \, (X_{t+1} + X_{t+2})$$

and the transfer function of the residual is found to be:

$$\zeta_R(\omega) = 0.8 - 0.8 \left(\cos \frac{3\omega}{2} \cos \frac{\omega}{2} \right)$$

which is plotted in Fig. 9.5. Providing that X_t is an independent random process, the trend-removal operation will create a pronounced oscillation at the frequency $\pi/2$. This is the Slutzky-Yule effect which was mentioned previously. Slutzky's theorem states that if a moving average of 2 is taken n times in a random series and if this is followed by the m-th difference of the result, then, if $n \to \infty$ in such a way that the ratio m/n tends towards some constant θ, the series will tend towards a sine wave with wavelength λ, where:

$$\lambda = \arccos \frac{1 - \theta}{1 + \theta}$$

If X_t is any other random process with the power spectrum $f_x(\omega)$, then the power spectrum of the residual will be:

$$f_R(\omega) = f_x(\omega)\, \zeta_R(\omega)$$

Although the trend removal does change the residual, the effect can be evaluated

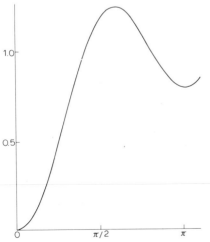

Fig. 9.5. Power spectrum of an independent random process after removing a moving average of 5.

9.6 SMOOTHING PROCEDURES

Whereas trend removal procedures attempt to find simple analytical functions to describe an overall change in the time series, the primary purpose of data smoothing is to eliminate irregular fluctuations. All smoothing methods try to maintain, or even enhance, the more systematic variations of the record. Whether the smoothing of stratigraphical data should be considered depends very much on the data and also on the purpose of the stratigraphical analysis. For example, if two sections are to be compared in order to establish a stratigraphical correlation, it is often useful to simplify the record somewhat by smoothing. On the other hand, if the ultimate aim is to understand the random processes which generated the section, then one might lose valuable information by eliminating random elements.

Again, the basic justification for smoothing procedures is an assumed composite structure of the series. One accepts that the stratigraphic record consists of a systematic element X_t together with a random element Z_t which is symbolically written as:

$$U_t = X_t + Z_t \qquad\qquad [9.18]$$

The term systematic is used here to describe a variation which can be represented by polynomials, at least locally. It means that the finite differences of X_t must be steady to a given order (Vistelius, 1967). This is best understood by considering the two short sequences $f(x)$ and $f(y)$ which are given in Table 9.1. This first finite difference for the first pair of x-values is defined as $x_1 - x_0 = 5$. This is written symbolically as Δ_1^1. The first difference for the second pair is written as $\Delta_2^1 = x_2 - x_1 = 3$ and so on. The whole sequence of first-order differences is decreasing monotonically and is, there-

TABLE 9.1

Finite differences of two observational series

$f(x)$	Δ^1	Δ^2	Δ^3	$f(y)$	Δ^1	Δ^2	Δ^3
0				0			
	+5				+5		
5		−2		5		−2	
	+3		−2		+3		−7
8		−4		8		−9	
	−1		−2		−6		+14
7		−6		2		+5	
	−7		−2		−1		−5
0		−8		1		0	
	−15				−1		
−15				0			

fore, steady. The same is true for the second differences $\Delta_1^2 = \Delta_2^1 - \Delta_1^1$. The third differences remain constant, which indicates that the series can be represented by a cubic polynomial. In contrast to this, the second series $f(y)$ already shows a turning point in the first differences and this series is said to be unsteady in the first degree.

If one now differences a series U_t of the type which was given in [9.18], two things will happen. First of all, it is obvious that the differencing will eliminate any polynomial term, providing that the order of the difference which is taken is higher than the polynomial order. Secondly, the mean of the random term will remain zero throughout the operation but the variance of the differences of a random series will increase. It can be shown (Kendall and Stuart, 1968) that this increase is given by:

$$\text{var } (\Delta_{z_t}^r) = \binom{2r}{r} \text{ var } (z_t) \qquad\qquad [9.19]$$

The so-called "variate-difference" method proceeds as follows. The series u_t is differenced and the variance of the first differences Δ^1 is calculated and divided by $\binom{2}{1} = 2$. The procedure is repeated up to the r-th difference and each time the resulting variance is divided by $\binom{2r}{r}$. In doing this, the variances of the differences should decline until all the polynomial terms are eliminated and then the value should remain constant, apart from sampling fluctuations.

The variate-difference methods, in common with other smoothing or trend-removal procedures, suffer from the practical impossibility of separating the "systematic" component cleanly from the random element. When the polynomial terms are cancelled by the differencing, they are invariably contaminated by random fluctuations which are incorporated into the systematic component. If the random element is large, this can lead to very inaccurate results. As an example of this, we can take the previously mentioned limestone percentages of the Glencar Limestone. (Carboniferous, NW Ireland.) The series consists of 440 observations and the variance of the original data was $\text{var}(u_t) = 1025$. The variance estimates of the random element after r difference operations, are given in Table 9.2. It may be seen

TABLE 9.2

Variance estimates after r differences (Glencar Limestone)

	Estimate
1	580.24
2	704.73
3	695.50
4	667.20
5	643.13
6	613.89
7	613.72

that the second variance is higher than the first one and that this must be attributed to some of the random variation being incorporated into the linear trend which led to the elimination of nearly 50% of the total variation. In addition, it can be seen that, after differences of the fifth to sixth order have been taken, the residual variance remains constant at around 600 and this must be the variation which is caused by the random element. It follows that a smoothing formula, using a fifth-order polynomial, should be sufficient to represent the systematic component. It has been noted before that the Spencer twenty-one-point formula reproduces the cubic polynomials accurately while approximating the higher orders fairly well and so it is a suitable smoothing procedure for the data.

If the process does have the structure of [9.18], then its variance can be written as:

$$\text{var } U = \text{var } X + \text{var } Z + 2 \text{ cov } XZ \qquad\qquad [9.20]$$

If Z, which is assumed to have zero mean, is independent from X, then the covariance term should vanish. This provides a test of how well a smoothing procedure has succeeded. If the Spencer twenty-one-point formula is applied to the Glencar Limestone data, then the smoothed series ought to be a graph of the systematic component X_t. The original data minus the Spencer data should be the residual Z_t; the variances of the Spencer series and the residual series should add up to the original variance. In fact, it is found that:

var (original) = 1025   var (Spencer) = 420

var (residual) = 525   2 cov (Spencer, residual) = 80

This indicates that, although the separation is fairly good, it is not complete. It may be noted that "var (residual)" is another estimate of the variance that is contributed by the random element Z_t which was found previously to be of the order of 600.

Smoothing can be achieved by fitting a trigonometric series to the observed data (Lanczos, 1957). It is obvious that a great number of Fourier coefficients must be calculated if considerable detail is required. The following may be written to represent the systematic component of a stratigraphic series:

$$X_t = A_0 + A_1 \cos t + \dots A_2 \cos it + B_1 \sin t + \dots B_2 \sin it \qquad [9.21]$$

The mean square deviation $(1/n) \Sigma (U_t - X_t)^2$ can be calculated from this series and its value must, of course, converge towards zero when i increases. The convergence is relatively rapid for continuous functions, but with the discontinuities and rapid changes which are typical of random data the convergence becomes slow.

To take an ideal situation, if one plots the mean-square deviation against i, which is defined as the number of computed Fourier coefficients, one finds that this function falls rapidly at first and then the decline becomes much slower at a distinct point. In practice, experiments with the Glencar Limestone series showed that this change of slope occurred at $i = 45$ approximately, but the change in the rate of decreasing error terms was not at all well marked. The variance of the residuals, after 45 trigonometric terms had been eliminated, was approximately 550. This is in agreement with the previous estimates. When the harmonic terms were estimated, part of the random component was mixed up with the systematic component and it was impossible to separate the two cleanly, which is what happened before. Since 440 observations were analysed for the Glencar data, it follows that the wavelength of the random disturbance must be shorter than $440/45 \approx 10$. The measurements were taken at 20-cm intervals and this means a length of 2 m. One must conclude for this particular set of data, that any stratigraphic variation which is shorter than 2 m will be masked completely by random fluctuations. These short fluctuations are eliminated by the smoothing process. The method is supposed to eliminate only the random component which may be regarded as disturbing or inessential. Unfortunately, it has been shown already that a clear separation between the systematic and the random components is impossible and any smoothing operation will affect both of the components. For this reason, smoothed data should be used with great care in stratigraphical analysis and it is often advantageous to compare the results of different smoothing procedures.

9.7 SOME GEOLOGICAL CONSIDERATIONS REGARDING STRATIGRAPHIC TREND

Some of the sedimentological aspects of trends must be discussed before the methods of trend analysis can be developed further. It has been shown that, although it is unsatisfactory, it is often necessary to base a stratigraphic analysis on a single variable. Under such circumstances, one will naturally choose a parameter which conveys the maximum amount of information about the environment in which the sediments formed. Many sediments have properties which are related directly to the energy of the depositional environment and measurements which are based on such properties provide particularly good indicators of the sedimentation conditions. For example, grain size in clastic sediments or the shale content in carbonates are recognized as good facies indicators. Other parameters such as faunal elements, colour, or various bedding forms, could be combined to give a single variable which might be used.

Because such variables are chosen especially for their good representation of the environment, they are usually very sensitive to changes in the rates of sedimentation. This is particularly well known in the case of grain size. High-energy environments produce coarse clastic sediments which have very much higher sedimentation rates than low-energy environments in which fine-grained material is deposited. This leads to the conclusion that a trend which is found by the study of stratigraphic variables very often goes together with a trend in the sedimentation rates.

In order to study the effects of a trend in sedimentation rates, let it be assumed that sedimentation is absolutely continuous and that a constant stream of particles arrives at the depositional area. If the number of particles per unit time is constant, then the rate of sedimentation will be strictly proportional to the grain size. This somewhat artificial assumption is made so that the following discussion is easier to understand. Consider a series of sandstones in which grain size increases steadily with continuous deposition. The variable grain size thus has a linear time trend. (See Fig. 9.6.) At regular intervals 1, 2, ... some time markers are formed which might be bedding planes which do not interfere with the constant sedimentation. The stratigraphic record will show the bedding planes spaced at exponential intervals because the sedimentation rates increase constantly and this is shown for the two examples A and B in Fig. 9.6. Clearly, if the grain size of the two sections is plotted against the stratigraphical position of the samples, a second-order curve will result. On examining the sections, one would find a quadratic trend which could be removed easily by the methods which have been discussed previously. However, this trend removal would not affect the positions of the bedding planes, which in the trend-free section would be still crowded together at the beginning of the section and more widely spaced towards the top. Fig. 9.6 shows that, when the time history is examined, the removal of the trend makes the two records A and B identical. In contrast to this, trend removal does not make the two sections identical in the case of the thickness record. This could be achieved only by making adjustments to the vertical scale.

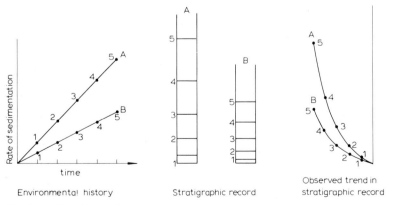

Environmental history   Stratigraphic record   Observed trend in stratigraphic record

Fig. 9.6. The effect of trend removal involving a change in sedimentation rates.

This rather simple example illustrates how the removal of a trend from stratigraphic data may be regarded as changing the various lithologies into an average lithology. This change should be closely linked with an adjustment of the vertical scale because of the very nature of sedimentation. Of course, this principle does not affect trend-like variation alone, but also operates on any systematic or random component.

The implications of this finding are quite considerable. Even if the trend in the time history affects only the mean values, and so could be removed, that same trend will cause non-stationarity of the stratigraphic series with respect to its variance and covariance properties. The analysis cannot proceed any further unless fairly precise data about the changes in the correlation structure of the section are available. If the records are very long, one may be able to estimate the covariance function in various parts of the section, but such long and continuous records are rare in geological work. Quite apart from this, even rather short sections can depart markedly from stationarity and it is then impossible to estimate the statistical properties of the subsections. The model concept is the only alternative approach. A theory is formulated as to how trend developed in a particular stratigraphic section and once the effects of such theoretical trends have been studied, one investigates whether the observed data are in agreement with the model. This model does not provide any proof for geological speculations but it can show the potentialities of some models or give positive evidence that certain theories are not practicable.

9.8 SOME SIMPLE MODELS FOR SEDIMENTARY TREND

As has been discussed at some length in Chapter 3, models can be either deterministic or stochastic. The classical concept of trend implies that one is dealing with a slowly moving variation which might be considered steady to a large degree and, therefore, predictable. Thus, the deterministic sedimentation models provide the logical explanations for the existence of a trend, but stochastic trend models will also be considered later.

It was shown in Chapter 3 that many deterministic models lead to an exponential relationship between time and the cumulative sediment thickness. The ubiquity of the exponential time—thickness relationship is caused by the simplicity of the chosen models which assume a linear rate of decay. This applies to the model of the eroded source area supplying sediment, as well as to the basin models where the water depth approaches an equilibrium which is dictated by a base level. Even the more elaborate models that are based on diffusion lead to time—thickness relationships which can be approximated by exponential curves to a reasonable degree. As will become apparent shortly, exponential trends seem to occur frequently in stratigraphic sections so that they are a natural choice for a trend function.

The following example is given to illustrate the effect of an exponential sedimentation trend. If Z_t is written for the thickness of the accumulated sediment, then using [3.4], which comes from the simple erosion—sedimentation model, one has the following thickness—time relationship:

$$Z_t = h_0(1 - e^{-at}) \qquad [9.22]$$

where h_0 is the ultimate thickness and a is a constant. It follows that:

$$t = \frac{1}{a} \ln \frac{h_0 - Z}{h_0} \qquad [9.23]$$

Next, let it be assumed that this sediment accumulates according to [9.22] and that simultaneously it records an environmental history which may, for example, determine the lithology X_t at any particular time. For instance, one may simplify the example by assuming that X_t is given by a sine wave, let us say:

$$X_t = A \cos \omega t \qquad [9.24]$$

In order to find the vertical change in the lithology X_t [9.12] must be substituted into [9.13] to obtain:

$$X_z = A \cos - \omega \frac{1}{a} \ln \frac{h_0 - z}{h_0} \qquad [9.25]$$

This function was plotted in Fig. 9.7 and it shows how the sine wave changes its frequency continuously and it becomes obvious that the stratigraphic record, which was produced by this mechanism, is no longer stationary. Strictly speaking, a sequence like this does not possess a correlogram or a power spectrum, but if the correlogram is calculated merely as an example, one finds that it changes with the length of the record. Thus, Fig. 9.8 shows the correlogram of the series that was plotted in Fig. 9.7 by using 50 and 100 observations, respectively. Both of the autocorrelation functions show a pronounced damping and they also differ in the wavelengths of the oscillations. As one would expect, the damping becomes more pronounced when

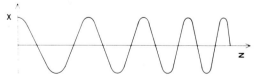

Fig. 9.7. The function of [9.25].

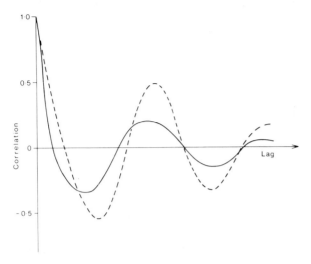

Fig. 9.8. Correlogram of the non-stationary function of [9.25] based on a sample of 50 (broken line) and based on a sample of 100 observations (solid line).

the length of the record is increased and, at the same time, the wavelength of the correlogram decreases. The example shows clearly that this type of non-stationarity, which was produced by an exponential trend, has serious consequences when the correlation structure of the series is interpreted.

It was assumed for the model in the previous paragraph, that the environmental history which determined the nature of the lithology did not affect the rates of sedimentation as well. It was assumed that this variation was just superimposed onto the trend of the accumulating sediment. In examining limestone sedimentation, and perhaps some clastic environments also, it may be of interest to study the situation where the lithological variation itself changes the rates of sedimentation. It has been noted that, in many limestone—shale sequences, the sedimentation is apparently faster during the formation of pure limestones and slows down when the shale content increases. It has been mentioned already that many of the bedding planes in these successions are marked by shaly partings which may represent sedimentation at a standstill, or even periods of erosion. A simple model which could account for this situation would assume that the shale-sedimentation rate remained constant at extremely low levels but that the limestone production fluctuated. At periods of high production, the limestone diluted the shale fraction to such an extent that its presence was masked completely by the limestone. A periodic environmental-time history can be investigated by assuming that the limestone production X_t is of the form:

$$X_t = h_0 - A \cos \omega t \qquad\qquad [9.26]$$

The maximum limestone production per unit time is denoted by h_0 and if the shale-sedimentation rate is low, this value can be taken to be near 100%. At first, it will be assumed that there is no long-term trend in this series and so one can calculate the cumulative thickness of the sediment as:

$$Z_t = \int_0^t h_0 - A \cos \omega t \, dt = h_0 t - A \sin t \qquad [9.27]$$

The function X_z is obtained most easily by using [9.26] and [9.27] in a parametric form and:

$$X_t = h_0 - A \cos \omega t$$

$$Z_t = h_0 t - A \sin \omega t \qquad [9.28]$$

are the equations of the curves known as cycloids, which are shown in Fig. 9.9. The function has the familiar geometrical interpretation of describing the path of a point on a rolling wheel. The constant h_0, which is the diameter of the wheel and which determines its progress, is a measure of the sedimentation rate. The constant A is the distance of the point from the centre of the wheel and it describes the amplitude or range of the lithological variation. Fig. 9.9 illustrates three possible situations. Either $A < h_0$, in which case sedimentation is continuous but retarded when the limestone percentage decreases, or $A = h_0$, in which case the sedimentation comes to a momentary halt at the limestone minimum. The third alternative $A > h_0$ leads to the cycloid becoming negative in part and this reverses the direction of the movement. The reversed movement in this context can be taken to

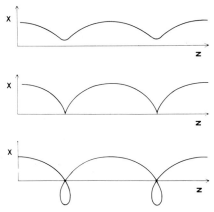

Fig. 9.9. Cycloids.

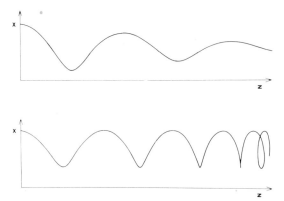

Fig. 9.10. Cycloids with linearly decreasing values of A (upper curve) and decreasing values of h_0 (lower curve).

mean that erosion occurs and that this is followed by renewed deposition which fills in the eroded gap and continues the stratigraphic record.

The cycloidal model can be modified in several ways by introducing trends. For example, the two constants A and h_0 could be changing during sedimentation and a linear decline of A and h_0 is shown in Fig. 9.10. A decrease in the value of A causes a non-stationarity which affects the mean values only. In extreme cases, there may be erosion with large values of A and sedimentation becomes more and more continuous as the amplitude of the lithological variation decreases. The decrease in the value of h_0 (lower curve in Fig. 9.10) produces not only a decrease in the amplitude but also a decrease in the wavelength of the oscillation and, at the same time, erosion becomes more likely. Both types of trend are likely to occur in real situations but there is the additional possibility of the thickness scale being controlled by outside influences, as was assumed in the exponential-decay model. This concept of an overall decline in the sedimentation rate was used to explain the variation in the limestone percentages for the middle Glencar Limestone of NW Ireland (Schwarzacher, 1967). The limestone percentages, which were actually observed for the lowest five "cycles" of this succession, are compared in Fig. 9.11 with a cycloid which is plotted on a logarithmic

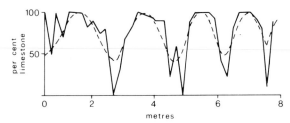

Fig. 9.11. Exponential cycloid model fitted to data from the Glencar Limestone.

horizontal scale. The method for actual fitting of the trend will be discussed in the next section. The figure shows that a very good agreement exists between the theoretical model and what was actually observed, up to the last "cycle". This last cycle alone is overdeveloped and it is followed by four more "cycles" which would fit perfectly into the pattern if the phase shift, which was introduced by "cycle five", was removed.

Finally, it should be mentioned that the cycloid model was first used by Barrel (1917, p. 791) when he discussed sedimentation in a constantly sinking basin with a superimposed fluctuation of the base level. The problem is very similar to the limestone sedimentation process and Barrel's model will be discussed in more detail later.

9.9 THE RECOGNITION OF SEDIMENTARY TRENDS

The discussion in Section 9.7 must have made it clear that the recognition of useful stratigraphic trends is, once more, linked to the thickness—time relationship. In order to establish a thickness—time relationship, one needs an interval scale for time. This can be taken to mean a scale which establishes the equality of the time steps but not necessarily the absolute value of the time units which are being used. Such a scale is hardly ever available. One can only substitute for it a scale which is based on steps which one has reason to believe are approximately equal. The natural choice for such intervals are the subdivisions in the stratigraphic section that are caused by repeated similar sedimentary processes, such as geological "cycles" and sedimentary bedding. Fortunately, it is not necessary for the assumed time-steps to be of precisely equal lengths and they may be used, provided they possess a common distribution giving a mean length.

Once a "time unit", such as sedimentary bedding, has been chosen, one can plot the cumulative thickness against the serial number of the bed. This "time"—thickness graph should be a straight line as long as the series is stationary. A trend in the series will produce a systematic departure from the straight line and, if one keeps to the simple models, this departure has to be either exponential or linear. An essential feature is the systematic nature of the departure from the straight line, which is unlikely to occur just by random fluctuations. One can fit an analytic expression to the graph because it has been accepted, for the time being, that the trend is a deterministic function. One can attempt to analyse the sedimentation—"time" relationship in terms of linear and exponential components. It cannot be stressed too strongly that any attempt at understanding the physical nature of the trend must be made in conjunction with an intensive sedimentological investigation of the stratigraphical record. This will be illustrated by the examples which follow.

The previous chapters used the Lower Carboniferous series of NW Ireland

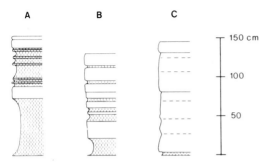

Fig. 9.12. Three types of sedimentary cycles in the Lower Carboniferous of NW Ireland.

frequently as an illustration of oscillating sedimentary conditions. The lime-stone percentages fluctuate fairly regularly and persistently, and the "cycles" which are observed as a result of this are an obvious choice for the approxi-mate time units. Although the details about the "cycles" change consider-ably when they are examined in the three formations of the Benbulbin Shale, the Glencar Limestone and the Dartry Limestone, the essential pattern and approximate thickness of the units remain the same throughout the 200 m of the exposed sediment. The oscillations are asymmetrical in all three for-mations, beginning with clastic shale sedimentation, which takes up more than half of the "cycle" thickness in the case of the Benbulbin Shale (see Fig. 9.12A). The initial clastic interval is less than a quarter of the "cycle" thickness in the middle Glencar Limestone (see Fig. 9.12B), and it is hardly more than a strongly developed bedding plane in the Dartry Limestone (see Fig. 9.12C). Nevertheless, the changes are gradual and it is reasonable to assume that the basic mechanism which produced the "cycle" remained unchanged.

Once this conclusion is reached, the cumulative "cycle" thickness can be plotted against the "cycle" number which was taken as the preliminary time scale. Fig. 9.13 shows this graph and there is an approximately linear rela-tionship between the cumulative thickness and the "cycle" number. This indicates that sedimentation was continuous and at an approximately con-stant rate. There are, however, deviations from this linear relationship which are regarded as significant because it is found that the changes in the sedi-mentation rates coincide almost exactly with the formational boundaries. In order to make this more obvious, the constant sedimentation rate of 230 cm/"cycle", which was found for the Benbulbin Shale, was continued as a straight line in Fig. 9.13 and it can be seen by comparison with this standard that the main deviations occur in the Glencar Limestone. The Dartry Lime-stone returns to a straight-line relationship with a slightly higher sedimentation rate of 280 cm/"cycle". The deviations in the Glencar Limestone can be examined in more detail. The measured cumulative thicknesses of the "cycles" were replotted on a much larger scale in Fig. 9.14 and, at the same time, a

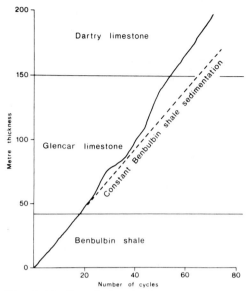

Fig. 9.13. Cumulative thicknesses of Lower Carboniferous limestone. Vertical scale in metres, horizontal scale successive numbers of sedimentary cycles.

constant rate of 150 cm/"cycle" was subtracted from the measured cumulative thicknesses, in order to counteract the linear trend. This simple procedure shows that the data can be fitted extremely well by two exponential curves. One curve covers the lower Glencar Limestone and the second exponential curve marks the beginning of the middle Glencar Limestone at point A in Fig. 9.14. The reasons for fitting these two exponential curves will be given in a later paragraph.

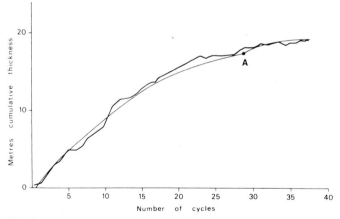

Fig. 9.14. Cumulative thickness of the lower and middle Glencar Limestone.

The purpose of this exercise was to find and recognize a deterministic trend in the stratigraphic record of the section which was examined. The graph which shows the cumulative thickness of the Glencar Limestone against the "cycle" number suggests strongly that this deterministic trend consists of an exponential and a linear component. It is important to realize at this stage, that the finding of such a trend depends in no way upon the assumption that the sedimentary oscillations, which were used as time units, are of precisely equal lengths. Indeed, if one were to choose independent random numbers with a common distribution to represent the assumed equal time steps, one would still obtain a straight-line relationship between cumulative thickness and the "cycle" number.

The chances of the deviations, being systematic and following an exponential law as it were by accident, can only be assessed if something is known about the mechanism that produced the trend. If it is accepted that the trend is caused by an essentially deterministic process, then the simple model of exponentially declining sedimentation rates leads to several possible interpretations of the sedimentary record. When the Glencar Limestone is considered, two possible ways in which the trend could have been generated suggest themselves. The first possibility is shown in Fig. 9.15A and attributes the linear trend of the cumulative curve to constant crustal sinking but the exponential components are interpreted as the effects of sudden subsidences which increase the sedimentation until a new base level is re-established. This is the model of sedimentary by-pass essentially, which was introduced by

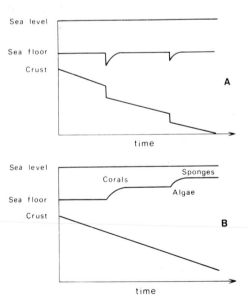

Fig. 9.15. Two hypotheses to explain the sedimentation rates in the Glencar Limestone.

Barrel (1917) and has been discussed already in Chapter 3. The second possibility is shown in Fig. 9.15B diagrammatically and consists of a steadily sinking crust but a sudden increase in sedimentation produces a shallowing of the sea. Base-level control is the effective factor which determines the amount of sedimentation in both of these models. This implies, automatically, that the crustal sinking is constant unless a link exists between tectonic activity and the sediment supply. More complicated situations like this are not considered in this particular case. The absolute water depths for the models are not known but the value of the shallowing or the sudden subsidence can be determined from Fig. 9.14 and is of the order of 5—20 m. Of the two alternatives, the second case (Fig. 9.15B) seems to be the more likely explanation for the Glencar Limestone because there is a definite facies change from the Benbulbin Shale to the Upper Glencar Limestone which, for various sedimentological reasons, has been interpreted as being the effect of a shallowing environment. For example, it is noticeable particularly that the benthonic fauna changes whenever there is a marked increase in the sedimentation rates. The first rapid increase in the sedimentation rate of the Lower Glencar Limestone coincides with the appearance of numerous caninide corals. In the second step, one finds a sudden abundance of lithostrotian corals, together with structures which have been interpreted as algal formations (Cotter, 1968). The Upper Glencar Limestone is regarded as the most shallow development and it contains numerous sponge colonies. It is possible that these faunal changes which alter the base level of the sedimentation system are, themselves, caused by changes in the supply of nutrients (Schwarzacher, 1968). A number of additional observations support this interpretation but the points which have been discussed are enough to show that the examination of a stratigraphic trend must not be separated from the palaeontological and petrographical examination of the section. A hypothesis based on the numerical analysis of a section becomes acceptable only if the hypothesis does not contradict the additional stratigraphical evidence, indeed a hypothesis should contribute to the general understanding of the sedimentation history of the series.

9.10 THE REMOVAL OF STRATIGRAPHICAL TRENDS

After a trend has been recognized, and if possible justified on geological grounds, one may attempt to remove it from the original section. It follows from the discussion in Section 9.7 that trend removal involves not only the subtraction of an analytical function from the observed data, but also an adjustment of the vertical scale to allow for the increase or decrease in the sedimentation rates. Inevitably, the estimation of sedimentation rates must be based on some hypothetical time scale. Although one assumes the existence of a steady and deterministic trend function in every case, the estimation of this will be blurred by either variations in the length of the assumed time intervals or by irregular fluctuations in the sedimentation rates. According to the theories which were developed in an earlier paragraph, the finer details of the sedimentary record consist of a number of discontinuous steps, so that any attempt at applying a continuous trend function to such data would lead to mistakes. Thus, trend removal will never isolate the residuals for detailed analysis as cleanly as one would wish. However, it is possible to reduce the non-stationarity of the observed geological sequences to such a degree that at least an approximate analysis of the data becomes feasible.

A short example, which was first given by Schwarzacher (1969), will illustrate the procedure. In examining the Pennsylvanian cyclothems of Kansas (U.S.A.), which will be discussed further in Chapter 10.8, it was noticed that the original data were highly non-stationary. The transition probability matrices, for example, which were calculated from the lower and upper halves of the section were significantly different. A graph of the cumulative cyclothem thicknesses was constructed in order to investigate this. Since each cyclothem consists of two lithological groups which are supposed to represent marine and non-marine environments, these were separated and the cumulative thickness of the marine group, which was called limestones for simplicity, was plotted separately from the so-called outside shales which represent the alleged non-marine interval. This graph is shown in Fig. 9.16. Once again, it can be seen that the thickness to cycle-number relationship can be approximated by a combination of straight lines and exponential curves. As was seen in a previous example (Fig. 9.14) the changes in the sedimentation rates occur at stratigraphic boundaries. The

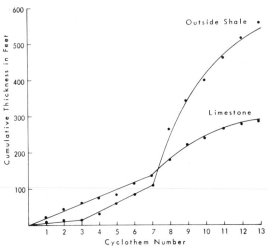

Fig. 9.16. Fitted trends for two lithological groups in the Upper Pennsylvanian of NE Kansas (after Schwarzacher, 1969).

boundary section between the Pedee and Douglass groups is noticeable particularly. It occurs at cyclothem No. 7. The clastic component indicates a marked break at cyclothem No. 3, which marks the boundary between the Kansas City and Lansing group. The equations of the straight lines and the exponentials were calculated by conventional regression methods and, once these were known, they were accepted as representing the true rates at every height in the section. The transformation of the vertical scale could now proceed and, in the programme which was used in this particular case, the following operations were carried out. Lithological units were read together with their stratigraphical thickness. Next, the position of the unit in the section was found and, depending on whether the particular unit was limestone or outside shale, the sedimentation rate which corresponded to that position was calculated from the equations of the regression lines. The sedimentation weight value was used for weighting the original stratigraphic thickness of the unit, which was then added to the transformed section. The weights were calculated in such a way that, if the observed cycle thicknesses had fallen precisely on the regression lines, each cyclothem would have consisted of twenty feet of shale and twenty feet of limestone. The transformation in the first instance resulted in a stationary sequence, and the transition probability matrices which were calculated from various parts in the new section were significantly the same. Furthermore, it was possible to analyse the series in terms of a strongly oscillating fourth-order process (see Section 10.8). It was concluded that the stratigraphic record, after adjusting the sedimentation-rate trend, showed strong geological cyclicity which one could not detect in the original data.

It is obvious that one could change any section into a series in which the assumed time units lead to precisely equal thickness intervals. This would indeed be the case in the present example if one took the observed cyclothem thicknesses of Fig. 9.16 as weights in the transformation procedure. However, this would imply that the sedimentation rates do not change according to a deterministic trend which can be represented by a simple analytical function, but by some sort of random process instead. The model of the deterministic trend has the advantage of simplicity, and in many instances a direct geological explanation for the trend can be given.

9.11 THE POTENTIALITIES OF STOCHASTIC TRENDS

Introducing a random process as a possible trend model appears, at first sight, to contradict the definition of a trend. Random processes are unpredictable and are associated with frequent changes in the direction in which they move. On the other hand, a trend is regarded as a slow variation which moves in one direction generally. However, this fluctuation of random processes is a question of scale and there are numerous stochastic processes which, on a larger scale, are either increasing monotonically or which move possibly in broad waves, once the shorter fluctuations have been removed by filtering. The unidirectional random walk, which was introduced as a simple sedimentation model in Chapter 4, is a typical example of a random process which increases on a large scale, providing that the probability of deposition exceeds the probability of erosion (see Fig. 4.1). It is of some interest that the unidirectional random walk can lead to a trend-like movement, even if the probabilities of the up and down steps are equal. This situation is well known from the theory of gambling and has been treated in some detail by Feller (1957), and simulation experiments, which can be performed easily with a random-number generator, give results which are often surprisingly similar to the so-called secular variations or trends which are frequently observed in time series.

The potentialities of stochastic trends lead to an interesting dilemma in the interpretation of stratigraphic records. The conventional approach, which was treated earlier, is to assume that the trend function is deterministic. One implies, by this assumption, that the trend is caused by a process which is physically different from the remaining variation which may consist of shorter oscillations and random fluctuations. Many examples of this type of thinking can also be found in the analysis of economic time series. Such series are often based on the prices or the quantity produced of some commodity over a period of time. Frequently they show a trend representing the variation over many years and, superimposed on this, shorter fluctuations which may be seasonal or due to other disturbances. The important step in analysing a series like this is the attribution of the trend to something different from whatever is causing the shorter variations. As an example, consider the monthly sales of ice-cream. It is not surprising that these show a strong seasonal variation and it is easy to explain why more ice-cream is sold during the hotter months of the year. However, the figures show a continuous increase over the years as well. This could be attributed to a general growth of the industry and to an increase in the demand from a more affluent society. In this hypothetical case, one might attempt to fit a growth curve to the trend and consider it as a deterministic process. In doing this all the random fluctuations which are suggested by the data must be explained by the variation of factors which are unrelated to the trend mechanism. For example, if there were no other factors involved then one might conclude that the temperature differences between the different months and years are irregular and introduce a stochastic element. If it is accepted, on the other hand, that the trend is not strictly deterministic and itself fluctuates, then it is obvious that less random variation is available for the seasonal temperature differences and for the differences from one year to the next. If a geologist was to examine, let us say, the fossil records of ice-cream sales, he would in all probability choose a deterministic interpretation of the trend. In this case, he would conclude that the climate was fluctuating quite considerably or he might accept the model of a stochastic trend and thus come to the conclusion that the temperatures were more even. Very similar considerations apply to stratigraphical records, where sedimentary "cycles" occur together with a trend and this was discussed in the previous section. One either assumes the trend to be deterministic, which may lead to fairly irregular "cycles", or one accepts stochastic elements in the trend, which leads to more regular "cycles".

Dividing a time series into two or more components may sometimes be seen as an artificial procedure but without such a division no analysis is possible. Indeed, the meaning of analysis is to recognize individual components; the recognition of components is simplified if they are defined clearly by a deterministic process. Therefore, there are situations where a certain amount of progress can be made by assuming that either the trend or the oscillating component is deterministic in nature, even if there are good reasons for believing that they both incorporate stochastic elements.

Chapter 10

Sedimentary cycles

10.1 INTRODUCTION

Cyclic or rhythmic sedimentation is one of the aspects of stratigraphy that has attracted many geologists' attention and has led to a considerable amount of controversy. Some geologists have attached a great deal of importance to the cyclic development of stratigraphic history at the risk of over-emphasizing the deterministic nature of the development of past events. Others may have shied away from anything which could be regarded as a regularity in the geological record and have adopted a hypercritical attitude towards the concept of sedimentary cycles. The reasons for this divergence of opinion are very often the lack of adequate definitions together with the great difficulty in formulating the concept of sedimentary cycles in quantitative terms.

Generally, the word "cycle" refers to a series of connected events which return to a starting point. When it is applied to a sedimentary sequence, the term is taken to mean a series of rock types that follow each other in a predictable pattern in which at least one rock type, which has been taken as the starting point, is repeated. The word "predictable" is an essential part of this definition since otherwise any sequence that contains the same lithology more than once would be called cyclic and this would make the whole concept superfluous. Every geologist knows that absolute predictability in rock sequences does not exist and it is necessary to qualify the term "predictable" to allow for the variations which always occur. In doing this, one treats the recognition of sedimentary cycles as a stochastic problem. The sedimentary models which were developed in Chapters 4—8 have a direct bearing on the subject, so that it will have been noted already that much of the quantitative work on stratigraphic sections originated from a study of cyclic sedimentation.

10.2 SEDIMENTARY CYCLES IN STRATIGRAPHY

Before reviewing the mathematical methods which can be used in the study of cyclic sediments, it is useful to consider some of the geological definitions and the motivations which have led to the study of the sedimentary cycles as such.

The term "cyclic" or "rhythmic" sedimentation was used by geologists for nearly a century (Duff et al., 1967) and many different interpretations of sedimentary cycles exist. In spite of this, it is recognized generally that a cyclic sequence is a sedimentary series in which the various rock types, A,B,C, ... are arranged in some recognizable pattern which is not random. Weller (1964, p. 613) writes, "The idea of a cycle involves repetition because a cycle can be recognized only if units are repeated in the same order." It follows that a cyclic sequence should be ordered into repetitive patterns like: ABCABC ... ABCBABC ... or CBACBA ... and so on. It is recognized that such a perfectly ordered sequence does not occur in nature and it is an abstraction which has been called an ideal cycle (Pearn, 1964). Most of the authors who used the ideal-cycle concept believed that it represented some, "law of nature" which would be followed by sedimentation, except for unforseeable disturbances.

The term "cycle" is sometimes used in the geological sense without any implication of a manifold repetition in one and the same section and so one speaks of a "cycle of erosion", "orogenic cycle", as well as others. These are all sequences of events that seem to follow a definite pattern which has been established by studying the whole geological column in many different parts of the world. Although such "cycles" may have a marked influence on sedimentation, they are excluded from the present discussion.

It may have been noted that the definitions of cycles which have been given so far, make no reference to the stratigraphical thickness of the repeated rock groups. Indeed, Duff and Walton (1962, p. 247) write: "In so far as the recognition and definition of cycles are dependent only on the order of rock units, the thickness of the units is an additional even superfluous feature." Certainly, cycles can be defined without any reference to the thickness of the sediments which are involved but it is doubtful whether they can be recognized in practice unless some scale factor is introduced. It is easy to show that the concept of rock-sequence cycles has only an extremely limited use in either stratigraphy or sedimentology.

The problem of sedimentary cycles has also been approached from a much more geometrical point of view (Sander, 1936; Schwarzacher, 1946). The general idea of space rhythmic fabrics was introduced by Sander (1948) and the term defines rock fabrics in which, at equal intervals in space, identical fabric elements may be found. When this concept is applied to the sedimentary succession, rhythmicity in space implies that at equal thickness intervals identical lithologies will be found. It is obvious once again that the repetition will not be absolutely precise and a statistical measure of space rhythmicity must be used (Schwarzacher, 1947). The type of cycle which recognizes a regularity of pattern in both lithology and thickness will be referred to as "geometrical cyclicity" in order to differentiate this fabric from the more simple "sequential cycles".

One can translate sedimentary thickness into time, at least in theory, and one of the most interesting aspects of sedimentary cycles is the search for the time history. A "time cyclic sequence" is a sedimentary series in which identical rock types have been formed at equal time intervals. This possibility has appealed to several geologists and it has been suggested (Moore, 1950) that the term "rhythm" should be reserved for such regular intervals while the term "cycle" should be reserved for regularities in rock pattern. Useful as this distinction would be, it has never become general practice and in the following paragraphs the term "time cyclicity" will be used for this type of regularity.

In order to understand why the three different aspects of sedimentary cycles — sequence, thickness, and time — have received such different attention from geologists, one has to look at the motives which led stratigraphers to the study of the subject. It is clear that the concept of cyclic sedimentation can lead to a considerable economy in the description of stratigraphic sections. If a sedimentary series consists of many repetitions of, let us say, sandstone, shale and limestone, then it is shorter to note that the cycle sandstone—shale—limestone is repeated x times, rather than giving the list of successive lithologies. This description is improved considerably if the average thickness of the cycle is given at the same time.

There is another and quite different motive which has encouraged the examination of sedimentary cycles. It is felt intuitively that if a certain pattern to sedimentation exists then it must have some cause. It is only right that the geologist should try to explain repetitive patterns of any kind. It is worth considering the following quotation from Hull (1862, p. 139) which is given by Weller (1964): "We cannot fail to have observed that the groups have a tendency to arrange themselves into three-fold divisions, the upper and lower being composed of sands or clay the middle of limestone ... Phenomena of so general a character cannot be accidental but must be in accordance with the system of nature." This reference to a "system of nature" is probably the main reason why studies in cyclic sedimentation are so intriguing. It is equally clear that the whole complex of cyclic sedimentation consists of many phenomena which operated on quite different time scales. On the one hand, there are the lithological cycles which are often complicated but were, stratigraphically speaking, formed instantaneously, for example, turbidity currents which represent such a short time interval that they cannot be taken as indicators of environmental changes. On the other hand, there are the sedimentary cycles which represent long time intervals, during which the environment may have changed quite considerably and, therefore, they represent an interesting portion of the geological history. Sedimentary cycles of the first type which represent stratigraphic time quanta, as discussed in Chapter 2, are of particular interest to the sedimentologist who has to find a sedimentation process to explain the sequence of the observed lithologies. Cycles which are composed of more than one sedimentation unit are of special interest to the stratigrapher since they are often a record of regularly changing environments. It is this possibility of finding some order or law in the evolution of geological history which is attractive and which has opened up a very wide field for speculations in theoretical stratigraphy.

The desire to find some order in a complex subject is, of course, not peculiar to geologists, and many others including meteorologists, astronomers, biologists, economists and even historians have often been preoccupied with the search for cyclic phenomena. It is well known that many of the so-called "cycles" that were discovered in these subjects were fictitious and the result of trying to find regularity where none existed. The use of quantitative methods and precise definitions is, therefore, essential.

10.3 SEQUENTIAL CYCLES

One is dealing with discrete states when studying lithological sequences and the Markov chain provides an ideal model. It was shown in Chapter 5 that the transition matrix of lithological states provides a very compact description of a section. It can also be used to predict future lithologies when the initial state, or the starting point of the section, is known. If one calculates the n-th power of the transition matrix $\mathbf{P}$, then the elements $P_{ij}^{(n)}$ give the probability of lithology j being encountered after n lithological units, providing that the starting point was lithology i. For this to be true, the sequence that is under investigation must be reduced to a simple Markov chain and, although this is always possible, it does need considerable care.

Consider the following numerical examples (Schwarzacher, 1969):

$$
\begin{array}{cccc}
a & b & c & d \\
\begin{array}{c}A\\B\\C\end{array}
\begin{bmatrix}
0.5 & 0.3 & 0.2 \\
0.5 & 0.3 & 0.2 \\
0.5 & 0.3 & 0.2
\end{bmatrix}
&
\begin{array}{c}A\\B\\C\end{array}
\begin{bmatrix}
0.8 & 0.1 & 0.1 \\
0.2 & 0.1 & 0.7 \\
0.4 & 0.5 & 0.1
\end{bmatrix}
&
\begin{array}{c}A\\B\\C\end{array}
\begin{bmatrix}
0.1 & 0.8 & 0.1 \\
0.1 & 0.2 & 0.7 \\
0.9 & 0 & 0.1
\end{bmatrix}
&
\begin{array}{c}A\\B\\C\end{array}
\begin{bmatrix}
0 & 1 & 0 \\
0 & 0 & 1 \\
1 & 0 & 0
\end{bmatrix}
\end{array}
$$

$$[10.1\ a\text{---}d]$$

The three states which represent lithologies are labelled A,B,C, in each of these matrices. If the first lithology in example [10.1a] happens to be A, there is a probability of 0.5 that the second lithology will be A also. The rows in this matrix are identical and it has been shown earlier that such a matrix does not change when higher powers are calculated. This means that lithology A will always be encountered with equal probability, regardless of the number of beds that have been laid down in between. In the second example, [10.1b], the first transition probability from A to A is 0.8 and the higher transition probabilities for the same state are shown in Fig. 10.1. The transition probabilities decline geometrically to the stable probability which in this case happens to be 0.6. A similar geometrical decline is seen in

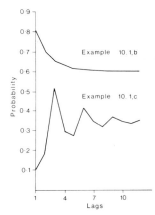

Fig. 10.1. Higher-order transition probabilities of state A in examples [10.1b and c] (after Schwarzacher, 1969).

example [10.1c] but in this case the stable probability is approached by a series of damped oscillations (Fig. 10.1). A powering of example [10.1d] will show that the probability of finding state A after three layers will remain constantly 1 and the higher transition probabilities for this state are 0,0,1,0,0,1 ... The four examples which are given in [10.1] represent the four types of probability pattern which are of basic interest in the analysis of sedimentary cycles. Example [10.1a] was identified as an independent random process in Chapter 5 and the arrangement of lithologies in such a section is what one would expect from complete random mixing. The case of a periodic chain ([5.34]) is easily recognized in example [10.1d]. The periodic chain is the discrete equivalent to the mathematical cycle which was defined in Chapter 7 (equation [7.52]). It is clear that example [10.1a] has nothing in common with the geological definition of sedimentary cycles. It is inevitable that such sequences contain repetitions of identical rock types but there is no regularity of pattern and the periods of repetition are quite unpredictable. In contrast to this [10.1d] does represent a deterministic process and it describes the pattern of an "ideal cycle" precisely. Any random disturbance will reduce the periodic chain of [10.1d] to a stochastic Markov chain of the type which was seen in [10.1c]. This will happen whether the random disturbance is part of the original generating mechanism or due to incomplete recording by the sediment. It can be seen without any calculations that a matrix such as:

$$\begin{bmatrix} 0.01 & 0.99 & 0 \\ 0 & 0.01 & 0.99 \\ 0.99 & 0.01 & 0 \end{bmatrix}$$

would behave almost like [10.1d] and would have strong recurrence maxima

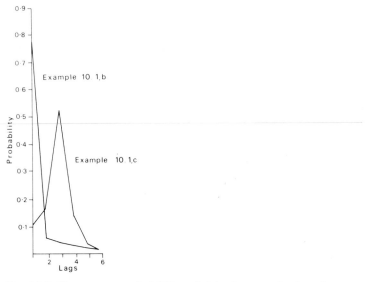

Fig. 10.2. Recurrence probabilities of state A, examples b and c.

at an interval of three beds. However, this matrix represents a stochastic process and the powered matrix $\mathbf{P}^n$, $n \to \infty$ will converge to a matrix which has identical rows. This means that any prediction for a large n will be the same as for an independent random process. Thus the example which was given above behaves like [10.1c] which, in this context, will be called an oscillating Markov chain.

The different behaviour of the four examples also may be seen by calculating the recurrence probabilities for the different states according to [5.16]. This has been done again for state A (Fig. 10.2) and the graph of [10.1c] shows a clear maximum after three beds have been deposited. The fact that there is a pronounced probability maximum after three acts of deposition makes A the most likely lithology in the third bed but others could turn up instead. A short series was simulated using [10.1c] and it gave the following:

ABCBCACABCABBCABC

Increased damping suppresses the oscillations and the chain then behaves like [10.1b]. In this case, the recurrence probability maximum lies at the first successive bed and declines geometrically from there. A series like this is not oscillating although it is not an independent random process; once a certain lithology occurs, the probability of its return decreases geometrically until the stable probability value is reached. The difference between [10.1b] and [10.1c] can be seen readily by generating a short series using [10.1b]. This gives the following sequence:

AAABABCAAACACBCBCAAB

There are matrices that belong to the oscillating type but which possess such strong damping that it is very difficult to recognize them just by graphing the higher transition probabilities. A test for oscillating behaviour can always be made by examining the eigenvalues of the transition matrix. It was found in Section 5.8 that, if a sequence has oscillating properties, then at least one eigenvalue of its transition matrix is either complex or negative. A negative eigenvalue in the transition matrix leads to the simple oscillation period of two, which produces the pattern ABAB ... in stratigraphical sections. This is commonly known as "banding" and it must be considered as a special case which will be investigated later. The damping of the oscillation is determined by the modulus of the complex eigenvalue. One is dealing with a strictly periodic case when the complex eigenvalue $|\lambda_c| = 1$, but with increasing damping, this value goes towards zero and so provides a useful measure of cyclicity. When there is no oscillating component, all the eigenvalues are real.

Doveton (1971) indicated a different but rather roundabout way by which the "degree of cyclicity" could be estimated. This involved calculating the mean recurrence time ([5.19]) and its variance. The mean recurrence time must be the same for any type of random or deterministic process model that has been fitted to an observed sequence, but this does not apply necessarily to the variance. Provided that a recurrence-time maximum exists that is different from step one, then the variance of the recurrence time distribution will be small for nearly periodic processes and large for highly damped processes. If this method is used it would be simpler, and in the case where more than one oscillation period is involved it would be more meaningful to use the probabilities of the recurrence-time maxima as easily computed measures of cyclicity.

The classification of time series into four types, which are independent random series, non-oscillating series, oscillating series and deterministic cyclic series, can be derived equally well from a study of autocorrelations and power spectra. These were considered in Chapters 7 and 8. The necessary diagnostic features are summarized in Table 10.1. The division of dependent stochastic processes into oscillating and non-oscillating ones was introduced especially to separate the two types of sedimentary sequences which geologists call "cyclic" and "noncyclic." A sedimentary cyclic section from now on will be regarded by definition as a sequence which possesses the properties of an oscillating or the properties of a periodic series as it is defined in Table 10.1.

The justification for this definition of cyclicity lies in a number of points but the most apparent one is that the definition corresponds precisely to the

TABLE 10.1

The identification of stratigraphic time series

Type of series	Transition matrix	Correlogram	Spectrum
Independent random process	Rows of transition probabilities are identical.	One at r_0. Zero otherwise.	Constant power over the whole frequency range.
Non-oscillating random series	Powers and recurrence probabilities exponentially approach a constant value.	Autocorrelation declines steadily from one to zero.	Maximum power at frequency zero.
Oscillating random series (geological cycles)	Powers approach stable probability in a series of damped oscillations. Recurrence probability maximum different from one.	Autocorrelation approaches zero in a series of damped oscillation.	Maximum at a frequency which is different from zero.
Deterministic process (mathematical cycles)	Powers of transition probability do not converge. Spike-like recurrence probability maxima.	Autocorrelation oscillates without damping.	Spike-like maxima.

general concept of sedimentary cycles as they are envisaged by most geologists. If the distinction between oscillating and non-oscillating is not made, then either one has to deny the existence of sedimentary cycles or accept that all sedimentary sections are cyclic. The concept of sedimentary cycles would be lost if either of the alternatives was to be chosen. The truth of this statement is borne out by experience. No quantitative analysis of stratigraphical sections to date has found either a purely random arrangement of rocks, as in [10.1a], or a perfectly ordered periodic sequence, as in [10.1d]. This shows that we are, in practice, dealing only with the oscillating and non-oscillating sequences and, if they are not differentiated, this leaves just one group.

Other attempts at defining a cycle have been made by Duff and Walton (1962) who write: "We define the term as a group of rock units which tend to occur in a certain order and which contains one unit which is repeated frequently throughout the succession." As it stands, this definition is applicable to any sedimentary section in which the rock types are repeated. However Duff and Walton also define what they call a "modal cycle" and this is defined as: "that group of rock types which occurs most frequently through any succession." Although this is not stated specifically in their paper, the authors obviously regard a cyclic sequence as a sedimentary series which contains a modal cycle. In order to show that this definition is quite inadequate, it is only necessary to show that an independent random process possesses a modal cycle. Consider example [10.1a] where the most commonly occurring lithology is given by state A. The most likely sequence in this series is A,A but this is not recognized as a "cycle"; the most common cycle must be of the form A*A which can be either ABA with the probability of 0.15, or ACA with the probability of 0.10. Therefore, ABA is clearly a modal cycle. To give the modal cycle of Duff and Walton any meaning, the existence of cyclicity as it is defined by Table 10.1 must be established first. This means that the series must be proved to be oscillating.

10.4 SEQUENTIAL CYCLES IN PRACTICAL STRATIGRAPHY

It was shown in the preceding section that theoretical criteria can be established for cyclic and non-cyclic sequential successions. However, it is equally important to realize that knowing whether or not a particular sedimentary section has got the property of sequential cyclicity contributes surprisingly little towards the interpretation of sedimentary cycles. How useless such information can turn out to be is seen by considering the following situation. Assume that an oceanographer has measured sea-level fluctuations carefully, and by doing this has established the sequential order of low and high water. Should this oceanographer have forgotten to note the time when each observation was made, it would be impossible to tell whether he has measured waves, tides or isostatic sea-level fluctuations. An analysis of such data can lead only to the trivial result that the sea level has changed. To draw any further conclusions about causal connections between high, medium and low sea levels would be obviously quite out of place because there is no information about the processes which are involved. This is exactly the situation that the stratigrapher faces when confronted with sequential analyses of sedimentary sections. The pattern may turn out to be cyclic or noncyclic but until he can relate the pattern to some physical process which generated it, such information must remain of little use.

This does not mean that sequential analysis is without any practical use. It has been mentioned already that sedimentologists attach particular importance to the establishment of associations of vertical rock sequences. These studies are similar to comparative facies studies which have shown that adjacent areas of deposition are genetically related. Vertical sequences which are not random show almost identical relationships. The very common upwards fining sequence of sand—silt—clay, for example, indicates a progressive decline of environmental energy and is a typical instance of a well established vertical association. The energy decline provides the link between the different lithologies just as the increasing water depth may provide a link between adjacent facies developments. It is evident that such associations can be studied in any type of sedimentary section and it is irrelevant whether the

sequence sand—silt—clay is followed again by sand which would complete
the cycle, or by an entirely different lithology. Thus, sedimentological information can be obtained from a section quite irrespective of whether the
section possesses the property of cyclicity or not. Nevertheless, the study of
vertical association is simplified if a sequential cycle exists and this aspect
will be dealt with in the next section.

Establishing sequential cyclicity is sometimes not as straight-forward as
one might think from the theoretical considerations in the previous section.
The section must be structured, as discussed in Chapter 5, in order to obtain
the transition matrix. There are clearly only two alternatives for the analysis
of sequential data. If the sections contain natural stratigraphic subdivisions
such as sedimentary beds, these may be used as the step units in the Markov
chain and the analysis can proceed as indicated. If there are no natural steps
and an equal-lithology method of structuring is employed, then it follows
automatically that the resulting matrix always will have the aspect of an
oscillating chain. It is clear that a process cannot be an independent random
process if by its very definition it is forbidden to enter the same state twice
in succession. This was discussed in Section 5.10. Since the leading diagonal
of the matrix is zero, the process cannot have a recurrence maximum at lag
one and this maximum must occur at lag two or at a higher interval. This
makes the chain oscillating, by definition. One may decide that the maximum at lag two is artificial and ought to be ignored so that only the recur-
rencies at the higher lags would be taken into account when one is trying to
find out whether a chain is oscillating. The difficulty which arises here is that
any recurrence maximum can reappear at multiple step intervals, just like the
harmonics in spectral analysis, and the maximum at two can produce maxima at four, six, eight ... step intervals. It is very likely that any higher
oscillating frequency is hidden by this phenomenon.

Fig. 10.3 gives an example to illustrate this. A lithological state recurring
plot was calculated from a 5 × 5 matrix in Table 10.2 which gives the
transitions of the "shaly group" in a coal-measure sequence in Ayrshire,
Scotland, which was studied by Doveton (1971). The solid line in Fig. 10.3
shows the recurrence probability of the rootlet beds (Y). The strong maximum at step two and its regular repetition at four, six and eight steps,
suggests that one is dealing with an artifact which was introduced by the
particular method of structuring. The example cannot be analysed any further if the artificial assumption is made that the chain must pass from any
state into a different state at each step. However, it could be assumed that
the zero diagonal in the original transition matrix is due to incomplete
observations, which means that the lithological states could have been subdivided but that the subdivisions were not observed. In this case it might be
possible to transform the transition matrix into one which has a non-zero
diagonal (see [5.48]). It was shown in Chapter 5 that this transformation
would require information about the stable probability vector of the missing

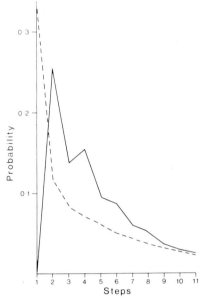

Fig. 10.3. Recurrence probability of rootlet beds in coal-measure data from Doveton (1971), indicated by a solid line. Recurrence probabilities of the same horizon after a transformation of the matrix (dashed line).

TABLE 10.2

Transition probabilities of "shaly group" in a coal-measure sequence of Ayrshire (after Doveton, 1971)

Lithological state matrix

A	M	B	C	Y
0	0	0.48	0.12	0.40
0	0	0.66	0.16	0.18
0.56	0.06	0	0.24	0.14
0.34	0.03	0.55	0	0.08
0.41	0.29	0.24	0.06	0

Transformed matrix

A	M	B	C	Y
0.25	0	0.36	0.09	0.3
0	0.50	0.33	0.08	0.09
0.42	0.05	0.24	0.18	0.11
0.17	0.02	0.28	0.49	0.04
0.21	0.15	0.12	0.22	0.30

A = barren shales, M = shales with non-marine bivalves, B = siltstones, C = sandstones, Y = rootlet beds.

matrix and so it could be done only by making arbitrary assumptions. The values $\alpha_1 \ldots \alpha_5$ in [5.47] were set as: 1.33, 2.0, 1.33, 2.0, 2.0, in order to illustrate the effect of such transformations. This produces only a slight change in the stable vector and the resulting transformed matrix is given in Table 10.2. The recurrence probability of the rootlet beds was calculated from the transformed matrix and is shown by a dashed line in Fig. 10.3. It can be seen that the oscillations at step two and its multiples have disappeared completely and that this matrix suggests a non-oscillating dependent random process. Similar results are obtained for all the other states except state M which represents shales with non-marine bivalves. This has a very weak but distinct recurrence maximum at step four and could, therefore, be regarded as cyclic. It should not come as a surprise that some states seem to indicate non-oscillating properties while others indicate the reverse. Admittedly, in a deterministic cycle, each state is dependent on all the other states so that if one of the states is cyclic, the remainder must be cyclic too. If this absolute dependency is relaxed, then finding cyclicity for one state does not necessarily imply that the remaining states must be cyclic.

10.5 THE MODAL CYCLE AND SIMILAR CONCEPTS

The term "modal cycle" was first used by Duff and Walton (1962) in an attempt to establish quantitative methods in the study of cyclic sedimentation. The modal cycle was defined originally as the most frequently observed sequence of rock types which can be found between two identical predetermined lithologies. Taking the following sequence as an example: ACBDCAB-CABDCABC ... one could designate C as the "cycle" boundary which divides the observed series into the groups: BDC, ABC, ABDC, and ABC. Since the ABC group occurs most frequently, it would be called the "modal cycle" of the series. The only objection that could be raised to the Duff and Walton approach has been mentioned; it is that independent random sequences can possess a "modal cycle" which is somewhat contradictory to the definition of a geological cycle. However, a slight modification of the original definition can remove this inconsistency and the modal cycle can be defined as "the most commonly observed rock sequence between two marker lithologies for which cyclicity has been established". In the example which was given above, one should first investigate the recurrence probability of C and, if this shows a maximum that is different from one, then cyclicity exists and one can proceed as before. Restricting the term 'modal cycle' to lithological groups which contain a definite number of lithological states, has the practical advantage of making the search for the modal cycle very much easier. For example, when it was found in Doveton's data (Table 10.2) that state M, which represents shales with non-marine bivalves, may have a recurring maximum at four steps, then one has to investigate only the possible combinations of groups M∗ ∗ ∗M, in which ∗ may be any lithology other than M. Any longer or shorter "cycle" is less likely, according to the recurrence probability graph. If one bases this search on the classification of lithological states which implies that no identical states must follow each other, one finds 6 × 4 = 24 combinations, in which each state occurs only once and 3 × 4 = 12 combinations in which the same state may occur twice. The search for the modal cycle is restricted in this way to one out of 36 possible arrangements. In the particular case of the coal-measure data, transitions from A to M and from M to A are impossible and this reduces the possible choice even further to 21. Calculating the probabilities of the possible combinations, one finds the group MBAYM, with a probability of 0.04273, is definitely the most likely sequence. It is followed by the group MBABM with p = 0.0106 and MBCBM with p = 0.00348. The recurrence maximum for state M with period four has a probability of 0.0914 and, therefore, the three given sequences account for 62% already of all the cases in which the cycle M∗ ∗ ∗M can be realized.

Duff and Walton did not use the transition matrix for calculating the modal cycle but they obtained it directly from the original data by an extensive sorting programme. It is more economical from the computing aspect to make use of the transition probability matrix which will have to be calculated anyhow. The choice of the marker horizon can influence the order of the modal cycle, of course.

A choice must be made between two policies. Either one chooses a geologically meaningful horizon as Duff and Walton did, or one calculates the modal cycle for each lithological state in the sequence, which is more informative.

In the example of the coal-measure sediments, the recurrence probability of the four-cycle following the state M had a probability of only 0.09. One may, therefore, ask which sequence is the most likely to follow state M in general. This sequence will be called the "modal sequence", it does not have to return necessarily to M. The modal sequence differs from the modal cycle because in theory, at least, it can be calculated for any length of group. The modal sequence can be read from the transition probability matrix directly by searching for the highest probabilities from each state. In Table 10.2, for example, the highest transition probability from M is to B with p = 0.66, from B to A with p = 0.56, and so on, giving a modal sequence of MBABA. However, the probability of obtaining this definite pattern is only 0.099. It is evident that, if the length of the sequence is increased, the number of possible permutations increases enormously and, although the modal sequence is still the most likely combination, it need not ever be realized in the limited length of the stratigraphic section. Short modal sequences can give valuable information about predominant lithological associations, just as relatively short modal cycles can.

Schwarzacher (1969) proposed an alternative to the modal sequence which was called the "maximum passage probability series". Instead of estimating the probability of a definite pattern occurring, one calculates the powers of the transition matrix and so one obtains the most likely state to occur after 1, 2, 3, ... beds have been deposited. One finds for the previous example M,B p = 0.66, M∗A p = 0.5, M∗ ∗B p = 0.42, M∗ ∗ ∗M p = 0.376. Combining this sequence one obtains, as before, MBABA but, in reading the series, it must be remembered that the probability estimate only refers to M∗ ∗ ∗A. The method is very easy to compute but otherwise is at a disadvantage when it is compared with the proper modal cycle.

Various methods which are related to the concept of a modal cycle have been proposed to construct so-called "facies relationship diagrams". The term was introduced first by De Raaf et al. (1965) but it was adapted by Selley (1970) to show some of the information which is contained in the tally matrix in graphical form. The diagram uses arrows to indicate, for each lithological state, the lithology which is most frequently found above and below it. It may be seen from the data of Table 10.2 that state C for example is most likely to be over and underlain by state B. State C, therefore, has an arrow pointing upwards and downwards towards B (see Fig. 10.4).

The complete facies relationship diagram may often turn out to be much more confusing than the original data matrix and, of course, it contains much less information. Selley proposed to simplify the relationship diagram further by subtracting a randomized matrix (see Section 5.7) from the original tally matrix which must have a non-zero diagonal. This difference matrix is supposed to show the number of transitions which occur more or less commonly than the number of transitions which would occur after the section has been submitted to complete mixing. Selley proceeds to draw diagrams which use the most frequently occurring transitions in the difference matrix. This procedure can be criticized on several points (Selley, 1970, p. 577). Selley does not try to establish the significance of a transition probability against the arbitrary hypothesis of random sedimentation, but he does attempt to measure this significance, admittedly in a very crude way. Selley justifies his procedure by stating that it leads to interesting results, which, of course, is a question of personal taste. As a quantitative method, this approach is confusing rather than illuminating.

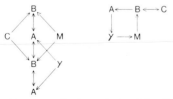

Fig. 10.4. Facies relationship diagrams for the data in Table 10.2. Basic data on the left and "significant" transitions on the right.

Doveton (1971) overcame some of the objections to Selley's method by simply testing the significance of the transition probabilities against a so-called "independent event matrix". Reference should be made to Section 5.10. In the coal-measure data of Table 10.2, Doveton found that only a limited number of the transitions was significant when they were tested against this hypothesis. Indicating significant positive tallies by a cross, the matrix simplifies to:

$$
\begin{array}{c}
A \\
M \\
B \\
C \\
Y
\end{array}
\left[
\begin{array}{ccccc}
* & * & * & * & + \\
* & * & + & * & * \\
+ & * & * & + & * \\
* & * & + & * & * \\
* & + & * & * & *
\end{array}
\right]
$$

which is almost cyclical and can be arranged into a facies relationship diagram which is also shown in Fig. 10.4. The basic cycle to emerge from this analysis is MBAYM and it is identical to the modal cycle of length four which was deduced previously.

The fundamental weakness, which is inherent in both this approach and that of Selley's, lies in the singling out of certain transition types which are regarded as significant. Since only some of the transitions are significant, the matrix containing such tallies is quite incomplete and can no longer be said to describe an observed section. Literally, one is looking at a section containing physical gaps where everything which is random has been removed. If the matrix of significant transitions was to be read as a transition matrix, then all the transitions except those to and from B would occur with deterministic certainty. Further account must be taken of the fact that the matrix against which the tests are carried out is not the transition of an independent random process (see Chapter 5). It must be concluded that the simplified facies relationship diagram is not a very objective way of describing a lithological sequence.

Other authors such as Read (1969, p. 209) have made no selection of transition probabilities and show the linkages graphically between the various states. This gives essentially the same information that is contained in the transition matrix. It is only the arrangement of the states in such diagrams which implies a preferred sequence of rocks. For example, the conclusion is reached (Read, 1969) that in coal-measure sedimentation there is a preferred path of coal → mudstone → siltstone → sandstone → rootlet beds → seat clay → coal. It is not clear why the rather vague term "preferred path" should be better than "modal sequence" or "cycle" as the case may be and it would help, in the geological interpretation, if it were to be stated clearly that the probability of this "preferred path" occurring is only 0.0103.

Duff and Walton's modal cycle has been criticized by several authors but it appears that once the distinction is made between a modal cycle and a modal sequence, the concept becomes useful. It carries considerably more information than the other schemes for developing facies relationship diagrams.

10.6 THE CYCLICITY OF ENVIRONMENTAL EVENTS

It was seen in the previous sections that reasonable criteria for geological cyclicity do exist. Indeed, if the proposed definitions are accepted, there can be no doubt that a variety of stratigraphical sections are composed of sedimentary cycles. The further development of the cycle concept and its extension to geometrical and time cycles is so closely linked to the interpretation of sedimentary cycles that it is useful at this point to consider the problem from a more genetical point of view. It is well known that there is a great variety of different sedimentary cycles and that their origin may have had many different causes but at this stage we are interested only in some very general features of cyclic sedimentation. The model concept will be used so that this generality can be achieved and it will be assumed that a dynamic system exists which generates an environmental model. It will be assumed that each lithology represents a definite environment, and so, to give an example, when coal is mentioned it will be implied that a coal swamp existed while this lithology was forming.

As soon as one introduces the term "dynamic" into the model, one is talking about processes which develop over a period of time; if the model is expected to explain an observed sequence of lithologies, it must explain the sequence of historical events in the environment too. It follows that time must be treated as a measurable quantity as far as the model is concerned. Chapter 2 referred to the relationship between the physical time of the process model and stratigraphical time as it is recorded by a sedimentary section. The relationship can be simulated by various sedimentation models, some of which were mentioned in Chapter 4. Any cycles which represent stratigraphic time quanta, and so could not be understood in terms of a time history, will be excluded from this discussion. Using the classification of time series which was developed previously, the dynamic models which determine the environmental history can be classified into one of the following groups: (1) deterministic cyclic models; (2) oscillating Markov models; (3) non-oscillating Markov models; and (4) independent random models.

The list could be elaborated to include a wide variety of non-Markovian stochastic models but this is unnecessary because these models can be reduced to Markov processes. At the same time it is useful to connect some actual geological theories with the various types of models and coal-measure sedimentation will be discussed for the purpose of illustration.

The Carboniferous cycles of America and Europe are, perhaps, the most frequently studied cycles in stratigraphy. The strong environmental changes which seem to be indicated by the coal-bearing strata make it fairly certain that their formation took place in a transitional, continental-marine environment. The very generalized sequence of "clastic sediments—coal—marine shales or limestones" has led to the picture of a delta on which a coal swamp develops and which is overrun in its turn by a marine transgression or at least some deepening of the depositional area. Comparative studies of various coal-measure cyclothems (Duff et al., 1967) have shown that the series "delta—swamp—marine sedimentation" is the exception, in fact, rather than the rule. In European coal-measure cycles, marine incursions occur relatively rarely and, in the marine-dominated cyclothems of the mid-continental basin of North America, the coals are either completely missing or they are of a very uncertain nature. However, the fact remains that most theories for explaining the mechanism of such cycles are concerned with explaining the environmental sequence of fluviatile-swamp and marine regimes. The present intention is to show how the existing theories may be fitted into the environmental model classification scheme, rather than to find a new solution to the cyclothem problem.

It must be stressed again that this discussion is concerned with process models and not with the properties of the stratigraphic record. Confusing these two concepts has sometimes led earlier workers to rather contradictory statements. On the one hand, it is known that the stratigraphic record contains the only observable data and that this shows a certain amount of disorder. Undoubtedly, this influenced geologists to assume that the generating process must have been some random process also. On the other hand, an "ideal" or a "theoretical" cycle is referred to frequently and indicates the belief that a definite order would exist in sedimentary cycles, "if only nature had behaved properly" as it were. This is indicated clearly by Pearn (1964) who writes that "an ideal cycle has the attribute of a natural law, real rocks represent this sequence only imperfectly". This implies that, if sedimentation followed a theoretical pattern, it would have to be generated by a deterministic model. No ideal cycle can exist if the new model of an oscillating random process is introduced because the theoretical sequence is replaced by a stochastic process.

It was argued in the previous section that the choice between a deterministic model and an oscillating random model can be based solely on the time histories of the environments which are generated by the models. While the deterministic model requires strict time cyclicity, the oscillating model can behave more irregularly. Most geologists reject the idea of strict time cyclicity largely because it is very difficult to think of any process which could lead to such behaviour on the scale that is required. It is probably characteristic that the only theories which do involve time periodicity stipulate that the cycle mechanism should have extraterrestrial causes (Jessen, 1956) and so are somewhat outside the sphere of our experience. The possibility of low-frequency astronomical cycles causing the sedimentary cycles, cannot be rejected outright and similarly the idea would be difficult to prove. What does seem certain is that a strictly deterministic model will always be a model of extreme simplification. Let us assume, for example, that a low-frequency cyclicity does exist in the orbital elements of the earth. It is possible that this would influence tectonic processes, the climate, sedimentation and biological developments. Each of these environmental factors is stochastically related to each other and the resulting environmental history would be oscillating, rather than strictly cyclic. It thus appears that, if a harmonic model is used to describe the environmental history, it has to be an approximation. Sometimes this approximation may provide a useful alternative to the possibly more complex stochastic models.

In contrast to this, it is quite impossible that an independent random process model should lead to sedimentary cycles and patterns of deposition as they are observed in coal-measure depositions. Independent processes have been considered as crude sedimentation models on a small scale, as for example, in Chapter 4, but if such a process had operated over stratigraphically measurable time, some dependence would have developed invariably. It will be seen later that an independent random sequence of events implies an independent random sequence of lithologies and that this excludes automatically such a process from being a possible process model. If some geologists have attempted to explain cyclicity as a purely random phenomenon, then this must be attributed to either ignorance or a very loose usage of the word "random."

The great majority of the theories which have been proposed to explain cyclothems can be reduced to processes which follow either the oscillating or the non-oscillating Markov model. The theories are customarily divided into three groups (Weller, 1964; Duff et al., 1967). These are: (1) the tectonic control theory; (2) the climatic or eustatic control theory; and (3) the sedimentation control theory.

Historically, the tectonic theories are the oldest ones and they range in scope from relatively elaborate models to simple statements like the following example: "a certain basin has undergone subsidence which occurred at irregular intervals." The so-called "arrested subsidence theory" is quite typical (Truemann, 1948). All that the theory specifies is that a relatively sudden subsidence occurred at irregular intervals which was followed by periods of no crustal activity. It is assumed that the subsidence caused a marine transgression but, eventually, continuous sedimentation of clastic material filled the depression to near sea level; a coal swamp developed which lasted until a renewed subsidence occurred. The following is found to apply when the model is analysed in terms of transition probabilities. Marine beds can occur with equal probabilities at any time because the subsidence that causes a marine transgression is an independent random process. This means that the transition probabilities from any state into the marine state must be constant. On the other hand, the model makes it impossible for sandstone (S) to follow coal (C) and neither can coal follow a marine bed (M). The transition matrix of the process must, therefore, have the following general appearance:

$$
\begin{array}{c}
S \\
C \\
M
\end{array}
\begin{bmatrix}
P_1 & P_2 & P_3 \\
0 & 1-P_3 & P_3 \\
1-P_3 & 0 & P_3
\end{bmatrix}
$$

The definition of the process implies that the recurrence times between the marine environments must be exponentially distributed and it follows that the process must be non-oscillating. The process does not lead to geological cyclicity in either rock sequence or in time sequence.

The difference between oscillating and non-oscillating models can be appreciated by comparing the arrested subsidence hypothesis with a much more elaborate tectonic control theory. When Bott (1964) developed a tectonic control theory, he argued that erosion in the source area leads to isostatic undercompensation which normally would lead to a rise in this area. A mountain uprising like this cannot occur unless mantle material replaces the immersed portion. This must lead to a flow of mantle material from the basin areas, which are subsiding as a result. The process is not a steady one but is controlled by intermittent movements along pre-existing fault lines. It is clear that this model is oscillating because a causal connection exists between the supply of sedimentary material and the subsidence in the basin area. Starting with a sudden subsidence, a marine transgression is inevitable but, in contrast to the previous hypothesis, this subsidence must be followed eventually by an increase in clastic sedimentation which will lead to a filling of the basin. Complete filling of the basin gives conditions for coal formation and the system is ready to start the cycle again. Bott has not specified the times at which the adjusting movements occur, but it is clear from the proposed mechanism that the sequence of sudden fault movements cannot be an independent event series. This is because it is unlikely that a number of subsiding movements would occur in rapid succession, just as it is unlikely that there would be prolonged periods without any activity at all. The intermittent movements along faults could be explained by a threshold stress which has to be exceeded and a time lag between subsidence and rising which is more or less fixed. This would ensure that the system is oscillating in time as well as in sequence. It is fairly certain that a time lag like this will occur because mantle material cannot be transferred instantly from the basin into the source area.

A further illustration of the difference between oscillating and non-oscillating process models is given by the so-called "sedimentation" theories. Several variants of this type of theory exist but generally it is assumed that a delta is built into a basin which is subsiding more or less constantly. As the delta progresses, its gradient will decline making it more likely that crevasse inter-distributary activity occurs. The main flow of clastic material is diverted to fill an inter-distributary basin and a subdelta may be started where there was no deposition previously. Vegetation is established on the abandoned parts of the delta and very little clastic material is introduced into the environment during peat formation. The environment must be almost at sea level by this time and so a marine invasion becomes likely. Meanwhile, the newly active subdelta will mature and then the main clastic sedimentation will be diverted to yet another centre.

The sequence of events is shown in a series of purely imaginary maps in Fig. 10.5. The theory is aptly called the "delta switching" theory (Duff et al., 1967) and it is evident that the proposed model is not based on an independent random process because, once a delta has occurred it is likely to stay in this position until maturity is reached. However, the process model does not provide any predictions about the chances of an abandoned delta becoming active again. Although the theory explains the sequence of sand—coal—shale, once the marine transgression has taken place, one cannot predict when the next terrestrial transgression will occur. It is clear that renewal process models of the electronic-counter type which were given in Chapter 6, can be adapted for this delta-switching problem. For example, the switching of a delta into an area could be said to constitute a renewal in time which is followed by a "dead time" which is either constant or variable but which is sufficiently long for the coal-measure "cycle" to be deposited. Once the deposition of the cyclothem is complete, the "counter" is open again and a new delta advance may occur at any time. It was shown in Chapter 6 that this model could be developed as a semi-Markov process in which a two-state embedded matrix controls the status of the environment and a waiting-time function, determined by the sedimentary process, determines the amount of the deposited sediment. The two states of the embedded matrix could be taken simply as: delta active and delta not active. In other words, this type

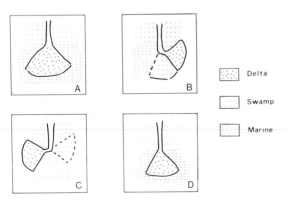

Fig. 10.5. The delta switch hypothesis.

of theory reduces the problem again to the trivial statement that deltas were either active or they were inactive, or that land and sea conditions alternated.

Oscillating models are obtained when the sedimentation theories are developed further, just as was found before. A cycle of processes will result if a causal connection exists between clastic sedimentation and coal formation and if, in addition, the termination of the coal-producing environment is found to lead automatically to renewed clastic sedimentation. This is shown in Fig. 10.6 in a much simplified form. It is assumed that there is a constant subsidence and a basin is filled with clastic material relatively suddenly, perhaps by crevasse formation in a distributary. The sand accumulation is rapid at first until a base level is reached, when it declines to keep pace with the subsidence. Vegetation develops gradually and this inhibits the distribution of sand (Robertson, 1952) but clastic material never enters the sedimentation during the maximum development of the swamp regime. The swamp conditions could be terminated by a number of circumstances (Duff et al., 1967, p. 154) and marine invasion has often been suggested as the cause. It has been suggested also that an ecological system as specialized as a peat-producing swamp may have only a limited life span, the system becoming extinct because of over-specialization or because of some changes in the supply of nutrients (Duff et al., 1967). In any case, once the swamp is declining, clastic sedimentation can start again. In Fig. 10.6 it has been assumed that there is a time lag before sand deposition and this allows the basin to sink below sea level. The time lag between the end of coal production and the beginning of the formation of a new crevasse may be caused by adjustments in the upstream distributary system.

The oscillating model which results from accepting the suggested causal connections could be elaborated considerably and it is certain that any real sedimentation process must have been far more complicated than any of the

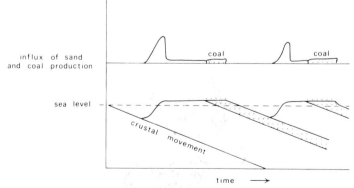

Fig. 10.6. The dependence of clastic sedimentation and coal formation in a coal-measure cyclothem.

models that have been discussed. It is also relatively easy to find flaws in the outlined theories but it would be useless to speculate blindly about alternative theories without considering all the available sedimentological evidence. Nevertheless, it is interesting to note that all the more elaborate theories lead towards oscillating processes. This is natural because the multiple causal connections between the environmental factors will lead to cycle-producing feed-back loops. For this reason, if a decision is made as to whether an oscillating or a non-oscillating model gives the better description, it is important that the choice should be accompanied by some indication of the complexity of the cycle-producing mechanism.

10.7 GEOMETRICAL AND TIME CYCLICITY

It was mentioned at the beginning of the previous section that process models deal with physical events in time and so every oscillating model will describe developments that are oscillating in time. This means that each repeated completion of a cycle takes an approximately equal time interval, because inevitably the processes that are repeated in a cycle are identical and, therefore, must take up similar amounts of time. Naturally, this does not imply that the time intervals between coal seams, for example, are precisely the same, but it does mean that there is a preferred time period which depends on the damping factor of the process model and which may vary from nearly random to nearly equal time periods. On the other hand, with a non-oscillating process, only a mean time exists between the recurrence of identical events but the actual recurrence times are distributed exponentially so that they do not have any preferred recurrence period. It is clear that the time history of the events which are recorded in the stratigraphical section must be known if a process model is to be fitted to some observed data. Usually, no time data are available and so stratigraphical thickness and estimated sedimentation rates must be substituted in order to obtain at least some approximation to this important parameter.

The necessity for this procedure may be demonstrated by comparing two imaginary sections. Let it be said that one section was generated by a non-oscillating delta switch model and that the other is imagined to have been generated by a more complex sedimentation process. Equal and constant sedimentation rates are assumed for all the lithologies in order to simplify the argument so that stratigraphical thickness can be substituted for time. Both of the models state that, once sand sedimentation is initiated, the deposition of coal and shale is along predetermined lines. The amount of sand and coal which is deposited in each cycle should be roughly constant for each type of model because the repeated processes will be identical. The two models differ only in the thickness distribution of their shales. The first model stipulates that the time that is taken up by the marine environment is an independent random variable and it follows that the shale thicknesses must be distributed exponentially. The second model stipulates a time interval which is approximately constant for the marine interval and leads to an approximately constant shale thickness.

Many geologists will object to this argument by refusing to make any time estimates from stratigraphic-thickness measurements when no other time estimate is available. The objections may be justifiable on practical grounds, but this does not alter the fact that the oscillating and non-oscillating models cannot be differentiated unless reference is made to thickness measurements. The relationship between sediment thickness and time in sedimentary sections will be discussed in more detail in Chapter 11 but attention must be drawn to the very obvious fact that, as long as there is some relationship between the amount of sedimentation and time, stratigraphical thickness must contain some time information.

The cycles which do not contain any time information have been called stratigraphical quanta and are excluded specifically from this discussion. Naturally, one of the greatest problems is the recognition of such cycles. For example, it could be the case that a delta-switching mechanism operated in coal-measure sedimentation to produce a cyclothem which was followed by a long and possibly quite irregular period of non-deposition. The time information in this cycle is insignificant and the cycle must be regarded as a stratigraphic quantum because the time that is represented by the sedimentary record is short when it is compared with the time period during which there was no deposition. This situation cannot be recognized by a geometrical analysis of the section and the likelihood of it occurring must be decided by geological reasoning that is based on detailed sedimentological work and a full three-dimensional stratigraphic analysis of the problem. Taking the specific example which was mentioned, one might ask whether long periods of non-deposition are likely in an area for which steady subsidence has been assumed and one may search for evidence of sedimentary breaks and investigate their areal distribution.

10.8 THE ANALYSIS OF GEOMETRICAL CYCLES

The definition of geometrical cyclicity in sedimentary rocks which was given previously was, "the ordering of strata in such a way that identical lithologies are repeated at equal intervals". The most obvious method for studying geometrical cycles would appear to involve measuring the distances between the repeated lithologies. However, this direct approach is impossible unless the cycles are relatively simple and clearly defined as, for example, in a study of bedding where the bedding planes could be regarded as the cycle boundaries (Schwarzacher, 1947). The more complex cycles contain a variety of lithologies and in many cases it is impossible to establish the "cycle boundary" in an objective way. A "marker horizon", which is repeated at identical intervals, is a feature of the precise mathematical cycles but, as it was pointed out previously, they are unlikely to be found in nature. The "marker horizon" may be out of sequential order or it may even be missing if an oscillating rather than a deterministic process model is chosen. Therefore, it is useful to introduce cycle measurements that are not linked to a definite lithological boundary.

The Markov chain approach could be used in this situation if the equal-interval method of structuring is adopted. In this method, which was discussed in Section 5.9, the section is divided into equal thickness intervals and the lithology is recorded at each artificially created step to give a sequence from which the transition matrices can be calculated. It has been pointed out already that, unless the assumed step structure can be related to some geological process, such a Markov chain is unlikely to give a useful model of the physical processes which produce cycles. This does not prevent the Markov chain model from being a valuable descriptive model for the analysis of the section. Care must be taken, however, that the states and the intervals at which the measurements are made should be chosen so that the resulting process does have a simple Markov chain structure. It is unlikely that this will be achieved at the first attempt, since a Markov chain may have to contain some lithological states which cannot be recognized by petrographic examination. The number of states to be used in describing the sequence depends on the measuring interval and the thickness of the cycles which are being examined: in many cases artificial states have to be created in order to make the measuring interval compatible with the suspected cyclicity in the section.

It cannot be overemphasized how important it is to examine the probability structure of the sequence; if this is neglected, the results are either meaningless or quite predictable. The following deterministic sequence: (12313, 12313, ...) illustrates the relationship that exists between the measuring interval and the oscillation period. Clearly, this sequence has a period of five. The numbers can be assumed to represent three different lithologies which were deposited in beds of 10 cm thick, let us say, and so the cycle is 50 cm thick.

When a transition probability matrix is prepared for the three states 1, 2, 3, at steps of 10 cm, one obtains:

$$\mathbf{P} = \begin{array}{c} \\ 1 \\ 2 \\ 3 \end{array} \begin{array}{ccc} 1 & 2 & 3 \\ \left[\begin{array}{ccc} 0 & 0.5 & 0.5 \\ 0 & 0 & 1 \\ 1 & 0 & 0 \end{array} \right] \end{array} \qquad [10.2]$$

This matrix represents a non-deterministic process with oscillation periods of 20 cm and 30 cm, which obviously is wrong. One needs at least five states in the transition matrix in order to show the full cycle of 50 cm. In general, a preferred recurrence period which is not a multiple of a fundamental oscillation cannot be longer than the total number of states. This fact becomes obvious when it is considered that the longest path to follow the most likely transition probabilities, is the path which visits each state in turn. Multiples of this period can occur but they are easy to recognize. This principle has been neglected frequently in the literature where Markov chains have been applied to sedimentary sections. Read (1969), for example, in examining the coal-measure sequence in Scotland chose a sample interval of 15 cm with five lithological states. The longest oscillation period that could be depicted by such a model is 90 cm when in fact the coal-measure cycles are of the order of 10—15 m. It is not surprising that the matrix, which was derived by Read, shows no oscillating properties.

The second problem that arises from using Markov chains which have been derived by the method of equal interval structuring concerns the thickness distribution of the individual states. It was shown in Chapter 5 that one of the properties of the Markov chain is that the time which the process spends in a certain state is distributed exponentially. This would imply that the thickness distributions of the lithological units should be exponential, which is obviously untrue, at least in most of the sections that have been observed. The semi-Markov models which were mentioned in Chapter 6, were developed to overcome this discrepancy. However, the semi-Markov models incorporate more information than is available from the equal interval structured data directly. It must be remembered that our concern here is to provide a descriptive model which does not necessarily have the physical meaning of a process model. With this in mind, one can use a method of stages like the one that was discussed in Chapters 6 and 7 and which can provide the required non-exponential thickness distribution of the lithological states. This is achieved by including states in the transition matrix that are not recognized necessarily by the petrographic description. This can be explained by considering a sequence which consists of only two rock types, let us say sand (S) and clay (C). The thickness distribution of the sand is not exponential. Obviously the following matrix cannot describe the section:

$$
\begin{array}{cc}
S & C \\
\begin{array}{c} S \\ C \end{array} \left[\begin{array}{cc} P_{ss} & P_{sc} \\ P_{cs} & P_{cc} \end{array} \right]
\end{array}
$$

However, if it is assumed that in reality there are two sands, let us say σ and s which are not recognized as such, one obtains the 3×3 matrix:

$$
\begin{array}{c}
\quad\ \ \sigma \quad\ \ \text{s} \quad\ \ \text{C} \\
\begin{array}{c}
\sigma \\
\text{s} \\
\text{C}
\end{array}
\left[
\begin{array}{ccc}
P_{\sigma\sigma} & P_{\sigma s} & P_{\sigma c} \\
P_{s\sigma} & P_{ss} & P_{sc} \\
P_{c\sigma} & P_{cs} & P_{cc}
\end{array}
\right]
\end{array}
$$

According to [6.1], the probability that the system stays for k steps in state σ is proportional to $p_{\sigma\sigma}^{k-1}$ and the probability that it stays in state s is proportional to p_{ss}^{k-1}. The probability that the process stays in the general sand state S is given by the convolution of the two geometrical distributions, which is a negative binomial and was discussed in Section 6.2. One can obtain various gamma distributions by increasing the number of additional states and decreasing the step size. It has been shown already that these gamma distributions provide excellent approximations to a variety of observed distributions.

The example which was given in [10.2] illustrates how the missing states for an expanded Markov matrix may be found. In the series 12313, 12313, ... one is dealing with two different types of state 3, one of which is preceded by state 2 and the other is preceded by state 1. If the states are relabelled according to their higher transition probabilities, one obtains the following matrix which is clearly cyclic with a period of five:

$$
\begin{array}{c}
\quad\ \ 123 \quad\ 231 \quad\ 313 \quad\ 131 \quad\ 312 \\
\begin{array}{c}
123 \\
231 \\
313 \\
131 \\
312
\end{array}
\left[
\begin{array}{ccccc}
0 & 1 & 0 & 0 & 0 \\
0 & 0 & 1 & 0 & 0 \\
0 & 0 & 0 & 1 & 0 \\
0 & 0 & 0 & 0 & 1 \\
1 & 0 & 0 & 0 & 0
\end{array}
\right]
\end{array}
$$

The rows in this matrix refer to the states of the system which were occupied at times $n-3$, $n-2$, $n-1$, and the columns refer to the states occupied at times $n-2$, $n-1$, and n. If only three states are used, as in [10.2] then the process is of the third order and requires knowledge of the states at times $n-3$, $n-2$ and $n-1$, in order to predict a state at time n. If the five state classes are used, the matrix becomes again a single-step matrix which is necessary for the Markov definition. The example shows that the state space increases automatically when the higher-order dependencies are removed and clearly such dependencies must be evident in the original record of the series. The following example is taken from a pilot study of the cyclic behaviour of the Upper Pennsylvanian sediments of Kansas in the U.S.A. (Schwarzacher,

1969). The series contains a great variety of lithological types (Davis and Cocke, 1972) which are arranged into sedimentary cycles ranging from 10 to 30 m in thickness. An average cycle thickness of 22 m is obtained when the section is subdivided into individual cycles on geological grounds only. The rock descriptions that were available for this investigation were only accurate enough for differentiating between two types of shale and silt that are referred to as S_i and S_o as well as one type of limestone (L) which gave no more than three states. It was decided to use a sampling interval of 1.51 m (5 feet) and this led to a 3×3 transition probability matrix which was derived from 400 sample points. The longest cycle which can be derived from this model is 5 m (15 feet) in length and this is quite insufficient. The higher-order dependencies were tested by calculating the statistic which was given in [5.45]. This was used to determine whether the probabilities $p_{ijk}^{(l)}$ are significant. The second-order transition probability that is used here is defined as:

$$p_{ijk}(l) = P\{X = i, \text{ at time } n | X = j, \text{ at time } n - 1 + l$$

$$X = k, \text{ at time } n - 1\}$$

where l is called the lag. The results of this calculation are shown in Fig. 10.7 where the likelihood criterion is graphed against l. The result shows clearly that the type of lithology which is encountered at a certain stratigraphical level is not independent of its past history. Particularly strong contributions come from the lithologies which are three and four steps below the observed level. The test suggests that possibly a second-order process of the following type would provide a good model.

$$p_{ijk} = P\{X = i, \text{ at } n | X = j, \text{ at } n - 3, X = k, \text{ at } n - 4\}$$

If this model is written as a single-step transition matrix, one gains six additional states giving a 9×9 matrix with the following states:

LL	S_iL	S_oL
LS_i	S_iS_i	S_oS_i
LS_o	S_iS_o	S_oS_o

The model that was fitted eventually was a fourth-order chain of lag 2 and the transition matrix was reduced to a 12×12 matrix which is sufficient to describe the cyclic behaviour of the section at the chosen sample interval. The thickness distributions that were generated by this model were not examined in detail because it was recognized that the model could not resolve the structure within the cyclothems. More data are needed for this.

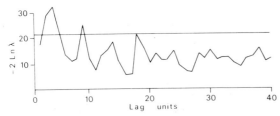

Fig. 10.7. Maximum-likelihood criterion for lags 2—40 for a trend-adjusted section from northeastern Kansas. 95% significance level is indicated. Lag interval 5 feet. (After Schwarzacher, 1969.)

No other study which uses the Markov chain approach for evaluating geometrical cyclicity has been made to date and it is possibly too early to assess its full potentialities. The methods of time-series analysis offer an alternative approach which has been taken by several authors and was referred to in Chapters 7 and 8. One may use either the methods of the correlogram or the power-spectral analysis. All the results refer to geometrical cyclicity because stratigraphical thickness has been substituted for time. In the univariate time-series analysis, a single variable is needed which is determined at discrete intervals along the section. The main difficulty lies in the choice of such a variable. This variable must be geologically significant and the measurements should be easy to carry out because a large amount of data is required. For example, Schwarzacher (1964) used shale/limestone ratios and average bed thickness within the sample interval; Carrs and Neidell (1966) and Carrs (1967) combined a number of observations to define two states only which were coded as +1 and —1. They calculated the correlogram from these data using the method which was discussed in Section 7.2. Mann (1967) used numerically coded data on the lithology, colour and bedding in addition to thickness measurements. Comparative studies must be made with a variety of parameters in order to find out which is the most suitable parameter to be used for each specific problem. One might venture to investigate more complicated and less obvious geometrical cycles in sedimentary sections if a more complex multivariate approach to time-series analysis could be developed.

10.9 CONCLUDING REMARKS

It is useful to summarize the ways in which the quantitative approach to the problem of cyclic sedimentation conforms to a certain extent with the more conventional stratigraphical approach as well as how it differs in emphasis from it. The basic concept of a sedimentary cycle is of an ordered sequence of lithologies which is repeated, although the orderliness of the sequence is never perfect. The idea of a process model leading to a sedimentary sequence shows that there are two basic problems to be considered. These are, the environmental history and the stratigraphical section which, one hopes, is a record of this history. Only if the cycles carry time information is there any relation between the environmental history and the section itself. Two types of sedimentary cycles have been recognized. The first of these does carry time information and can be resolved stratigraphically. The second kind of sedimentary cycle forms instantaneously, as it were, and, therefore, carries no time information. The two types of cycle will be referred to as type 1 and type 2. The classical theory of varve formation provides an example of a type-1 cycle which represents the summer and winter of a full year; an example of the type-2 cycle is the graded bed which is produced by a turbidity current. Sometimes it may be very difficult to differentiate between these two types but this should not prevent it from being recognized that they do represent two genetically different models. Type-1 cyclicity requires a mechanism which dispenses sedimentary material over equal time intervals and type-2 cycles require that the amount of sedimentary material should be measured out in equal portions.

Once it is recognized that the ordering in a cyclic sequence can and usually does depart from the perfect mathematical cycle, these departures can be explained in different ways. With type-1 cycles, the departures from the perfect mathematical cycle can be attributed to irregularities in the sedimentation rates, in which case one accepts a deterministic process model and believes in the existence of an ideal cycle. Alternatively, it can be assumed that the process model itself is of a stochastic nature, in which case it is almost impossible to define the concept of an ideal cycle. If one is dealing with type-2 cycles, it is meaningless to talk about sedimentation rates, and whether it should be linked with either a deterministic oscillating stochastic or a non-oscillating process model cannot be determined from any geometric analysis of the section. The nature of the process model cannot be found unless its time history can be reconstructed. The only links that exist between time history and type-1 cycles are the sedimentation rates and the geometrical cyclicity of the stratigraphic sequence. The latter can be assessed objectively by the various methods which were discussed in this chapter. The effects of various sedimentation rates will be discussed at a later stage, in Chapter 11.

The quantitative approach to the problem of cyclic sedimentation presents a strong case to consider seriously geometrical and possibly even time cyclicity of processes. This is different from the conventional approach where the sequential structure of lithologies has been emphasized. This departure from the relatively safe procedures of a sequential analysis is absolutely necessary if one is trying to construct process models. It is quite clear from a study of the literature that essentially trivial results have been obtained from the sequential study of rock types. Indeed, Duff et al. (1967, p. 241) conclude their monograph on sedimentary cycles with the sentence: "It is still arguable that the attempt to discern a pattern of repetitive sedimentation in successions has led to more misrepresentation and difficulties than it has enlightened". This appears to be a natural consequence of ignoring both the importance of an exact definition of cyclicity and the time aspect of sedimentary cycles.

Undoubtedly, the time estimates which are derived from sedimentation figures are unreliable. Duff et al. (1967, p. 250) show that the time estimates for the length of the coal-measure cycle range from 20,000 to 350,000 years, when the results from various authors from different formations and in different continents are compared. The difference between the extreme estimates is 94% of the maximum and Duff et al. "feel provoked to the reflection that such exercises may be unprofitable". However, in the facies analysis of a single limestone in the Permian of Kansas, U.S.A. the estimates of the water depth during its formation, which were based on palaeontological evidence, range from 60 m to 10 m (Elias, 1964) which is a difference of 83% of the maximum. These large discrepancies are common to most geological deductions and we will feel provoked to improve them and, if this is not possible, to find out why such large variations exist.

Chapter 11

Sedimentary cycles and time

11.1 INTRODUCTION

The importance of time in sedimentological and stratigraphical problems is one of the central themes of this book. It has been noted repeatedly that no physical explanation can be given to a process unless something is known about the speed with which it has developed; the fact that stratigraphy is supposed to supply the time scale for historical geology speaks for itself.

In spite of this urgent need for more precise time measurements, only very approximate methods for estimating absolute time intervals are available. It was seen in Chapter 2 that, apart from radiometric age measurements, stratigraphic time determinations rely almost entirely on estimations of sedimentation rates. An alternative to stratigraphic dating, which is unfortunately rare, is provided by varve analysis. Varves are sedimentary cycles, which are used as time units, and one of the main purposes of this chapter will be to investigate to what extent other cycles which are not annual ones can be used in a similar way. From what has been learnt in the preceding chapter, it is unlikely that any sedimentary cycle can ever approach the true varve in time accuracy. Thus, the main problem in using sedimentary cycles as stratigraphic units is that of assessing their time accuracy and, once again, the knowledge of the relationship between the deposited sediment and time is necessary for this. Theoretical models will be used to examine this relationship, but it must be emphasized at this stage, that an analysis like this, which is based on strict and formal logic, is somewhat limited in its scope. The practical problems in stratigraphy involve so much geological information which cannot be quantified and which is often accepted without proper proof, that it would be impossible to use mathematical methods exclusively. An attempt will be made, therefore, to combine formal mathematics with geological experience and, possibly, even intuition. It is typical that the basic assumption of a functional relationship between sedimentation and time can be assessed only on geological grounds.

It was seen earlier that any stratigraphic record consists of what have been called "time quanta" or steps, which do not carry any time information. The size of the timeless intervals in the stratigraphical record is relative, and sedimentary cycles have been discussed which may contain such long gaps of non-deposition that their stratigraphical thicknesses show no relationship to time at all. Sedimentation within such units is discontinuous and, in order to obtain some relationship between the deposited sediment and time, one needs, at the very least, some statistical continuity of sedimentation. This implies that the smallest thickness interval of a stratigraphic section must be long enough to contain a sufficient number of steps to permit one to speak of a mean rate of sedimentation. The length of this interval depends on the sedimentary environment. Questions as to whether there has been continuous subsidence, or whether there has been a continuous supply of sediment, or whether intermittent erosion has occurred, will have to be answered, as well as other questions. In other words, any analysis which is based on the assumption of continuous sedimentation can only be made after a detailed facies analysis has been undertaken on a regional scale, taking into account not only all the sedimentary evidence but also everything which is known about the tectonic framework of the ancient environment.

11.2 THE STRATIGRAPHIC TIME SCALE

Every measurement which is made involves the use of a scale and so, if stratigraphy is to measure time, some scale must be agreed upon. It is common practice to talk about "absolute age" if a stratigraphic time interval has been measured in years, or in any other standard physical time unit. In contrast, "relative age" refers to the time which has been established by using the geological time scale.

The geological time scale is a time scale of the type which has been called an "ordinal scale" (see Krumbein and Graybill, 1965); events have been ordered into a sequence, starting with the oldest and proceeding to the youngest. The steps on such a scale can be quite unequal and, indeed, the lengths of the individual steps are not known. Clearly, any thickness measurement in a single section can be regarded as a time measurement using an ordinal scale; this follows directly from the principle of superposition, which states that stratigraphically higher strata are younger.

The absolute age scale can be the type of scale which Krumbein and Graybill (1965) call an "equal-interval scale" or it can be a "ratio scale". Both of these scales use equally spaced units but the interval scale has no absolute null point, whereas the ratio scale has either a natural or an assumed fixed origin to the scale. In the study of a single section, the difference between these two scales is of little consequence, because an arbitrary zero can always be chosen at the base of the section. However, once the problems of stratigraphical correlation are introduced, a ratio scale is essential, since different sections must refer to the same zero point. The equal-interval scales may be classified further into the type of scales in which the steps are known in absolute time units such as years, and into the scales in which the units are known only to be equal. Obviously, an "absolute equal-interval scale" is more valuable but a number of stratigraphical problems can be solved if an equal-interval scale is used with unknown units. When it comes to assessing the accuracy of the stratigraphical time scales, two questions arise which can be treated independently. These concern the accuracy with which the scale represents equal time intervals and, secondly, the accuracy with which it represents absolute time intervals. If it can be shown that the time intervals of a stratigraphical scale are equal, then the step towards an absolute time scale can be taken at any time and improvements can be made when better estimates of sedimentation become available.

The concept of an inaccurate equal interval scale introduces a third type of scale, which is not normally used in physical measurements. This is the "statistical equal-interval scale" which is intermediate between the nominal scale and the equal-interval scale. Although the size of each individual step is unknown, the random distribution of the steps could be known. For instance, the steps on a scale might be normally distributed with mean μ and variance s^2. Any measurement which is taken on such a scale will contain an observational error. If the measurement R has the value of r variable steps, that is $R = x_1 + x_2...x_r$ then its most likely value will be given by the mean:

$$\overline{R} = r\mu \tag{11.1}$$

and its error by the variance:

$$s_R^2 = rs^2 \tag{11.2}$$

Situations where repeated measurements of R suggest that it consists of a constant number r, of subdivisions, will be discussed later. The mean value of the step units remains the same under this hypothesis, but the variance of the steps is reduced. Therefore:

$$\overline{x} = \mu \quad \text{and var } x = \frac{1}{r} s^2 \tag{11.3}$$

Use will be made of this result when discussing composite cycles (see Section 11.6).

To postulate an equal interval, or at least a statistical equal-interval time scale, is a perfectly logical procedure and does not cause any conceptual difficulties. Such scales are valueless, unless the thickness increments which correspond to the time units can be recognized. A considerable amount of speculation must enter here. The time—sedimentation relationship is always hypothetical, and this may be seen by using the Carboniferous limestone series of NW Ireland as an example. Fig. 11.1 illustrates three different attempts at constructing an equal-interval scale for the Benbulbin Shale and the Lower Glencar Limestone. The three scales are shown along the horizontal axis of the graph. The vertical axis represents the stratigraphical thickness and so the three curves give the sedimentation—time relationship, as deduced from the different scales.

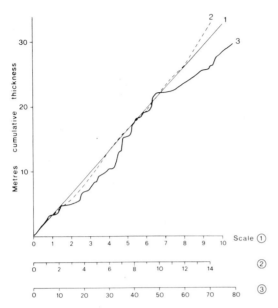

Fig. 11.1. Three equal-interval time scales for the Lower Carboniferous limestone of NW Ireland.

Scale 1 divides the investigated section into ten arbitrary units and the scale is used under the hypothesis that the sedimentation rates were constant. This leads to a straight-line relationship between the cumulative sedimentation and time.

Scale 2 is based on the sedimentary cycles, which occur in this section and which were discussed earlier. There are fourteen divisions corresponding to the number of cycles in this part of the section. The lower ten represent the Benbulbin Shale, and cycles eleven to fourteen belong to the Lower Glencar Limestone. The lower cycles belong to the type which is shown in Fig. 9.12A and the upper cycles are similar to the type shown in Fig. 9.12B. The base of the shale has been taken as the cycle boundary in each case and the cumulative thickness of the cycles has been plotted against the time units. The sedimentation—time relationship uses the cycle scale and follows the straight line fairly closely, and what is more, the deviations from the constant sedimentation rate are of a systematic nature. It should be noted that the sedimentation rates which are given by the second curve, are average values for complete cycles, ignoring differences between the limestone and shale rates of sedimentation. In common with all the cases which are being considered here, sedimentation is, at least, statistically continuous.

A comparison between the hypothesis of equal cycle lengths and the constant-rate hypothesis can be made by trying to divide the complete section into two, in such a way that each part represents the same time interval.

The length of the total section is 33.14 m and therefore the half time, according to scale 1, occurs at the thickness of 16.57 m which is equal to 5 units. The half time is at seven units, according to the second scale, and this corresponds to a thickness of 15.55 m. If the thickness values are expressed as a fraction of the total thickness, of course, one obtains 0.50 for scale 1 and 0.43 for scale 2. This result may be expressed in yet another way. Assuming that the hypothesis of equal sedimentation rates is correct, then the result which was obtained from the equal-cycle scale is wrong by 6%. On the other hand, if scale 2 was correct, then scale 1 contains an error of 6%.

A third scale can be based on the hypothesis that each limestone bed, together with its underlying shale, represents a time unit. "Limestone bed" is taken in this context to mean the smallest stratigraphic unit that was used during field work and, although laboratory investigation showed that such units can be composite, this was ignored in order to obain a third scale, which can be derived directly from field data. It may be seen that the cumulative thickness curve is much more irregular for these units. The step-like deviations from the straight line are noticeable in particular and these correspond roughly with the cycle boundaries of scale 2. This behaviour is not unexpected since it is very likely that sedimentation rates vary within the individual cycles. However, even if this variation is ignored, it is found that scale 3 differs quite considerably from scale 2. This is reflected also by the half time which is obtained with this scale: the thickness corresponding to 44 units is 19.20 m or, expressed as a ratio of the total thickness, it occurs at 0.58. If the constant-sedimentation hypothesis is correct, scale 3 has, at the half time, an error of 16% and if the equal-cycle hypothesis is accepted, the error becomes 22%.

Common sense would suggest the rejection of scale 3 as a likely candidate for a time-interval scale but there is actually very little reason for doing this, unless additional evidence is taken into account. It may be noted that the largest deviation from the straight line occurs from unit 50 onwards. This point coincides with the Benbulbin Shale—Glencar Limestone junction and the graph shows that the number of limestone beds per cycle and per unit thickness has apparently increased in this formation. Detailed stratigraphical work in the area shows that if, for example, the lowest cycle of the Glencar limestone is traced laterally in a northeasterly direction, it becomes progressively more shaly and, at about 3 km from the section which is shown in Fig. 11.1, the lowest Glencar Limestone cycle is developed in the Benbulbin Shale facies. This means that the number of beds in this cycle decreases when the cycle is traced to the northeast, or, in other words, the Benbulbin Shale facies migrates upwards when it is traced laterally. The number of beds per time unit is, therefore, facies-controlled and so it is not a reliable basis for an equal-time-interval scale. It may be mentioned at this stage that there is plenty of evidence that the limestone beds in this sequence are primary sedimentation features and are not the effect of "rhythmic unmixing" during diagenesis — a possibility which has been suggested by Duff et al. (1967). The primary nature of the beds can be demonstrated by their deposition fabric, the differences in fossil content, and by well-defined burrow horizons. These are all criteria which can be recognized with even an elementary knowledge of petrology in spite of the diagenetic changes which do occur. The changing number of beds which is observed laterally within the same time unit as well as vertically within a unit thickness or within a cycle can be explained quite naturally by different sedimentation conditions and this will be discussed later.

Additional geological considerations must be discussed so that a definite choice can be made between the three different time scales. One of the basic arguments which favours sedimentary cycles as stratigraphic time units will be developed in the next section.

11.3 SANDER'S RULE OF RHYTHMIC RECORDS

Sander (1936) derived the following rule in examining the relationship between cyclic events in time and the resulting stratigraphical record. "Cyclicity in space (which means stratigraphical thickness) indicates cyclicity in time but the absence of cyclicity in the stratigraphical record does not indicate the absence of time cyclicity." Sander assumed, quite naturally, that it is very unlikely that a cyclic sequence (which he regarded as an ordered pattern) would result, as it were, by accident from disorder. Consequently, a regular stratigraphic record indicates with all probability that it resulted from a regular time history. Clearly, Sander regarded sedimentation as a process which introduces a stochastic element into the stratigraphic record. Otherwise, the second half of his rule would have been superfluous. Sander understood cyclicity (rhythmicity in his terminology) to mean something approaching the regularity of a deterministic cycle. No attempt was made at this time to differentiate between mathematical and geological "cycles", as has been attempted in Chapter 10. Keeping this in mind, one can write Sander's rule, without changing its meaning, in the following way:

(1) Equally spaced events can be recorded in a stratigraphic record as:

(a) equally spaced lithological events, or

(b) unequally spaced lithological events.

(2) Unequally spaced time events can be recorded in a stratigraphical section as:

(a) unequally spaced lithological events, or

(b) equally spaced lithological events, but this is unlikely.

The following arguments show that Sander's rule is consistent with the model of random sedimentation but that it requires certain assumptions to be made before it can be applied in practice.

Let it be assumed that sedimentation proceeds by a random walk of unit steps, which occur with probability p, and of non-deposition occurring with probability q. The probability that a certain thickness θ is reached after precisely n steps is given by the binomial distribution with mean $n\mu = n(p - q)$ and the variance $ns^2 = n[p + q - (p - q)^2]$ (see [4:14]). When n becomes large, the distribution can be approximated by the normal distribution (see [4.15]) and so one can write that the most likely thickness value after a fixed time n will occur with the probability:

$$P(\theta = n\mu) = (2\pi ns^2)^{-1/2} \tag{11.4}$$

Assuming that Sander's rule operates under condition 1a, which means that precisely the same amount of sediment is deposited at equal time intervals, then the probability of this occurring twice under the most favourable conditions is $(2\pi ns^2)^{-1}$ or, of occurring r times $(2\pi ns^2)^{-r/2}$. The alternative, which is that Sander's rule operates under conditions 1b, has the probability of $1 - (2\pi ns)^{-r/2}$ and this increases rapidly with the number of repetitions.

The second situation, where it is specified that unequal time intervals lead to equal thicknesses, is more difficult to assess. If it is accepted that the time intervals leading to a cyclic stratigraphic record are normally distributed, then one may argue in the following way. In order to reach a definite predetermined thickness, the random walk may be regarded as having an absorbing boundary at θ. The absorption times again approach the normal distribution with a mean of θ/μ (see [4.21]) when the number of steps becomes large. The probability with which the mean of the time distribution occurs can be written as:

$$P(T = n) = \mu(2\pi ns^2)^{-1/2} \tag{11.5}$$

and, since $\mu = p - q$ and consequently $\mu < 1$, the probability given in [11.5] is always smaller than the probability of [11.4]. Once again, the mean value is the one that is the most likely to occur in the absorption time distribution. For condition 2 of Sander's rule to operate, it is necessary that the same thickness should be produced at least twice, but during different times. Thus, the mean value can occur only once and the probability for condition 2b must be smaller than $\mu(2\pi ns^2)^{-1}$. A numerical example illustrates these results. If sedimentation proceeds as a random walk with the parameters $p = 0.8$, $q = 0.2$, then the probability of condition 1a occurring twice under the first hypothesis is $(0.111)^2$ or smaller. The probability of condition 2b occurring under the second hypothesis is always smaller than $(0.067)^2$.

The analysis shows clearly that, under condition 1, it is possible and indeed, if the accuracy of the thickness measurements is not too great, it is quite likely that condition 1a will operate and a cyclic record will result. On the other hand, it is less probable that sedimentary cycles are produced under condition 2. This probability decreases rapidly if the absorption times vary widely. Sander's rule can therefore be regarded as proved but it must be noted that the two conditions of the rule are based on two exclusive hypotheses. Therefore, it is impossible to apply the rule to practical problems, unless something is known about the likelihood with which the two conditions occur in nature. If a cyclic record is observed, one cannot conclude that this is most likely to have originated from a time cyclic history without making any further assumptions, even if the condition of steady sedimentation is known to have been fulfilled. Sander was aware of this and he justifies the application of the rule by his belief that time cyclic processes in nature are much more likely to occur than non-cyclic processes. This assumption is supported by the study of sedimentation processes, providing that "time cyclic" is not applied in the strict mathematical sense but refers to oscillating processes. Such processes do arise very commonly from the interaction of environmental factors. The analysis of the sedimentation—time relationship becomes somewhat more complicated once it is admitted that stochastic processes are active in the environmental history as well as during sedimentation, and this will be seen in the next section. However, the essential conclusions which arise from such an analysis are the same as the ones that were reached by Sander.

11.4 THE SEDIMENTATION—TIME RELATIONSHIP OF RANDOM MODELS

It was suggested in Section 10.6 that the relationship between sedimentation and time can be investigated by introducing definite sedimentation models. Such an approach is based on three separate stages. Firstly, there is the environmental history, which is a function of time (T). Secondly, there is the mechanism of sedimentation, which is regarded as a random process that is turned on and off by the environmental history but is otherwise independent from it. Thirdly, there is the stratigraphic record, which is produced by both the environmental history and the sedimentation, and this is a function of the stratigraphic thickness θ. In order to understand the model more easily, one may think of a vessel which is being filled with a liquid. A tap is opened and closed at certain intervals, T. It is assumed that the liquid flows irregularly, following a certain random pattern which models the sedimentation process. Whenever the tap has been closed for a short time, a reading of the liquid level θ is taken and these records correspond to the stratigraphic column. The problem involves either predicting the various liquid levels from a known time pattern of the tap closures or, alternatively, it involves reconstructing this time pattern from the liquid levels. Obviously, the weakness of the model lies in the assumed independence of the sedimentation process from the environmental history and the only justification for this is the great simplification that can be achieved by the assumption. However, the model is well justified in the analysis of the environmental events which leave their mark on the stratigraphic column but which do not influence local sedimentation conditions. This situation arises whenever the environmental history can be resolved into a number of point events such as, for example, the change from deltaic into marine sedimentation, or the change from a high-energy environment to a low-energy environment. It is assumed in each case that, once the switch to a new environment has occurred, sediment production is governed by a definite random process which will be called the "sedimentation model", together with the controlling environmental history which is itself a stochastic process.

The introduction of a stochastic sedimentation model of the type that was discussed in Chapters 4 and 6 permits one to express the sedimentation—time relationship as a conditional probability. One can write:

$$\text{Sedimentation model} = P(\theta = z | T = t) \qquad [11.6]$$

which stands for the probability that the stratigraphic thickness θ takes on a certain value z, provided that the sedimentation process was active over a definite time T, of length t. Nothing is known about the environmental history other than that it is a function of time. However, a possible classification of histories which are based on the probability structure of time series was proposed in the preceding chapter. It was seen that the time of occurrence between events may have been at precisely equal intervals, which led to the model of a deterministic cycle. Alternatively, there may have been a preferred period of recurrence which provided a model for oscillating sedimentation. As the third possibility, the recurrence times may have been distributed exponentially, which leads to the independent random event model. It should be noted that the recurrence pattern of a non-oscillating Markov process is the same as that of an independent random process and so it will be treated as such. The recurrence pattern of events will be described by the function q_T and this is a frequency distribution of the events occurring in the history of the environment. It was pointed out that, essentially, all histories can be treated by one of three mathematical models. If q_T is a negative exponential distribution, the events will follow each other in an independent random sequence. If it is certain that the time between events is constant, q_T becomes unity and one is dealing with a determiristic time history. Any case which lies between the deterministic and the random distribution can be approximated conveniently by a gamma distribution. The three models are shown in the first column of Table 11.1. In order to calculate the stratigraphic sequence, it is only necessary to choose an environmental model together with a sedimentation process model. This procedure will be discussed in detail, using the case of random sedimentation as an example.

Let the sedimentation process be represented by a simple random walk in which positive unit steps occur at exponential time intervals ρ. The sedimen-

TABLE 11.1

The time—sedimentation relationship of random walk models

Environmental history q_T	Var (q_T)	Number of sedimentation steps N_T	Var (N_T)	Thickness distribution f_θ
1 $P(T=t) = \lambda e^{-\lambda t}$	$\dfrac{1}{\lambda^2}$	$\dfrac{\lambda}{\lambda+\rho}\left(\dfrac{\rho}{\lambda+\rho}\right)^t$	$\rho(\lambda+\rho)\,\mathrm{var}(q_T)$	exponential
2 $P(T=t) = \lambda\dfrac{(\lambda t)^{k-1}}{(k-1)!}e^{-\lambda t}$	$\dfrac{k}{\lambda^2}$	$\left(\begin{array}{c}-k\\t\end{array}\right)\left(\dfrac{\lambda}{\lambda+\rho}\right)^t\left(-\dfrac{\rho}{\lambda+\rho}\right)^t$	$\rho(\lambda+\rho)\,\mathrm{var}(q_T)$	compound-Poisson
3 $P(T=C)=1$	0	$\dfrac{(\rho c)^t}{t!}e^{-\rho c}$	ρ	compound-Poisson

TABLE 11.2

Joint probabilities of independent random sedimentation

θ	$T=1$	$T=2$	•	$\sum\limits_{T=0}^{\infty}$
0	$\lambda e^{-(\lambda+\rho)}$	$\lambda e^{-2(\lambda+\rho)}$	•	$\dfrac{\lambda}{\lambda+\rho}$
1	$\lambda\rho e^{-(\lambda+\rho)}$	•	•	$\dfrac{\lambda}{\lambda+\rho}\dfrac{\rho}{\lambda+\rho}$
2	$\dfrac{(\lambda\rho)^2}{2!}e^{-(\lambda+\rho)}$	•	•	$\dfrac{\lambda}{\lambda+\rho}\left(\dfrac{\rho}{\lambda+\rho}\right)^2$
•	•	•	•	•
•	•	•	•	•
$\sum\limits_{\theta=0}^{\infty}$	λe^{-1}	λe^{-2}	•	•

tation mechanism is a Poisson process (see Chapter 4) and the conditional probability of [11.6] can be written as:

$$P(\theta = z | T = t) = e^{-\rho t} \frac{(\rho t)^2}{z!} \quad (z = 1, 2, ...) \qquad [11.7]$$

Next, assume that the events which are recorded by this sedimentation are also distributed as independent random events. This implies that the time history can be written as:

$$q_T = P(T = t) = \lambda e^{-\lambda t} \qquad [11.8]$$

Knowing [11.7] and [11.8] permits one to write the joint probability $P(\theta, T)$ from the relation:

$$P(\theta, T) = P(\theta | T \cdot q_T) \qquad [11.9]$$

This can be done in the form of a table (see Table 11.2) which contains any combination of θ and T-values as well as their associated probabilities. The stratigraphical record consists of units which represent the amount of sediment which is deposited between environmental events. Such units are called "beds" for simplicity. The bed thicknesses have a probability distribution f_θ which can be obtained from Table 11.2. If the θ-value is kept constant and let it be said, equal to k, then one obtains:

$$f_\theta(k) = P(\theta = k) = P(k, t_1) + P(k, t_2) + ... \qquad [11.10]$$

Expressed in words, this means that the probability of obtaining a certain thickness in the stratigraphic record is given by the appropriate row sum of the joint probability table. By integrating each row in Table 11.2, one obtains:

$$f_\theta = P(\theta = z) = \frac{\lambda}{\lambda + \rho} \left(\frac{\rho}{\lambda + \rho} \right)^z \quad (z = 1, 2,...) \qquad [11.11]$$

Naturally, by summing the columns of Table 11.2, one obtains again the distribution q_T, which is the model of the environmental time events. The two distribution functions f_θ and q_T are known as the marginal probabilities of the joint probability $P(\theta, T)$. The sedimentation—time relationship of two stochastic systems may be described fully by the joint probability $P(\theta, T)$. It links the two marginal distributions of events in time q_T and the stratigraphic bed thickness f_θ. The rules by which the marginal probabilities are linked are given by the conditional probability $P(\theta | T)$ which has been called the sedimentation model.

It can be seen from [11.9] that the joint probability $p(\theta, T)$ cannot be calculated from the two marginal probabilities without knowing the conditional probability. Therefore, any solution to the sedimentation—time problem can be based only on a hypothesis of a sedimentation model. Field data provide only two types of measurements. The bed-thickness distribution can be estimated from a sufficiently long section which is taken at a single locality. Thus, the marginal distribution f_θ will be known with some confidence. A second set of data may be found from the thickness distribution of a single bed which has been measured at different localities. Such a bed would have taken an identical time everywhere for its formation but it would vary in thickness because of the random sedimentation mechanism. It can be seen from Table 11.2 that, in the case of random sedimentation, the thickness of a single bed should be Poisson-distributed and so the sedimentation model could be checked in this way. However, this argument requires not only accurate time correlation but it also requires the thicknesses of sediment which are deposited in different localities to be completely independent of any regional sedimentation pattern. This is a situation which is very unlikely in realistic situations, as will be discussed later.

The problem of calculating the thickness of the sediment which is produced during a random time interval has been treated already in Chapter 6, and a comparison of [11.11] with [6.17] shows that the results which were obtained by the different methods are identical. The use of moment-generating functions in Chapter 6 showed further, that the number of steps which a random walk performs in an exponentially distributed time interval, is always an independent random number itself. This is true for any renewal process which may constitute the sedimentation model. This is an important geological result, because it gives clear guidance on the interpretation of the stratigraphic record. One can state positively, that an environmental history consisting of exponentially distributed time events, will lead to a stratigraphic record in which the thicknesses of the beds are exponentially distributed. The reverse and even more important result for the geologist, that an exponentially distributed stratigraphic record can only originate from exponentially distributed environmental histories, can be seen by examining the alternative models of deterministic cycles and gamma-distributed recurrence patterns. The constant time interval in the deterministic cycle model, leads to Poisson-distributed renewal steps. If these are assumed to be unity, then the counting distribution N_T equals the thickness distribution of the beds f_θ. Any other sedimentation process which is based on random walks will lead to compound Poisson distributions (see Chapter 6), and the only conditions under which these may degenerate into the negative exponential distribution will be discussed shortly. The second alternative model of a gamma-distributed counting interval, was treated in [6.18], and it was found that the number of steps follows the negative binomial distribution law, which is again a compound Poisson distribution (see Section 6.4).

The results of this analysis are summarized in Table 11.1, which shows the hypothesis that was made about the environment in the first column. Model 1 assumes a random recurrence pattern, model 2 assumes that the times between the recorded events have a preferred length which is different from zero, and model 3 assumes a deterministic cycle in the time events. The properties of the resulting stratigraphic record can be seen from columns 3 and 5. If the sedimentation mechanism is some renewal process with density ρ, then the number of renewals N_T are given in the third column. Thus, N_T is the thickness distribution resulting from a random walk with unit steps but if variable sedimentation steps are assumed (see Chapter 6) then the resulting thickness distribution will be of the type which is shown in column 5 of the table. One can see immediately that the first model can be recognized from the bed-thickness distribution f_θ which is specific for the exponential q_T distribution; models 2 and 3 will be much more difficult to separate.

In order to investigate the difference between models 2 and 3, one may examine the dispersion of the time distribution q_T (column 2) and that of N_T (column 4) which is taken as the thickness distribution of the beds, for simplicity. Considering model 3 first, which represents constant time steps, shows that the dispersion of the thickness measurements always must be larger than the dispersion of the time events, which is zero of course. The shape of the bed-thickness distribution is controlled entirely by the parameter ρ and only if $\rho < 1$ will its maximum occur at step zero. This is the only case where it will be difficult to separate this distribution from the negative exponential. The situation where $\rho < 1$ implies, however, that the mean of a sedimentation step which occurs within the time interval T is less than unity and that there is no guarantee that the environmental events are recorded. Such a situation leads to an incomplete sedimentary record. This is precisely the situation that was discussed in the introduction to this chapter, where it was pointed out that, if no relationship exists between sedimentation and time, an analysis becomes impossible.

Very similar considerations about the continuity of sedimentation apply to models 1 and 2, where it is accepted that the counting interval is variable. If λ is the density with which environmental events occur, then a statistical continuity of sedimentation will obtain when $\rho > \lambda$. Since the time scale is an arbitrary equal-interval scale, one can choose its units in such a way that $\lambda = 1$, and continuous sedimentation therefore means $\rho \geqslant 1$ as before. Choosing $\lambda = 1$ is the same as setting the mean time between environmental events equal to unity. Thus, if $\rho = 1$, a step of the random walk occurs during each time unit on average, or, if $\rho > 1$, then more than one sedimentation step occurs in the same time interval. Under these conditions of statistically continuous sedimentation, the variances of the time history and of the stratigraphic record can be compared. It may be seen from Table 11.1 that, if $\rho > 1$, the variance of N_T is always larger than the variance of q_T.

This is what one would expect on common-sense grounds. If an equal time interval is registered repeatedly by some faulty recording mechanism, then one expects an error. Furthermore, if the times between the events are variable in themselves, then this error will be increased. The reverse conclusion is important, however, and is frequently forgotten. If a regular pattern is found in the thickness measurements of a stratigraphic record, then a correspondingly regular pattern must exist in the distribution of the time events. Indeed, the regularity of the time pattern will be higher generally than the one that is observed in the sedimentary record.

These conclusions are essentially the content of Sander's rule (see Section 11.3). The rule could be formulated in the following way if it is adapted to random processes. "A strictly periodic time history will lead to an oscillating stratigraphic record; if the time history is an oscillating random process then its record in the stratigraphic column will be oscillating also, but its damping will have increased." Vice versa, one could state that "an oscillating stratigraphic record can result from either a deterministic cycle or from an oscillating time history". The rules are subject to the condition of statistically continuous sedimentation. The more general rule that "if the time intervals follow an independent random pattern, then this will be observed in the stratigraphic record", is not subject to this condition. An oscillating stratigraphic record cannot result from a time history which follows an independent random process.

11.5 THE APPLICATION OF MODEL ANALYSIS TO PRACTICAL EXAMPLES

It was seen in the previous section that the model approach to the sedimentation—time problem does provide some general results but because the parameters of such models, as well as the models themselves, are often purely hypothetical, one cannot expect the specific solutions that are needed for constructing absolute time scales. Nevertheless, to illustrate how the stratigrapher can benefit from these model studies, some further aspects of the limestone—shale sedimentation of the Carboniferous Glencar Limestone will be discussed.

The Glencar Limestone was used in Section 6.5 to illustrate how the analysis of bed-thickness distributions may yield information about the origin of the beds. It was argued that the sedimentation in this limestone proceeded in jumps, each step representing a fairly short interval. Fitting a gamma distribution to the bed-thickness data indicated that a sedimentary bed consists of between two to three steps, on average. A possible mechanism for this type of sedimentation was suggested and it is illustrated in Fig. 6.4B. The figure shows how the biological relief of the growing benthos could determine the height of a sedimentation step by acting as a baffle during a short incident of deposition which might be caused by a storm. If such a process operates, one

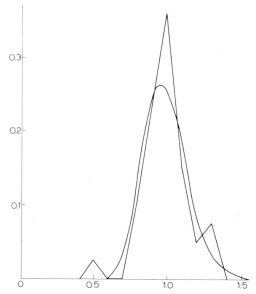

Fig. 11.2. Thickness distribution of a single bed measured at 1-m intervals along its strike.

would expect the thickness distribution of the steps to approach the normal distribution and, as long as there are no changes in the density and height of the benthonic growth, the average steps should be fairly constant.

The bed-thickness distribution of the middle Glencar Limestone (see Fig. 6.10) indicates that the mean step thickness is approximately 3 cm and a single bed of approximately that thickness was traced, therefore, in a cliff exposure over a distance of 39 m. The bed was measured at 1-m intervals and the mean thickness was found to be 3.25 cm. The data were standardized to give a mean value of 1 and the frequency distribution of the thicknesses is shown in Fig. 11.2. It can be seen from this figure that this distribution is indeed fairly symmetrical. Fitting a gamma distribution to the data gives the parameter values $\lambda = 36.1$ and $k = 30$. Providing that the assumed sedimentation model is correct and providing that the data do not incorporate a regional trend, these measurements should give an estimate for the distribution of sedimentation steps because they provide samples which were taken from a constant time interval, according to the accepted hypothesis.

A preliminary hypothesis can be formulated now. Let it be assumed that each bed represents a constant time interval (model 3 of Table 11.1), and that the sedimentation steps are gamma-distributed. The bed-thickness distribution may then be calculated, using [6.30]. The results of such calculations are shown in Fig. 6.8 and these are compared with the bed-thickness distribution which is actually observed in Fig. 11.3. Calculating bed thicknesses by the expression which is given in [6.30] is limited by the computer's

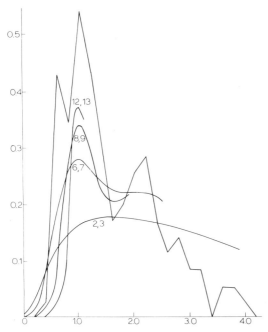

Fig. 11.3. Theoretical and observed bed-thickness distributions from the Glencar Lime-stone.

capacity for large numbers which arise from the factorials in the denomina-tor of the sum and so the calculated curves are incomplete. Nevertheless, it can be seen clearly from Fig. 11.3 how the model would behave if the parameters of the step distribution were to be increased until the maxima of the calculated curves coincided within the maxima of the observed bed thicknesses. With increasing height, the peaks of the calculated curves be-come much sharper and their dispersions become much smaller than those of the thickness distributions that are observed in the field. Indeed, the mea-surements which were taken on a single bed support this finding, because they too suggest that the variability of the step thickness is relatively small.

It is clear that the hypothesis that beds represent a constant time interval must be modified and, to account for the observed dispersion, one can assume that the time that is taken up for the formation of a single bed is, itself, variable. This is model 2 of Table 11.1 and the number of steps which occur in each bed should be negative binomially distributed. Simulation techniques were used to obtain these bed-thickness distributions. First of all, bed thicknesses were simulated using a constant time interval, and by varying the following parameters: ρ, which is the step density and λ and k which refer to the thickness distribution of the steps. The best results were ob-tained by setting $\rho = 1$, $\lambda = 20$ and $k = 21$. Secondly, simulations were

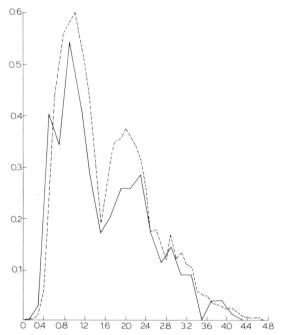

Fig. 11.4. Observed (solid line) and stimulated (broken line) bed-thickness distributions from the Glencar Limestone.

repeated with these parameters by sampling from a negative binomial distributed random array, simulating the condition of a variable counting interval in this way. Good results were obtained by giving the counting interval a standard deviation of between 0.3 and 0.5 and Fig. 11.4 shows one result of such an experiment.

The analysis gave the following results. It was found that each bed was formed by a small number of sedimentation steps. The distribution of the step thicknesses was approximately normal with a mean of 3 cm and a standard deviation of 0.456 cm. It was argued that the time it took to complete a single bed was variable. If the mean time for completing a bed was taken as unity, then the standard deviation of the time distribution was of the order 0.3—0.5. This is a relatively high variation and indicates that it would be very inaccurate to base an absolute age determination on a count of bedding planes. This finding is in agreement with the earlier inference that a time scale which is based on the number of bedding planes is unsatisfactory (see Section 11.2).

The conclusions which are given above are based, in the first instance, on a specific sedimentation model which is hypothetical and, in the second instance, on the good agreement that was obtained between the observed bed-thickness distributions and the simulated distributions using the model.

At this point, it may very well be asked whether such a good agreement could have been obtained by introducing other quite different models, or perhaps by using a combination of different sedimentation models, together with a different time history. This is an important question for the sedimentologist, since he might expect, quite justifiably, that an analysis on these lines could be used to prove the correctness of a certain sedimentation model. The prospects of such an approach are not very promising. It was seen in Chapter 6 that a variety of different sedimentation processes can all lead to the same mathematical model of a compound Poisson distribution. Such distributions always possess two or more parameters which are unknown generally and which have to be obtained from simulation experiments. It is very easy indeed to set up models for fairly complex sedimentation processes. These models can be realized by computer simulation and one can demonstrate that different models lead to the same result. Therefore, one cannot conclude that a certain well-behaved sedimentation model provides a unique solution to the problem of what happened actually during the formation of a sediment. It was concluded from the analysis of the Glencar Limestone that bed formation does not represent absolutely equal time intervals but there are models which could operate under such a condition. Models like these would involve much more complicated sedimentation processes than the ones that were postulated in this analysis. Thus, the conclusions which are drawn here are based on the simplicity of the model but they cannot claim to be proved beyond all doubt.

11.6 THE NUMBER OF BEDS IN A SEDIMENTARY CYCLE

Most sedimentary cycles are composed of a number of beds and one can investigate the time interval that is represented by a cycle by taking the average time of bed formation as a time unit. The mean duration of a cycle is obtained by counting the number of beds in a sufficiently large sample of sedimentary cycles. The frequency distribution which could result from such counts can be deduced in the following way. The first assumption to be made is that sedimentary cycles represent precisely equal time intervals of, let it be said, length T. Let the time t that it takes to produce a single bed be normally distributed with mean μ and standard deviation s. The random variable S_r is the sum of the times that it takes to produce r beds.

$$S_r = t_1 + t_2 + \dots t_r \qquad\qquad [11.12]$$

will be normally distributed as:

$$P(S_r < T) \sim N\left(\frac{T - \mu r}{s\sqrt{r}}\right) \qquad\qquad [11.13]$$

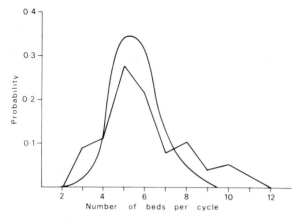

Fig. 11.5. Theoretical and observed number of beds per cycle in the Glencar Limestone.

where N is the normal distribution function. The probability of obtaining r beds in the interval T can be regarded as the number of r renewals in this time, and this can be calculated from [6.5].

A count of the number of beds per cycle in the middle Glencar Limestone gave the frequency distribution that is shown in Fig. 11.5. The average number of beds per cycle was found to be 5.89 and so each cycle should represent a time interval of five to six units. If the theoretical distribution on $T = 5$ is calculated, then a curve is obtained which agrees very well with the observed data; only some of the observed cycles seem to contain too high a number of beds when compared with the theory (see Fig. 11.5) and if this can be explained, then there is no reason for rejecting the deterministic cycle hypothesis.

One of the major difficulties in this type of work lies in the precise definition of a "bed". The continuity of sedimentation can be recognized at different levels and this has been discussed repeatedly. It is important to emphasize that the data which are shown in Fig. 11.5 are based on counting beds which were recognized as field units. It was mentioned in Chapter 6 that by sharpening the methods of observation, using hand specimens and possibly thin sections, a more detailed stratigraphy can be developed. In the Glencar Limestone, for example, the "step" structure of sedimentation can be recognized when peels are examined under the binocular microscope. Field observations are carried out in natural and artificial exposures of all kinds and the amount of details which can be observed depends very much on the type and amount of weathering which the exposure has undergone. It is quite a familiar situation for the field geologist to find that the same series of rocks may look quite different in, let us say, a cliff exposure and a quarry face. This difference in the natural preparation of rock surfaces can have a very pronounced influence on the number of observable beds.

Fig. 11.6. The field aspect of some Carboniferous, Triassic and Jurrasic limestone cycles (drawn after photographs).

A very good illustration of the weathering effect became evident during a study of sedimentary cycles in the Alpine Trias of Lofer (Austria). The pattern of cyclic sedimentation in this limestone dolomite sequence is similar to that of the Carboniferous limestone in many respects, although it occurs on a very much larger scale (see Fig. 11.6). The limestone beds are grouped together approximately in sets of five (Schwarzacher, 1954), and are separated by dolomitic beds. Counting the number of beds per cycle in twenty successive cyclothems gave an average of 4.56 for the south-facing exposures and an average of 7.12 beds for the north-facing exposures. The cyclothems are the same stratigraphic units in both cases but the north-facing exposures in this area show much stronger weathering, largely because they are covered by snow for much of the year. The variability in the number of beds is caused partly by this difference in weathering, but it is also influenced by the very loose definition of a bed which was used in these counts. This particular study was carried out largely by counting "beds" on telephotographs of cliff exposures which could not be reached easily for closer examination. Fischer (1964) measured a 120 m long section in much greater detail and he was able to subdivide the sequence, on lithological grounds, into two distinct facies. These are the massive limestone beds which were interpreted as subtidal deposits, and an intertidal facies which is dolomitic and gives rise to the bedding planes. Intertidal facies were recognizable in close-up examination, even when they were not expressed as a morphological feature as a result of weathering. Fischer found six, seven and eight intertidal horizons per cycle in the three cyclothems which he examined. It is unfortunate that Fischer's section includes only three cycles (megacycles, see later) and it is too short for deriving an average number of beds per cycle, but there are clear indications that such an average would be higher than the one which was derived from the relatively crude analysis of photographic records. The situation is exactly analogous to the examination of the Carboniferous limestones, where more stratigraphic subdivisions were found on closer investigation.

Recognizing that the definition of a "bed" depends not only on the variation of lithologies but also on the conditions under which the bed is observed makes it clear that the categories of the stratigraphic subdivisions cannot be as sharp as they were assumed to be in the previously discussed models. The cycles in the Carboniferous limestone were subdivided into beds and each bed itself was subdivided into a variable number of sedimentation steps. The petrographic examination shows that the bed and step structures are different but that there are gradations between the two. The limestone beds are well separated from each other by a layer of shale which can be 2—3 cm thick, particularly at the base of a cycle. This shale decreases in thickness towards the centre of the cycle and often there is only a very thin film of clay between the beds, a separation which can disappear laterally. A change in fabric is frequently the only indication of a step structure within the limestone beds and the top of a step shows an orientation of skeletal fragments which lie parallel to the sedimentary s-plane. Sometimes there are zones of higher dolomite content as well, and these can be slightly shaly, in turn. It becomes impossible in situations like these to differentiate objectively between a bed and a sedimentary step. In fact, the only way that one might justify any distinction being made between the bed and the step formations, is that the transitions from well developed beds to less distinct units can be seen, both containing the sedimentation steps. Chance weathering can sometimes accentuate a step boundary rather than a bedding plane, and this can lead to an apparent increase in the number of beds per cycle.

It is self-evident that a detailed examination of a section will lead to a better definition of the stratigraphic units than an examination which is based on the chance effects that weathering might produce. Unfortunately, there are often practical reasons which make the perfect examination of sections impossible. In this case, an attempt may be made to eliminate some of the effects of observational variability by examining the same section in different localities and under different conditions. A study like this can lead to extra information and this can be illustrated by using the Triassic as an example.

The large cycles (megacycles) of the Lofer limestone consist of five or possibly more "beds". Each bed, which may constitute a cycle itself, consists of a short intertidal facies which is followed by a subtidal development. The grouping into megacycles was caused by a prolonged intertidal regime which took place after a series of beds had been laid down. The groups of beds, which are 25—30 m in thickness, can be traced over considerable distances and they can be used locally for stratigraphic correlation. The boundaries of the large cycles are, therefore, master bedding planes (see Chapter 2) which means that they can always be recognized under any condition of weathering. The individual beds or cycles in the group are not necessarily persistent, and one can find massive developments of limestone which represent a megacycle with no apparent subdivisions into individual beds. If such a group is traced laterally, however, the individual beds may appear and the same megacyclothem can present a variety of field aspects which depend on the weath-

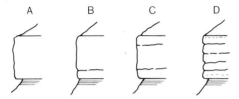

Fig. 11.7. Four different field aspects of the "Lofer" megacyclothem.

ering conditions, and perhaps on regional variation also. Some of these field aspects are shown in Fig. 11.7. The following procedure can be adopted in order to obtain some quantitative data on this variability. Groups are measured in a variety of different exposures and from many different levels in the section. The group or megacycle R is taken to be of unit thickness, and each measurement within the group is standardized by dividing it by R. Next, the position of any bedding plane that is inside the group is measured and plotted as a frequency distribution (see Fig. 11.8). The practical method is quite simple (Schwarzacher, 1954). If a group R contains one bedding plane, then it must consist of two beds with the thicknesses, let us say, of r_1 and r_2. Thus, r_1/R is calculated for the position of the bedding plane; if a group contains three bedding planes, for example, then it must contain four beds, and so one calculates: r_1/R, $(r_1 + r_2)/R$, and $(r_1 + r_2 + r_3)/R$ for the positions of the three bedding planes. Fig. 11.8 shows a frequency diagram which is based on 78 cycles, and it contains five maxima which would suggest a composite cycle consisting of six beds. This can be compared with

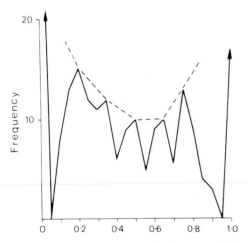

Fig. 11.8. The positions of minor cycle boundaries in the Lofer megacyclothem (after Schwarzacher, 1954).

an average of 5.84 for the previously given counts. It is interesting to note that by far the most common boundary occurs at 0.2, which is in the first fifth of the cycle, and this explains the frequent occurrence of the cycles which are similar to type B, in Fig. 11.7. The next most commonly occurring boundary is near 0.8, that is at 4/5 and this leads to the development of type-C cycle in Fig. 11.7. Type D is less common and it is shown in the same figure. It may be concluded that, whatever the agent was that was responsible for subdividing the megacycle into subcycles, it must have varied in intensity in the way that is shown by the dashed line in Fig. 11.8. Fischer's (1964) short but detailed section appears to support this conclusion because it shows that the intertidal development is considerably better developed near the megacycle boundaries.

11.7 COMPOSITE CYCLES

The term "composite cycle" was introduced by Barrel (1917) who applied it to the successions in which cycles of different wavelengths are superimposed upon each other. This type of sedimentation pattern was noted by many geologists and it led to the term "megacyclothem" (Moore, 1936) which is defined as a "cycle of a cycle". It follows from the discussion in the previous section that, since a sedimentary bed may be regarded as a geological cycle, cycles which consist of several beds can be regarded as megacycles or composite cycles.

Two fundamentally different situations can be found in the analysis of composite cycles and they are shown diagrammatically in Fig. 11.9. Two or more cyclic processes, which are quite independent of each other and of different wavelength, can determine the properties of a sediment. For example, the longer cycle R in Fig. 11.9A could control a chemical component like colour and the shorter cycle r could control a different parameter, such as grain size. The stratigraphic record of such a system will show evidence of both cycles which would not interfere with each other. The second situation is shown in Fig. 11.9B, where both cyclic processes interfere with each other and influence the same environmental factor. For example, R could represent an annual temperature variation and r a daily cycle. Let it be assumed that the sediment which forms under these conditions is temperature-sensitive and let it be said that a different mineral is formed above a critical temperature. The stratigraphic record would contain a pattern like the one which is shown in Fig. 11.9B. Knowing how closely the environmental factors are interrelated, makes it easy to understand that complex cycles of type A are uncommon and cycles of the second type are more commonly found. The Triassic cycle was discussed in the previous section and it is a typical example of this. If Fischer's (1964) interpretation is correct, then water depth was the factor which determined whether subtidal or intertidal

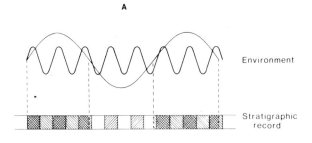

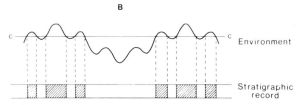

Fig. 11.9. Two different types of composite cycles. A. Two independent cycles. B. Superimposed cycles.

sediments were formed. Each short cycle, which in the previous section was referred to as a bed, commences with a horizon of reworking which is followed by algal mat developments and these in turn give way to massive calcarenites. The individual cycles are grouped together into a complex cycle which is caused by a thicker development of the intertidal deposits near the megacyclothem boundaries and by a decrease in the thickness of the calcarenite beds. The frequency distribution of the cycle boundaries (Fig. 11.8) illustrates the weaker development of the intertidal horizons towards the centre of the megacycle.

All the previous remarks which were made about geological cycles apply also to composite cycles. In particular, one can differentiate between strictly deterministic composite cycles and cycles in which either one or both of the cyclic components are replaced by oscillating stochastic processes. In the deterministic models, the phase between cycles of different wavelength remains constant and, therefore, if cycle R is, for example, three times larger than cycle r, then each cycle R will always consist of precisely three cycles r. This means that the number of cycles per megacycle remains constant (Schwarzacher, 1947). If either R or r or both cycles are replaced by an oscillating stochastic process, then the number of cycles per megacycle will be a random variable which is similar in concept to the number of beds per cycle which was treated in Section 11.6. However, it should be noted that, in the calculation which was made about the numbers of beds per cycle, it was assumed that successive bed thicknesses were stochastically independent. If a composite cycle is a type-B one (Fig. 11.9), then obviously this assumption cannot be made. The development of minor cycles would be expected to depend on their position within the major cycle. This may very well affect the sedimentation rates and so give much more complicated stratigraphic sections. These complications are in addition to the problems of defining cycle boundaries, and of their variable field appearance, which have been discussed already.

In spite of the difficulties in analysing complex cycles, one of their most intriguing aspects must be mentioned. This is the possibility of recognizing similar patterns in sedimentation, not only in different localities, but also in sections from different environments and from different parts of the geological column. Attention has been drawn already to the apparent similarity between the cyclic pattern of the Carboniferous limestones of Ireland and the Triassic limestones of the northern calcareous Alps. A third example of a carbonate sequence which is taken from the Jurassic (Malm β) of South Germany has been added to Fig. 11.6 and, once again, this bedded series is remarkably similar to the other two. One can make out a clear grouping of the sedimentary beds in each of the three examples. This may be interpreted as a complex cycle, the beds representing the short cycles (r) which are controlled by the longer cycle (R). Counts of the number of beds per R cycle give the very similar results of 5.89 for the Carboniferous limestone, 5.84 for the Triassic limestone and 5.24 for the Jurassic limestone. Whether such a close agreement is accidental is difficult to assess as yet. The average numbers were derived by counting between thirty and fifty individual cycles in each case. The agreement would be statistically highly significant if these could be regarded as random samples but the recognition of minor cycles as well as weathering and possibly other dependencies may introduce bias which cannot yet be analysed. Detailed petrographic examination of long sections is needed to test the reliability of the results. Complex cycles of a possibly similar nature have been reported by Van Houten (1964) from the non-marine Triassic Lokatong Formation (U.S.A.) and by Elliott (1961) from the English Keuper. In both cases, a complex cycle was found which is based approximately on the multiples-of-five principle.

The three examples may be compared further by considering the average thicknesses of the cycles R. The Carboniferous limestone cycle from NW Ireland has an average thickness of 2.8 m, the Triassic limestone dolomite cycle in the Lofer area has an average thickness of 21.5 m and the Jurassic cycle of SW Germany has an average thickness of 2.2 m. The last figure is based on only five direct measurements and it is much less representative than the other two values. It appears at first sight that the thicknesses of the Jurassic and Carboniferous sediments are beyond comparison with the Triassic sediments. This apparent discrepancy can be explained when it is remembered that the latter are geosynclinal sediments which were deposited in a basin of maximum downwarp and that the former are shelf sediments. The ratio of the maximum sedimentation in the Trias to the shelf sedimentation in the Carboniferous is found to be 10.7/1 (from Table 2.3). The ratio of the maximum "effective" sedimentation in the Trias to the shelf sedimentation in the Carboniferous is 5.1/1. A ratio of 7.7/1 is obtained when the thicknesses of the Triassic and the Carboniferous cycles are compared with each other, and this is well within the limits of what would be expected if the cycles in both of the formations represent the same time interval precisely. Less convincing results are obtained if the equivalent values are worked out for the Jurassic sedimentation data. The observed ratio of the thickness of the Triassic cycles, to that of the Jurassic cycles, is 9.8/1. Table 2.3 gives the ratios of the maximum sedimentation to shelf sedimentation as 4.2/1 and the observed Jurassic cycles, therefore, seem to be shorter than the Triassic ones. However, one must not attach too much importance to either of these comparisons because the relatively small stratigraphic interval which is represented by the cycles may contain quite appreciable gaps of non-deposition or periods of reduced sedimentation. This is quite apart from the fact that the data in Table 2.3 are not infallible and give values for average conditions only.

It should be noted that no reference has been made to either the lithological sequences or the inferred environmental histories in this comparison of different cycles. This is an unconventional procedure. Any attempts which have been made to compare different cycles have been restricted, so far, to sediments of similar geological age which were produced in similar environments. For example, Merriam and Sneath (1967), Merriam (1970) and Read and Merriam (1972) tried to compare British and American Carboniferous cycles which were deposited in approximately similar environments. The methods which they used were essentially correlation techniques which will be considered later (see Chapter 12). However, there are many unanswered questions connected with cyclic sedimentation, asking for a comparison of cycles from entirely different environments. Thus Fischer (1964) states that, if the cycles of the Alpine Trias are eustatically controlled, one would expect evidence of a similar cyclicity in other shallow-water deposits of similar age. If the observed cycles are climatically controlled, then one would indeed expect some comparable evidence of cyclic sedimentation in both marine and non-marine environments. The identity of such cycles can be established only by proving absolute, or as a second best, relative time equivalence. Relative timing is made possible by the study of composite cycles because one cycle may be taken as a time scale which is common to different environments. Additional cycles must show the same relationship to the standard, everywhere. Unfortunately, very few studies have been made with this ultimate aim in mind and so it will be necessary to collect considerably more information on the distribution and occurrence of composite cycles.

11.8 THE CAUSES OF SEDIMENTARY CYCLES

Attention was drawn in Chapter 10 to the many speculations and arguments which have been devoted to the explanation of sedimentary cycles. It appears useful at this stage to review briefly the various types of information which may help in the interpretation of cyclic phenomena and this can be done under five separate headings which can be listed as follows:

(1) The reconstruction of the environmental history of a cycle.
(2) The cyclicity, as discussed in Chapter 10.
(3) Problems arising from composite cycles.
(4) The time duration of a sedimentary cycle.
(5) The persistency of sedimentary cycles in space and time.

11.8.1 The reconstruction of the environmental history of a cycle

The reconstruction of sedimentary environments is the conventional method for analysing geological cycles, and this is certainly the most obvious way of attacking the problem. If there is a regularity in a sedimentary sequence, it must be caused by certain environmental factors, and if they can be identified, then the causes of cyclicity may become more obvious. For example, if the stratigraphic record indicates oscillating water depths, then one may interpret the cycle as being caused either by tectonic processes or by some mechanism which produces eustatic sea-level fluctuations. If the sedimentary record indicates, let us say, temperature fluctuations, then one might be more inclined to think in terms of climatic variations. Unfortunately, sedimentology cannot always supply as many details about the environmental history as one would wish, and some very important environmental factors are notoriously difficult to establish. A typical example of this is water depth, which may be judged from fossil assemblages, the energy state of the environment, light penetration, or temperature. Usually, but not exclusively, all these factors are determined by water depth. Indeed there are situations where it may become difficult to decide whether an environment is marine or non-marine, yet most of the cycles which have been studied happen to come from transitional environments which lie between the marine and the continental regimes. Clearly, such environments will provide the most spectacular lithological changes when the sedimentation conditions vary. Unfortunately, transitional environments are also particularly prone to periods of erosion and non-deposition, and so it is possible that many of the stratigraphic records which come from the deltaic environments in which coal-measure cycles were formed are relatively incomplete and are not as ideally suited to the study of cyclic sedimentation as one might think. More reliable results are to be expected to come from environments which can be regarded as being entirely marine but, in this case, the interpretation of the various facies which compose the cycle becomes correspondingly more difficult.

11.8.2 Cyclicity (as discussed in Chapter 10)

The so-called "cyclicity" of the sedimentary record was examined in some detail in the previous chapter and it was argued in Sections 11.3 and 11.4 that, providing that sedimentation was statistically continuous, stratigraphical cycles must correspond to some cyclicity in the history of environmental events. The implications of either oscillating or strictly cyclical developments in the environmental history were discussed in Section 10.6, in the light of the various theories which have been put forward to explain coal-measure cyclothems. Such speculations can be generalized so that they include any cyclic sedimentation process and its corresponding process model, which must belong to the type that was discussed in Section 7.5. It has been stated that, in order to produce oscillations or harmonic movements, a physical system must contain a feed-back mechanism together with an appropriate time lag. The true explanation of a cycle will be capable of identifying these factors in the system. It is known that environmental processes will often lead to feed-back loops, but of course it is difficult to choose the environmental factors which are genuinely responsible for one particular cycle. If the "degree of cyclicity" of a process is known, this may sometimes help in the choice of a model.

A biological cycle which is based on the predator—prey relationship, for example, will almost certainly be more irregular than a climatic cycle which is based on the nearly deterministic pattern of seasonal variation. Alternatively, it would be reasonable to assume that long astronomical cycles depending on the solar system for their generation, are more regular than "tectonic" cycles which originate within one structural basin. Naturally, such arguments are speculative but they must be taken into account, nevertheless.

11.8.3 Problems arising from composite cycles

The study of composite cycles introduces new aspects which are of direct concern when it comes to the interpretation of cyclic sedimentation. If a sedimentary cycle must, as has just been argued, be associated with an oscillating physical system, then the existence of more than one cycle implies that more than one physical system contributes to the formation of the sediment. Thus, the rotation of the earth around its axis produces day and night, and the rotation of the earth around the sun produces the seasons. The two physical processes are completely different, but they control the same environmental factor: namely the amount of light which reaches the surface. In a similar way, it is often found that complex sedimentary cycles are controlled by the same environmental factor. For example, in the Triassic limestone of Lofer, Fischer (1964) concluded that changes of water depth are the primary cause for the cyclic sedimentation. The cyclothem records three phases. The massive limestone contains a well-developed and partly benthonic fauna which proves it to be definitely marine. This is followed by a relatively thin horizon of re-working, together with the deposition of red or green clays which are interpreted as being the remains of soil and which, therefore, indicate immersion. Lastly, there follows a series of algal mat sediments containing dolomites, mud cracks and desiccation pores which Fischer believes to be intertidal. The latter facies in particular, is in accordance to a remarkable extent with the intertidal and indeed, the supratidal sediments of Florida and the Bahamas. The fluctuation of water depth is, therefore, well established by following the accepted rules of facies interpretation. Three processes must be accounted for in order to explain the sedimentation pattern of the series. The basin of deposition must have undergone continuous subsidence to accommodate the 800 m of Upper Triassic limestones. This subsidence would have caused a continuous deepening of the basin. Fischer assumed this subsidence to be intermittent in order to explain the megacycle, so that, by combining sedimentation with crustal movement, a cyclic variation of water depth could have resulted. Having used up, as it were, the structural explanation for the major cycle, a different cause had to be found for the minor cycles, and Fischer proposed that eustatic sea-level fluctuations were responsible for the latter. Therefore, this interpretation follows the principle that different orders of cyclic variation are associated with different physical processes. This is not meant to imply that it is wrong to postulate that cycles of several orders are caused by the same agent, such as the tectonic movement for example; but if such a complex movement exists, it must result from several physical causes. Too little is known about cyclic sedimentation in general to make a-priori statements about the likely causes. Bubnoff's contention (1947), which has also been followed by Fiege (1952), that all major cycles are tectonic in origin and all minor cycles are climatic in origin is a generalization which cannot be accepted without further evidence. The relative merits of structural and climatic cycles will be discussed further in the next section.

11.8.4 The time duration of a sedimentary cycle

The absolute time span that is represented by a sedimentary cycle would seem to be an important factor in its explanation, but there are only a few situations where it is helpful to know the time period very accurately. Clearly, if it is known that a cycle represents one year precisely, then there will be no hesitation in connecting its causes with some seasonal variation. Similarly, a cyclicity of about eleven years or multiples of this will be a strong indication that one is dealing with some process that is connected with solar activity. When it comes to longer cycles where their duration is measured in thousands or tens of thousands of years, our knowledge becomes considerably less certain. Either nothing is known about the mechanism of such long cycles, or it is not at all clear whether such cycles can determine the sedimentation pattern, let alone how they might. The latter is the case with the so-called astronomical cycles which have been appealed to repeatedly by geologists to explain cyclic sedimentation. Astronomical cycles are concerned with the variations of the orbital elements of the earth and these may have some effects on the terrestrial environment. It is relatively well known that changes in the inclination of the earth's axis and changes in the distance from the sun affect the so-called insolation which is the total solar energy which reaches the outer atmosphere of the earth. The amount of solar energy reaching the earth's surface can be calculated and such data were used by Milankowitch to explain the occurrence and distribution of ice ages in the Pleistocene (see Emiliani and Geiss, 1957). Three different cycles turn out to be important in this theory and these are:

the precession of the equinoxes	21,000 years;
the obliquity of the ecliptic	41,000 years; and
the eccentricity of the orbit	92,000 years.

The precession of the equinoxes is the result of two astronomical cycles which are: the precession of the earth's axis which involves a period of 25,800 years and a slow rotation of the perihelion which is completed in 110,000 years. It must be emphasized that the cycle lengths which have been given are approximate values. Not only are the cycle lengths influenced by the remaining planets and their moons but they are also influenced by the possible changes of mass distributions in these planets, so that they are not known with sufficient accuracy to be extrapolated indefinitely backwards into the geological past. Milankovitch, for example, calculated radiation values for the last 600,000 years only, and beyond the age of approximately one million years no details are known about the solar radiation. There is, however, no reason to believe that the astronomical cycles did not exist and, indeed, have varied much from the present ones, in earlier geological ages. It is possible that the cycle of equinoxes may have been slightly shorter when the earth's rotational speed around its axis was slightly higher.

The Milankovitch theory of ice ages has been criticized heavily by geologists and meteorologists, not because anyone doubted the existence of the astronomical cycles, but because there is very little agreement as to the effects which variable insolation can have on climate. The problem is very complicated. According to the theory, the insolation averaged over a year remains constant, but the winter and summer insolations vary in opposite ways. The maxima of, let us say, the summer insolation on the northern hemisphere occur at different times from the ones on the southern hemisphere, and it is beyond our theoretical understanding of meteorological processes to predict how this could affect the climate, particularly when it is remembered that the configuration of land and sea masses may have changed considerably in the geological past. The arguments against the Milankovitch theory as an explanation of the ice ages have been discussed in detail by Öpik (1967) who believes that insolation changes only cause minor fluctuations in the climate. However, there is relatively reliable evidence of a 40,000-year period in the palaeotemperature records of Pleistocene deep-sea cores (Emiliani, 1955) and it is worth noting that an environment which is determined by threshold values can be regulated by very minute fluctuations.

Almost nothing is known as yet about the possible effects which astronomical cycles could have on processes in the interior of the earth. It is almost certain that changes in the inclination of the earth's axis will affect processes in the mantle and core, but the connection between tectonic activity and astronomical cycles can only be suggested at this stage (see Heirtzler, 1968).

It was seen in Section 10.9 that absolute time estimates for the duration of cycles lead to highly variable results, particularly when such estimates are based on sedimentation rates that were averaged from composite sections representing a complete geological system. It has not yet been possible to count the number of cycles throughout a system and, at the best, such data have only been obtained for individual series. The length of the epoch must be estimated from the geological period for which the dates are known approximately from radioactive dating. The discrepancies which may arise from such a procedure can be illustrated by the Triassic limestone example which was discussed earlier. Fischer (1964) argues as follows. The Triassic period lasted for $45 \cdot 10^6$ years and in the Alpine stratigraphy it is subdivided into six epochs. The Dachstein Formation in which the cycles occur extends through two epochs and, therefore, lasted for $45/3 \cdot 10^6$ = 15 million years. Fischer estimated that there are 300 cycles approximately in the formation and, therefore, a minor cycle lasted for 50,000 years. A megacycle would then last for 262,000 years, using the average of 5.24 minor cycles per megacycle. An independent estimate which was made prior to Fischer's estimate involved calculating the thickness ratio of the Dachstein Formation to the total Trias from composite sections, and gives a value of 1/3.6. Thus, the duration of the Dachstein Formation was estimated to be $12.5 \cdot 10^6$ years. The counts of cycles in the Dachstein Formation gave 390 minor cycles, which is slightly higher than Fischer's estimate. It now follows that a minor cycle lasted for 32,000 years and that a megacycle lasted for 168,000 years. Fischer writes that his value of 50,000 years is not close enough to the 21,000-year periodicity of equinoxes, but that it is sufficiently close to the 41,000-year cycle of obliquity to suggest a causal connection between this and the observed sedimentary cycle. It has been seen, however, that a very small change in the estimation method brings the time value much closer to the 21,000-year cycle, and the truth is that neither estimate is accurate enough for deciding which hypothesis is correct. The estimates are good enough, however, for one to be able to state that the cycle length is of the order of tens of thousands of years.

A comparable result was obtained for the Carboniferous limestone cycle of NW Ireland, which has an average thickness of 2.8 m. Using the average rate of shelf sedimentation for the Lower Carboniferous, which is 33 B (see Table 2.3), one finds the duration of a cycle to be 84,000 years. In an earlier attempt (Schwarzacher, 1964) the same cycle was estimated to have a duration of 27,000—150,000 years. Taking the figure of 5.89 to be the average number of beds (minor cycles) per cycle, gives a time value of 14,000 years for the minor cycle. It may or it may not be fortuitous that the ratio of the eccentricity of the orbit cycle to the precession of equinoxes is 4.4 and that the ratio of the perihelion cycle to the equinox cycle equals 5.23, which comes close to the ratio of megacycles to minor cycles.

Further consideration will be given to the absolute length of sedimentary cycles in the next section. Problems which are related to the persistency of cycles in space and time are essential for this and will be discussed first.

11.8.5 The persistency of sedimentary cycles in space and time

If it were possible to follow a single sedimentary cycle over its entire development, a good deal could be learnt about its origin. Even if this is impossible in practice, it is useful to have a conceptional classification of cycles caused by strictly localized factors, against environmental factors acting on a large scale which might be world wide.

Localized cycles such as microclimatic variations, biological ecosystems, or tectonic processes, affecting a basin or source area, can originate from oscillating systems which are restricted in their geographical extent. Global cycles will be caused by climatic variations on a large scale, global tectonics and anything which is controlled by extra-terrestrial causes. This distinction becomes important in the classical example of the differentiation between the so-called tectonic and eustatic cycles. The latter should occur, by definition, on a world-wide scale but the difficulties of establishing a world-wide cycle are manifold and it has not yet been achieved. Consider, for instance, a eustatic sea-level fluctuation of moderate extent, perhaps of a few metres. This will clearly have some pronounced effect in the shallow marine and the littoral environments, but it will not be registered on the continents or in the deep-sea environments. This means that there is no continuous record which could be traced, even theoretically, and to prove a synchronous rise and fall of sea levels would require intercontinental correlation of an accuracy that is impossible at present. Further complications arise when it is remembered that most environmental factors are interrelated. Eustatic sea-level fluctuations, for example, can change the terrestrial climate and, through this, the biological conditions. In general, the effects of such a cycle will be quite unpredictable in different environments.

Studies involving the lateral variation of cycles can contribute considerably towards the understanding of cyclic sedimentation, in spite of the difficulties that have been mentioned. The bigger the area over which synchronous cycles can be established, the less likely it becomes that the sedimentation is controlled by localized sedimentation processes. For example, it would be quite difficult to formulate a sedimentary control theory (see Chapter 10) for the Trias limestone dolomite cycles which were considered earlier because they cover an area of several thousand square kilometers.

The persistency of cycles in time can be used in very similar arguments. If it can be established that the same type of cycle persists throughout long time spans and is independent of the local facies at any one time, then its cause must be fairly universal. In the Carboniferous limestone of NW Ireland, for example, the same cycle was found in the Benbulbin Shale (which contains only 25% limestone) as well as in the Dartry Limestone (which contains 94% limestone). Clearly, the factors that are involved in the process of cycle formation are independent of the more localized factors determining the facies change which occurs throughout this section.

A great deal more work is needed in mapping the distribution of sedimentary cycles in space and time, and there is every hope that this will lead to much better results than the exclusive studies of sequential properties in the sedimentary record.

11.9 SOME REMARKS ON TECTONIC CYCLES

Duff et al. (1967) conducted an interesting geological opinion poll among some European authors about the origin of coal-bearing cycles. Out of a total of 51 papers, 64% of the geologists favour a tectonic origin, 17% of the geologists favoured a sedimentation-control theory and 5% favoured a climatic-control theory.* Considering that practically nothing is known about the processes which could produce oscillating tectonic movements, this explanation appears to be astonishingly popular. One might almost suspect that the cause of sedimentary cycles is blamed on that part of the environmental history about which least is known.

Most geologists will agree that the basins in which large amounts of sediments are deposited are actively subsiding. Such movements of the crust or lithosphere must be caused by movements in the mantle which are driven by either thermal convection, or by chemical differentiation leading to density differences. Thanks to the relatively recent discovery of ocean-floor spreading, at least some information is available about the horizontal movements of

---

*It is difficult to assess how reliable these statistics are, since the present author was quoted as being one of the geologists who were in favour of a climatic control of coal-bearing cycles, although he has not, in fact, published any work on such formations.

mantle material. Two facts have emerged. Firstly, that the movement is uniform and has been relatively constant, at least for several million years. Secondly, the speed of movement is of the order of 2—4 cm/year (2000—4000 B). The idea of either thermal or chemical convection in the mantle provides for ascending and descending movements, and it is reasonable to connect subsiding basins with such descending motions. The rates of subsidence (200—30 B) are usually very much lower than the horizontal transport rates, and possibly this is determined by the angle of descent. Trend studies of sedimentary sequences have shown, however, that most crustal sinking can be regarded as being constant for millions of years, and this is in good agreement with the picture that is emerging from the study of sea-floor spreading. It looks as if the tectonic component is fully explained by the sedimentary trend.

Geologists who appeal to tectonic forces as generating cycles usually assume that the subsidence is intermittent. For example, in Fischer's (1964) interpretation of the Triassic megacycle, it was assumed that a subsidence of 70—100 Bubnoff units was slowed down or halted at intervals of 100,000—200,000 years and the question that arises is whether one can think of any mechanism which produces such a phenomenon.

One might consider some process which follows, to a certain extent, the principles of the elastic rebound theory of earthquakes, where stress is accumulated either in the crust or in the mantle and, when the stress is suddenly released, it produces movement. Stacey (1969) has discussed the strength of the mantle compared with that of the crust and found the mantle's strength to be considerably lower. It follows that, when stress is accumulated, the mantle will always fail before the crust and the latter is, therefore, best suited for storing stress. It is known from observations on shallow earthquakes that the periods of stress relief within the crust are irregular and certainly very much shorter. For example, the time between major earthquakes in the California area is estimated to be of the order of a hundred years, while in Japan and Chile shorter periods are likely (Stacey, 1969). Assuming the similar rates of mantle flow which are suggested by the recent data, it is quite out of the question that stress can be accumulated over tens of thousands of years.

An alternative to the stress-storage hypothesis may be developed by considering a subsiding basin in connection with a rising mountain area which acts as the source of the sediment. Bott's (1964) mantle-flow hypothesis was discussed briefly in Section 10.6 and is an example of a model which provides an oscillating system of this type. It is suggested in this theory that the subsidence of the depositional basin is caused by the withdrawal of mantle material which flows towards the mountain areas which are rising isostatically. In itself, this system should show a continuous movement, but if it is disturbed stochastically by small shock movements being generated, possibly on faults, then a preferred period of oscillation could develop. The "periodicity" is caused by the lag which is approximately constant and which is introduced by the time that it takes for a disturbance to travel from the basin to the mountain area. If one makes the assumption that the rate of mantle flow is of the order of 2—4 cm/year, then one can attempt to estimate the physical size of the oscillating system. If the oscillating period is about 100,000—200,000 years, then the distance over which any disturbance travels is only 2—8 km. These are very small and local systems, and obviously they are not large enough to explain the cyclothems which are developed over hundreds of square kilometers or more. One would have to accept considerably faster rates of mantle flow to account for such systems under a similar hypothesis and this is very unlikely.

In this connection, it may be of interest that some geologists have proposed tectonic cycles of much longer duration (cf. Stille, 1924; Umbgrove, 1947). In more recent work, Sloss (1964, 1972) attempts to show that the geological history of North America since Cambrian times may be divided into six cycles, each lasting for approximately one hundred million years. Taking this value to be the order of magnitude for long cycles and accepting a mantle flow rate of 4,000 B, one finds that the transport distances are some 4,000 km. This distance is, at least, in a scale that is comparable to the size of continents, and it is perhaps of the same order of magnitude as that of convection cells which may exist in the mantle.

One can apply these same arguments to evaluate the likelihood of climatic or sedimentation-controlled environmental systems. In such processes, the time lag can only be provided by some physical transport of materials such as ocean currents, the advance of a delta, or the migration of a vegetation belt, that is, by any mass transfer which can produce feed-back in the environment whether physical, chemical or biological. The physical size of the system will then be determined by the oscillation period and the rate of lateral transport. An attempt was made in Chapter 2 to estimate the rates of various exogenous geological agencies and it was seen that these rates are of the order of 10^7—10^{14} Bubnoff units (see Table 2.5). This means transportation distances of the order of 10^4—10^{11} km, when expanded to a period of 100,000 years, which suggests systems which are at least of world-wide dimensions. Thus, localized sedimentation control cannot possibly provide cycles of the required length but some world-wide climatic processes might do so.

It is possible that there are processes within the inner mantle or the core about which nothing is known yet, but it seems unlikely that any process in the interior of the earth will lead to localized effects on the surface. It is much more likely that tectonic activity of this type will lead to world-wide phenomena, perhaps to eustatic sea-level changes.

Undoubtedly, one of the reasons why tectonic causes for cyclic sedimentation have been favoured by so many geologists is that so many cycles appear to indicate a change of water depth. Quite apart from the previously mentioned difficulty of establishing water depth from the fossil record, the following must be kept in mind. If the general subsidence of a basin is compensated for by the sediment which is supplied during a cycle as is usually assumed, then it is clear that the change of water depth can only equal the compacted thickness of the cycle when an arrested downward movement is assumed. In many of the sedimentary cycles that have been discussed, the thickness is only of the order of meters or even centimeters. There are very few environments in which water-depth changes of such a small order would be recorded. If stratigraphical analysis indicates changes of depth which are larger than this, then one is driven to assume that the crust moved actively up and down. Such movements must be superimposed on to the general subsidence and no convincing mechanism for such an activity is known, and it is regarded by many geologists as being unlikely.

A good example of changing water depth is provided by Fischer's (1964) analysis of the Triassic cyclothems. The cyclothem in Lofer (Austria) has an average thickness of 3 m and, within this thickness, the environment changes from intertidal algal mats to a marine environment. The depth of the marine phase is regarded as being very shallow, say 5 m, and it is thought that a relative uplift brings the sediments to about 10 m above sea level. The latter is deduced from the solution cavities which exist up to 10 m below a surface of slight erosion, and presumably fresh water must have percolated to this depth. Some of the solution cavities are filled with a reddish material which Fischer interprets as being the remains of fossil soil. One has to conclude that the amplitude of the sea-level fluctuations was approximately 15 m. Clearly, the arrested subsidence theory could only provide 3 m and Fischer prefers a eustatic sea-level control, rather than accepting active uplift. The proof of emergence in this case rests on a number of observations and perhaps the most convincing evidence is the occurrence of a variety of shrinkage cracks which were most likely to have developed under sub-aerial conditions. It is not intended to challenge this interpretation in any way, but it is useful to keep in mind that even such a complete analysis as Fischer's contains features which are not completely explained or which could be interpreted differently. Thus, Fischer attributed the solution cavities below the disconformity to fresh-water diagenesis of the limestone and yet the actual surface which was exposed to the atmosphere shows no karst phenomena nor any signs of progressive lithification. The red and green sediments which are associated with the erosion surface, and which were interpreted as being modified soil formations, occur in precisely the same position in the contemporaneous Hallstatt facies which is generally accepted as being entirely marine. Perhaps the most puzzling feature of the environment, however, is the absence of any channelling in either the "intertidal" or the "immersed" part of the sequence, in spite of the fact that many of these beds are continuously exposed over several kilometers.

These few remarks may serve to show that the study of cyclic sedimentation can still lead to challenging problems in sedimentology and they may also illustrate that the analysis of stratigraphic sections comprises a wide range of different problems.

11.10 CONCLUDING REMARKS

It was pointed out in Chapter 9 that separating the trend from the oscillating components of a record is an arbitrary procedure which depends on the length of the record itself. In spite of this, an attempt was made in this and the preceding chapter to demonstrate that there may be physical differences between the trend and the oscillating part of the stratigraphic record. The latter is associated with a physical system which incorporates a feed-back mechanism and a time lag. The admittedly speculative considerations in the previous section suggest that this time lag or "geological memory" is limited by the processes which are involved and is linked to the physical size of the oscillating system. Very long cycles affecting whole continents and lasting for hundreds of millions of years may be of tectonic origin and be associated with convection processes in the mantle. These might be the magnacycles of Merriam (1963) or the transgression—regression cycles of many earlier geologists (cf. Suess, 1892). Cycles with periods ranging from tens of thousands to hundreds of thousands of years could be of climatic origin or be caused by unknown tectonic factors. In either case, it is suggested that they are of world-wide effect and it is possible that such cycles are controlled ultimately by extraterrestrial causes.

With our present knowledge of the stratigraphic column, it is difficult to assess whether the "long" cycles can be regarded as being part of an oscillating system. Sloss (1964) could not find any common pattern in the repeated "cycles" of North America, but he attempted to demonstrate (Sloss, 1972) that the cratonic activity of the U.S.A. runs synchronously with that of the Russian platform which leads him to the suggestion that different plates oscillate in phase. Whether this is an oscillation in the sense which has been defined here, must be regarded as being unknown. However, there are indications that the crustal movement is relatively constant during periods of basin subsidence which means that it behaves like a trend over shorter time periods (epochs). In the few examples which have been considered here, such trends could be treated as linear but there is no reason why tectonic trends should not be exponential and even more complex when, for example, tilting movements are involved.

The major problem about the "shorter" or proper sedimentary cycles remains the question of whether or not they are associated with an oscillating system. This is, in fact, implicit in the decision to classify them as type-1 and type-2 cycles (see Section 10.9). Only type-1 cycles contain time information. The practical question of whether such time information is good enough for one to be able to regard sedimentary cycles as time-stratigraphic units has been discussed by several authors. Vella (1965) is very optimistic. However, he stresses the world-wide distribution of cycles rather than their equal length of duration. The attitude taken by Duff et al. (1967) may be judged from the following remarks which they made about the work of Vella who "has even recommended the adoption of cyclothems as time-strati-graphic units. When it is realized that it is often not possible in America to correlate Carboniferous cycles between say Kansas and Illinois or even within each state and that in the British Coal Measure there are often correlation problems between Sheffield and Nottingham a distance of about 30 miles, we find it difficult to take Vella's suggestion with any seriousness". Considering that the same authors write in the next paragraph: "Lithologies in cyclic successions are obviously controlled by the environment of deposition" one finds it hard not to come to the conclusion that their inability to correlate different areas is due to lateral changes of the environment. Why this should prove the "cycles" to be unacceptable as time-stratigraphic units is somewhat obscure. It would be extremely naïve to expect that correlations can be made from one locality to the next by simply counting the number of cycles, particularly in environments which, by their very nature, must contain rapid lateral changes. The quantitative examination of some Carboniferous cycles in Kansas (Schwarzacher, 1969) and some coal-measure sequences from Scotland (Doveton, 1971) has shown that the oscillating behaviour of these sequences is uncertain unless definite assumptions are made about the sedimentary trend (see Section 10.4) and it is, therefore, not yet fully demonstrated whether the examples which were chosen by Duff et al. (1967) are, in fact, type-1 cycles which may be used as time units. Indeed, as far as the coal-measure cycles are concerned, Elliott (1968) claims that they are not cyclic in any sense. It is only certain that the typical coal-measure environment will always lead to predictable difficulties in correlation.

The various limestone—shale and limestone—dolomite cycles, as well as some evaporite cycles, appear to be much better candidates for cycles which carry time information. Serious attempts at using these for wide-ranging correlations have not yet been made. However, attempts to map cycles of this type within restricted areas have been highly successful and the working hypothesis that they represent equal time intervals can lead to geologically interesting results (cf. Fischer, 1964; Schwarzacher, 1964, 1969).

Chapter 12

Stratigraphical Correlation

12.1 INTRODUCTION

Much attention was given in the preceding chapters to the analysis of single vertical sections. However, stratigraphy in general has much wider aims and the historical geologist is interested in the development of large areas rather than a single geographical location. Indeed, his ultimate aim might be said to be the reconstruction of ancient environments on a world-wide basis. Even if one is satisfied with the study of a small area, the interpretation of stratigraphic records almost invariably requires the study of the lateral sedimentary relationships and a complete stratigraphical analysis becomes much more revealing when it is considered as a three-dimensional problem. The regional approach to stratigraphic problems seems only natural when it is remembered that most environmental factors are linked by processes which operate within a system which has no absolute boundaries, as discussed in Chapter 1.

Under these circumstances, it might well be asked why more attention has not been given to the three-dimensional aspect of stratigraphical analysis in the previous treatment and this criticism could be made, as indeed it has been made (Jacod and Joathon, 1972), about the more recent quantitative approaches to stratigraphy in general. A moment's consideration shows that most workers in this field must have been extremely well aware of this shortcoming because of the very nature of their training. The artificial one-dimensional approach to the analysis of sedimentary sections has been governed by necessity rather than by ignorance and three-dimensional analysis has lagged behind the development of single-section analysis simply because of the lack of suitable data. It is comparatively easy to collect a continuous and complete stratigraphical record from a vertical section where data can be sampled within well defined boundaries, but it is much more difficult to expand such sampling into a lateral variation study because of the scales that are involved. In Chapter 2, it was estimated that sedimentary transport, that is to say the lateral accretion rates, are 10^6-10^7 times faster than the vertical sedimentation rates. This means that studying a vertical interval of 1 mm or 1 cm in thickness is equivalent to a horizontal distance of 1—10 km. Certainly, the continuous exposures of sediments which would be required for a proper three-dimensional study, hardly ever exist. The difficulties arising from subsequent erosion and structural deformation are known only too well.

To facilitate the handling of the large volume of sediments which are used in stratigraphical analysis, stratigraphical subdivisions are necessary. The use of such units has been formalized by various national and international commissions on stratigraphic classification (Reports of the International Subcommission on Stratigraphic Classification, Montreal, 1971; Code of Stratigraphic Nomenclature, American Association of Petroleum Geologists, 1970). It is now generally agreed that the three types of stratigraphic subdivision which may be used are: lithostratigraphic, biostratigraphic and chronostratigraphic units. The latter is only rarely used in practical field mapping and the normal procedure for studying an area is to erect initially some reference sections which are subdivided into either lithostratigraphic or biostratigraphic units or perhaps, into a combination of both. Once the preliminary system is established and the section has been structured, individual units can be traced laterally and established in other localities. The recognition of an identical stratigraphic unit in two geographically different locations is known as stratigraphical correlation and since correlation is one of the basic operations in stratigraphical analysis, it must be discussed in some detail.

12.2 THE CORRELATION OF LITHOSTRATIGRAPHIC UNITS

Corresponding to the three types of stratigraphic units, one can distinguish between lithological, palaeontological and time correlation. There can be no doubt that it is an ultimate aim of stratigraphic analysis to establish all three types of correlation. Time correlation in particular, is absolutely essential for an interpretation of the environments so that it is often thought that palaeontological and even more so, lithological correlations, are inferior substitutes for the latter. This is true to a certain extent but the three are fundamentally different and they would provide different answers if absolute accuracy could be achieved. Time correlation should establish the reference system of isochronous surfaces in the sediment, whereas lithological correlation should establish the three-dimensional distribution of rock types or to put if differently, the shape and size of the lithosome. Palaeontological correlation aims to show the three-dimensional distribution of the various fossil assemblages which are called biosomes for short (Wheeler, 1958).

Lithological correlation is clearly an integral part of the mapping procedure and once the lithological units have been defined, the only difficulties which arise are those caused by insufficient exposures or by the fact that no geologist can spend a limitless time in the field. It helps to see the lithological correlation problem in its right setting if it is remembered that if one had a mountain range made of transparent plastic in which each lithological unit was marked by a different colour, then there would be no difficulty in correlating different sections. One can speak of more or less accurate correlations according to the rank of the lithostratigraphic units, whether they are bed or member or formation or group but the correlation itself is either correct or wrong within a defined interval. This means that it would be meaningless to introduce a degree of correlation in an attempt to quantify correlation procedures. On the other hand, one might hope that it would be possible to associate a probability with each correlation to indicate the likelihood of the operation being correct. It will be seen shortly that there are great difficulties in formulating such a measure of reliability.

A further discussion of lithostratigraphic units is needed before any quantitative statement can be made about correlation procedures. In the Preliminary Report on lithostratigraphic units, the International Commission on Stratigraphic Classification gives the following definition. "A lithostratigraphic unit is a stratigraphic unit which is unified by consisting dominantly of a certain lithological type or combination of lithological types or by possessing other impressive and unifying lithological features . . . However the many lithological variations manifested by rock strata offer a wide variety of choice in drawing the boundaries of specific units . . . ''. There should be no difficulty in mapping and therefore correlating rock units if continuous exposures are available, as has been mentioned already, but if the exposures are intermittent, then problems can arise from two kinds of circumstance. Firstly, the shape of the rock body can be quite complicated. It may be discontinuous which means that it would be present in one section and absent in an adjacent area, or it may consist of several tongues in which case it can be found as a single unit in one section and as several units in a section close by. Secondly, the lithological units cannot be regarded as unique features in a section, meaning that a group of rocks will usually contain several types of limestones, sandstones and so on. It would be quite possible for example, for a limestone 1 to be mistakenly correlated with a limestone 2 in a second exposure.

The complications which arise from shape variation are shown diagrammatically in Fig. 12.1A—D and it may be seen that a straight forward correla-

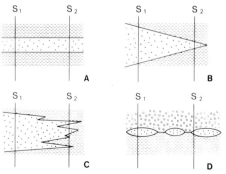

Fig. 12.1. The correlation of various lithological units. A. Parallel bedding. B. Wedge-shaped unit. C. Interdigitation. D. Syntaxial lenses.

tion of two sections, S_1 and S_2 is only possible in example 12.1A. The situation shown in Fig. 12.1D needs special explanation. Here, the rock body consists of several isolated lenses which may or may not form a single lithostratigraphic unit. The following recommendations are given by the international code. If such an isolated rock body has sufficient size and is distinctive, then it should be regarded as a separate lithostratigraphic unit. However, if there is a whole series of isolated bodies occupying the same stratigraphical position, then they can be unified as a common stratigraphic unit. The latter procedure is not quite logical if lithostratigraphical correlation is viewed essentially as leading to accurate mapping but it may well be more economical and therefore permissible, particularly when the mapping need not be very accurate.

It is useful at this stage to make a clear distinction between stratigraphic correlation, which was defined in Section 12.1, and the matching of two sections. Matching refers to two or more sets of serial data, such as are provided by the lithological logs of stratigraphic sections. Two sections are said to be perfectly matched if for each stratigraphic unit in one section, a correponding unit can be found in a second section. Matching may either refer to lithological types only, or it may include the thickness of the units. Thus the two sections in Fig. 12.1A are perfectly matched, both in thickness and lithology but the sections in example 12.1B only match in lithology. Clearly in the example of Fig. 12.1C, the two sections cannot be matched although they can be stratigraphically correlated. A perfect match can be achieved in the example which is shown in Fig. 12.1D but if the lenses are regarded as being different lithological units, no correlation is possible. This difference between matching and correlating becomes especially important when it is conceded that lithologically defined units are not necessarily unique and Fig. 12.2A gives a particularly obvious example of this. The rock units are repeated cyclically and consequently, if the sections were of indefinite length, there would be as many perfect match positions as there are cycles. Naturally, there can be only one stratigraphical correlation.

Once it is recognized that stratigraphical correlation and matching are two different concepts, it follows that although statistical tests can be designed for matching, this is impossible for stratigraphical correlation. The perfect match can be defined as being the complete correspondence between two

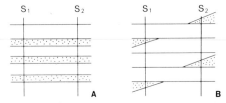

Fig. 12.2. Correlation problems involving repeated lithologies.

strings of data, whether they are repeated cyclically or not. This would be called the null hypothesis in statistical terms and one can now proceed to explain mis-matches by introducing error terms which change the lithology from one type to another. Of course something must be known about the probability with which such lateral errors can occur and the collection of such data involves correct stratigraphical correlation which may be possible, at least in some restricted areas. Once such information becomes available, significance tests for matching the results can be set up. The situation is quite different for stratigraphical correlation. The only criterion for correct correlation, that is the null hypothesis, can be that an identical stratigraphical unit was recognized in two geographically separated localities. This can only be verified by the continuous tracing of the same unit from one place to the next (a method which is known to geologists as "walking out") or by establishing that the unit which is being correlated is absolutely unique in the section. In the latter case, a match would also imply a correct correlation. One might consider basing confidence tests for the correct correlation on the "probability" of a certain marker horizon being unique. Thus, if one found, let us say, an orange coloured rock with blue and red spots it would be regarded as being highly unusual but there is no statistical probability measure for something being unique. In order to establish uniqueness one needs a knowledge of all the rocks in every part of the world. In this context, the biostratigrapher is in a very much better position because the first appearance of a new species is a unique event, at least in theory. Admittedly, difficulties might arise in defining a species or finding the genuine point of first appearance but if these could be overcome, then the matching of such events would provide a correct biostratigraphical correlation.

Let it be assumed next that a stratigraphic series contains only one lithology that is not repeated in the section. Such a horizon is rare obviously, and will thus be a suitable marker horizon. Bentonites, boulder beds and similar rare lithologies may occur once only in a given stratigraphical interval. Given this knowledge, can one predict the probability of a wrong correlation being made? Consider for example, section 1 containing one hundred beds, of which only one bed is a bentonite, let us say. This is to be correlated with section 2 consisting of one hundred beds that also contain one bentonite bed. If the second hundred beds really do correspond to the standard section, then a wrong correlation can only have come about in the following way. Some lithology other than bentonite must have formed in section 2 when bentonite ought to have developed. The probability of this occurring (assuming that the error could occur at any time) is equal to 0.99. For section 2 to have one bentonite in a wrong position, it is also necessary for a bentonite to have formed somewhere else where there should have been no bentonite deposited. The probability for this is 0.01. The probability that a wrong correlation results is therefore $0.01 \cdot 0.99 = 0.0099$ and conversely, the probability that the two bentonites were correlated correctly is 0.9901.

However, this apparently encouraging result depends on the a-priori knowledge that the two sections can be correlated bed by bed and it also depends on the entirely unrealistic assumption that the sequence of the deposited rocks is an independent random series. It is well known that difficulties in correlating lithological units do not arise from random changes in the actual depositional pattern but largely from the thinning out or splitting of units, or from the gradual facies changes which may occur. This is not meant to imply that relatively rare marker horizons are not a considerable help in lithological correlation, but it does appear that the reliability of such beds cannot be judged directly in quantitative terms. Geological experience of a non-quantitative nature is often the main basis for accepting lithostratigraphic correlations and the history of many field investigations shows that mistakes can occur.

12.3 QUANTITATIVE APPROACHES TO LITHOSTRATIGRAPHIC MATCHING AND CORRELATION

The previous discussion indicated that it is more profitable to concentrate on the matching of stratigraphic sections rather than on their correlation, in any attempt at designing quantitative or automatic procedures for comparing two sections. Geological judgement has to be used in deciding whether such results are valid correlations or not. Matching in general involves establishing the similarity of two sections and a section must be structured into convenient units so that such a comparison can be made. The problems of structuring are the same as the ones that were discussed in Chapter 5 and equal thickness intervals or intervals of lithological similarity can be used for the matching procedures. If the structuring is based on lithological characters and if the information comes in the form of continuous variables as are provided by well log data for example, then several analytical methods might be considered. For example, the stratigraphic log could be approximated by a trigonometric series ([9.3]) and the maxima and minima of this analytical expression could be used to define subdivisions of the sequence that is to be structured by this method. A rough correlation of two sections may be achieved by aligning the maxima (Vistelius, 1961). The series will only have one maximum if the first-order harmonic is fitted. Its position is given by the phase angle tg $\phi = b_1/a_1$, where b_1 and a_1 are the first-order Fourier coefficients ([9.3]). As an example, the second-order harmonic trends of limestone percentages have been calculated for the three different sections from the Glencar Limestone and are shown in Fig. 12.3. The three sections lie roughly on a line running northeast to southwest, and the horizontal distance between section 1 and section 4 is 5.6 km while the distance from section 4 to section 7 is 5.1 km. The maxima which falls into the middle Glencar Limestone could be used to align the three sections and so provide a

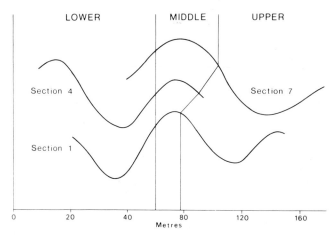

Fig. 12.3. Correlation by harmonic trends.

rough match. In Fig. 12.3, the base of the middle Glencar Limestone was used as an independent reference for the relative positions of the sections and the lower boundary of the upper Glencar Limestone is also shown. Two observations can be made about this example. Firstly, there is a considerable thickening of the sequence when it is traced from the northeast (section 1) to the southwest (section 7). Secondly, the maximum value for the limestone percentages shifts stratigraphically downwards in the same direction. In section 1, the maximum occurs very close to the top of the middle Glencar Limestone, whereas in locality 7 it occurs approximately in the middle of the series. The shift is systematic and can be observed in other sections of this area when moving from the margin of the depositional basin (section 1) towards the centre (section 7). This is clearly a facies change and the limestone-rich development appears to occur earlier in the central part of the basin than it does towards the margins. Approximating the lithological variation by a low-order trigonometric expression is a convenient technique but it gives a very artificial picture of the lithological trend. The positions of the maxima which are used for segmenting the series are very much constrained by the harmonic nature of the process. Attempts have been made to provide a more realistic segmentation of sections and Kulinkovich et al. (1966) used numerical differentiation of well log data to pick out the points where the lithological character of a sequence changes relatively rapidly. Hawkins and Merriam (1973) describe a search procedure which subdivides the section into segments within which the lithological variation is at a minimum. In this way, the entire sequence is divided into lithologically homogeneous segments and so this procedure corresponds very closely with the conventional way in which sections are divided into lithological units. The segments must have a certain minimum length, as was the case for the previously mentioned methods, and the segmentation cannot be carried on indefinitely. The use of

standard well logging methods in conjunction with any of the analysis procedures that have been discussed already, is unlikely to yield lithological correlations that could be any more detailed than, let us say, the formational
level. More experience must be accumulated before the potentialities of
these methods can be assessed fully.

Different matching methods can be used if it is possible to obtain a
detailed stratigraphical classification by working with exposures on the surface or with continuous cores. Consider the types of data in which each
individual bed is coded by its lithology so that each section consists of a
string of symbols such as A, B, C, D, E, F . . . where each letter represents a
lithology. A might stand for sandstone, B for shale and so on. If a section is
going to be compared with a second section, it is only necessary to count the
number of matches or the number of positions in the two chains where the
two letters are identical. The similarity between the sections may then be
expressed by the so-called matching coefficient S_m (Sackin et al., 1965)
which is defined as:

$$S_m = \frac{\text{number of matches}}{\text{total number of comparisons}} \qquad [12.1]$$

In searching for the best matching position, one section is kept stationary
and compared with the second section which is moved in unit steps. The
matching coefficient can be calculated at each position to give a complete
cross-association diagram for the two sections. The procedure is illustrated in
Table 12.1. If the length of the first series is M and that of the second series

TABLE 12.1

The calculation of cross-association diagrams

Position	Number of comparisons	Number of matches	Matching coefficient
ABCDE ABCGK	1	0	0
ABCDE ABCGK	2	0	0
ABCDE ABCGK	3	0	0
ABCDE ABCGK	4	0	0
ABCDE ABCGK	5	3	3/5
ABCDE ABCGK	1	0	0

N, then there will be $M + N - 1$ match positions. A useful feature of this procedure is that it can deal with sections containing gaps. If certain lithological units are not exposed for some reason, or they cannot be identified, they can be coded as blanks. In such a case, they will not be counted as comparisons. Clearly, if the sections contain sequential sedimentary cycles there will be a series of maxima corresponding to the positions in which the sedimentary cycles mesh and the method can be used to demonstrate this type of stratigraphical behaviour. The computer programme written by Sackin et al. (1965) contains several refinements which might be useful sometimes. For example, provisions are made for inverting one section and running it against the standard in reverse order. One can therefore test the matching of a section that was inverted tectonically or one can test for symmetry in sedimentary cycles which, if it exists, should give the same cross-association diagram in an inverted and non-inverted run. In addition, it is possible to use a multiple classification for the lithological units. Characters such as composition, presence of fossils, colour and so on can be coded for each unit in the same order as that represented by three or more letters in the data string. When the comparison for the cross-association diagram is made, the second section is moved against the standard section in steps which correspond to the number of descriptive characters and once again, the matching coefficient is calculated for each position. Sackin et al. (1965) introduced a further quantity which is called the similarity index S_L. Essentially, this is the proportion of the two sections which can be paired off as matching sequences. The similarity index is defined as:

$$S_L = \frac{\sum\limits_{r=1}^{M} \sum\limits_{s=1}^{M} (n_{r,s} - 1)}{\max (M, N) - 1} \qquad [12.2]$$

where M and N are the lengths of the two sections and $n_{r,s}$ is the number of matches for any position, subject to the conditions that $n_{r,s} > 2$ and any element which has been matched already in a subchain, cannot be matched again. The similarity index is 1 if the two sections are identical and it is 0 if there is no matching sequence of more than two units. Sackin et al. (1965) tested the matching procedure which has been outlined on a number of stratigraphical sections from the Pennsylvanian rocks in east-central Kansas, southern Kansas and northern Oklahoma (U.S.A.). These are all "cyclic" deposits and therefore yielded multiple matching positions. In no case was it found that the best matching position coincided with the best stratigraphical correlation but the two always came close to each other. Merriam (1970) used the similarity index S_L for comparing a number of sedimentary cycles of Carboniferous age from the United States and Great Britain. A precise stratigraphical correlation was not the aim of this study but instead, a search was made for similarities in the sequential pattern of coal-measure cycles. Assuming that data are available from n localities, one can calculate S_L (1,1) which is the similarity between locality 1 and itself, S_L (1,2) which is the similarity between locality 1 and locality 2 and so forth. These data can be arranged in matrix form and used like a correlation matrix in some numerical classification scheme. By using clustering techniques, Merriam (1970) was able to show that different sequential types correspond to different environments of deposition.

An attempt was made to introduce a statistical test of significance for the matching coefficient in the original study of stratigraphical matching (Sackin et al., 1965). It was argued that if a standard section is matched against a section containing the same characters but with the characters mixed in a completely random fashion, then the probability of obtaining a specified number of chance matches will be given by a multinomial distribution. With the aim of simplifying the test, the multinomial distribution was approximated by a binomial with the mean P which is given by:

$$P = \frac{\text{total number of matches summed over all match positions}}{\text{total number of comparisons summed over all match positions}}$$

[12.3]

Each coefficient S_m, can be expressed in units of the standard deviation which is appropriate for a binomial distribution having the mean given in [12.3]. The sampling distribution of standardized coefficients is approximately normal. The significance test is based on the hypothesis that the lithologies of a section that has no similarity with a standard section are in random order. This is quite unrealistic and the testing procedure which was proposed by Sackin et al. (1965) has very little practical use, for this reason. It was seen earlier that sedimentation systems invariably produce successive lithologies that are correlated in sequence and this correlation structure must be known if the significance of a similarity coefficient is to be tested.

It was shown previously that the stratigraphic record can often be expressed by one or more continuous variables which are measured either at constant intervals of thickness or for succeeding sedimentary beds. Cross-correlation coefficients can be calculated if such data are available. The two sections are represented by the two series of measurements $x_0, x_1...x_n$ and $y_0, y_1....y_n$. The correlation coefficient that is calculated between $x_0 y_0$, $x_1 y_1...x_n y_n$ is written as r_{00}, where the subscript denotes that the two sections have been compared without any shifting, using the formula:

$$r = \frac{\text{cov } xy}{\sqrt{\text{var } x \text{ var } y}}$$

[12.4]

The calculation is repeated for finding $r_{0,1}$, which is the comparison of the values $x_0 y_1$, $x_1 y_2 ... x_{n-1} y_n$ and in a similar way, the correlation between $x_1 y_0$, $x_2 y_1 ... x_n y_{n-1}$ is calculated and is written as $r_{1,0}$. It may be seen that when calculating $r_{0,1}$, the series x is leading by one step; when calculating $r_{1,0}$, the series y is leading. The calculations can be continued to find $r_{h,0}$ to $r_{0,h}$, where h is the lag by which one section is left behind and clearly, $h < n$. As well as being similar to the search procedures which were used in the previously discussed cross-association programmes, this sliding of one section backwards and forwards against its standard is similar to the methods used in calculating the autocorrelation of a section (see [7.8]). In fact, if one considers two sections for which the cross-correlation is calculated, then the similarity between the sections will be expected to increase with decreasing distance between the two localities, and when the physical distance between the sections becomes zero, the cross-correlation will be identical with the autocorrelation. Just as with the correlogram (Section 7.2), the autocorrelation coefficient is plotted as a function of the lag, so one can show the cross-correlation structure in the form of a diagram (see Fig. 12.4). This cross-correlogram is shown as a two-sided function which if $r_{0,0}$ corresponds to the correct stratigraphical correlation, should be symmetrical around this point. It is unnecessary to show both sides of the correlogram in the auto-correlation diagram, where the position of r_0 is always known but in cross-correlation problems, one is largely interested in finding the point of best correlation and therefore the search has to be extended to both sides of r_{00} which was chosen arbitrarily.

The data for Fig. 12.4 were obtained from two sections which were 37 m in length from the Dartry Limestone in the Lower Carboniferous, NW Ireland, where the two localities were 1,600 m apart. In this particular case, a well developed marker horizon permitted the two sections to be correlated stratigraphically and the position of r_{00} was chosen accordingly. As one

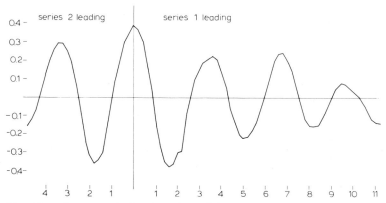

Fig. 12.4. Cross-correlation function of the Dartry Limestone (after Schwarzacher, 1964).

Fig. 12.5. Cross-correlation function of the Glencar Limestone (after Schwarzacher, 1964).

might hope, r_{00} turns out to be the highest correlation coefficient, with a value of +0.3901 but this value is not much higher than the neighbouring peaks which have values around +0.2500. In fact, experiments with cross-correlations are very often disappointing. An example from the Glencar Limestone of the same area (see Fig. 12.5), where the localities were only 500 m apart, gave a cross-correlogram in which the "correct" stratigraphical correlation is hardly higher than the alternative maxima. In several cases, it was observed that the correlation which had been established stratigraphically did not coincide with the best values suggested by the cross-correlation coefficients.

Some of the limitations that are inherent in cross-correlation methods become more obvious when one considers the difficulties which are involved in testing the significance of correlation coefficients between the two time series. It is easy to see that a comparison between the two sections can only be made after the trend component has been removed. If this is not done, the chance of obtaining so-called nonsense correlations are high. Let it be assumed that within a certain area, the stratigraphic parameter which is measured increases with an approximately linear trend. Any subsection that is taken from such a sequence will have this property of linear increase and therefore it will have a positive correlation with any other subsection that is taken from the same sequence. If the trend is not removed, then the values which are to be compared will form an ordered sequence and are not the independent random samples which are required by the classical theory of regression. A similar effect is observed when the time series contains a non-zero autocorrelation and one has to make allowances for the correlation structure of the series, as had to be done in the problem of matching by cross-association that was considered earlier.

The situation might be understood best by considering the following example. Let it be assumed that two sections are observed that are in close proximity and that the sequence consists of various beds with different lithologies. If there is a precise correspondence between the rock types, the stratigrapher will correlate without hesitation, whether this is done by matching the lithologies or by a measured parameter. Let it be assumed next that the measurements are taken at much more closely spaced intervals, so that several measuring points fall into the same bed and therefore give the same result. The correlation between the two sections has not changed but the number of observations has increased considerably and gives the correlation coefficient between the two sections an apparently much higher statistical significance. In fact, the additional measuring points are not independent and so they fail to provide any extra information. This is precisely the type of situation one gets when the subsequent values in a time series are correlated with each other. Therefore, in order to assess the significance of a correlation coefficient, one must estimate the number of independent observations which enter into its calculation. This can be done approximately by using a formula which is due to Bartlett and which was adapted for this purpose by Quenouille (1952). If the autocorrelation coefficients of two series of N terms are r_1, r_2 ... and r_1', r_2', ... then the effective number of independent observations which are used in testing the correlation coefficient is given approximately by:

$$N_e = N/(1 + 2r_1 r_1' + 2r_2 r_2' + ...)$$
[12.5]

The formula was applied to the Dartry correlation experiment in which $r_{0,0}$ was estimated from 185 observations. A value of 7.39 was obtained by summing the first 30 autocorrelation coefficients according to [12.5]. The effective number of observations $N_e = 185/7.39 - 25$. This is a very much smaller figure than the number of field observations, making the coefficient of +0.39 only just significant on the 5%-level but statistically insignificant on a 1%-level. It may be seen from Table 7.1 that the autocorrelation coefficients which were obtained from the Dartry limestone show very little tendency to damp out. If one increases the value of 7.39 (which was obtained from thirty lags) to 8.00, then the correlation is no longer significant, even on the 5%-level of significance. It was mentioned in Chapter 7.5 that the failure of damping to occur in a correlogram may be due to two reasons. Either the section from which the data were derived was too short and disturbed too much by random noise, or alternatively, the process which was responsible for the sedimentation was of a harmonic nature which means that it incorporated partly deterministic cycles. In both cases, it is interesting to find that formula [12.5] indicates that a correlation is impossible under such circumstances, which is obviously as it should be; if the data are inadequate, one should not obtain a significant correlation. If the sequence is cyclical, then each cycle should be equivalent to the next so that again, there should be no singular correlation between the two sections.

The method of cross-correlation would be suitable for sections approaching randomness or for sections which have rapidly damping correlograms, such as the simple Markov process. It is essential to know the correlation structure of a sequence before any correlation can be attempted but unfortunately, this is not sufficient in itself. In order to apply the methods in practice, one must know in addition to this, the trend and possibly the oscillating component which a stratigraphic unit may possess when it is traced in a horizontal direction. This is a problem that must be considered later.

Finally, a method of correlating two sections which was introduced by Burnaby (1953) will be discussed briefly. The method is very simple and of historical interest because it was the first attempt at quantitative correlation as well as being probably the first geological paper in the English literature to make reference to time-series analysis.

Burnaby's method was designed to correlate varve sections which are given by a series of thickness measurements: x_1, x_2 ... and y_1, y_2, ..., from two different localities. A cross-difference product is defined as:

$$E_{ss} = (x_s - y_{s+1})(y_s - x_{s+1}) \qquad [12.6]$$

The mean, which is defined as:

$$\overline{E} = \frac{1}{n}(E_{1,2} + E_{3,4} + ...) \qquad [12.7]$$

is positive if the corresponding x's and y's differ less from each other than they do from their neighbours in the series. A rapid χ^2-test can be applied to the number of positive and negative signs of the cross-difference products. In random matching, one would expect them to occur in equal proportions. The crosswise comparison as it is defined in [12.6], eliminates the effect of the first-order autoregression and it would be easy to adapt the method to compensate for higher-order serial regressions as well. In practice, one would use the method as a search procedure which would again involve sliding one section against a fixed standard. To facilitate this, a computer programme was written by Davis and Sampson (1967).

12.4 FURTHER PROBLEMS OF QUANTITATIVE CORRELATION

The methods which were discussed in the previous section deal essentially with straightforward matching procedures that are applicable to simple stratigraphical correlation problems like the ones that are shown in Fig.

12.1A. When it comes to the more complicated problems such as those that are indicated in Fig. 12.1B—D, neither the method of cross-association nor the method of cross-correlation are very satisfactory in their present form. The straightforward matching of lithological states which was used by Sackin et al. (1965) appears to be a logical approach at first sight because the method is not affected by thickness changes and the only discrepancies which can arise between the two sections must be caused either by the thinning out of some strata or by the splitting of one horizon into two or more (see Fig. 12.1C). Unfortunately, both are very common occurrences in almost any sedimentary basin. The cross-correlation which was first used by Schwarzacher (1964) is handled more easily because it deals with only one variable. It might prove useful as a method in theoretical studies but, from a practical point of view, no reasonable results were obtained. Thickness changes, the disappearance of strata, or extra insertions can cause a damping of the cross-correlogram which cannot readily be interpreted unless a detailed stratigraphical correlation is known beforehand!

Perhaps the most pertinent criticism of any of the methods which have been discussed so far, is that the results, if indeed they can be obtained, are of a type which is of relatively little interest to the geologist. For example, unless a rock-type matching procedure establishes a perfect match, it finds the best compromise solution which is the relative position of the two sections with the maximum number of identical rock strata being brought into juxtaposition. This implies invariably that a large number of stratigraphical units are out of phase, even if only slightly. Such units are wrongly correlated from the geologist's point of view. For example, in the experiments which were carried out by Sackin et al. (1965), two sections from the Pennsylvanian of northern Oklahoma and southern Kansas were compared and the standard section consisted of 36 lithological units. The best matching position that could be obtained while making full use of the 36 units, contained only eleven matches. Even if it could be proved that the matches are valid stratigraphic correlations, the procedure only succeeded in correlating 30% of the available section, in spite of the fact that the two localities are relatively close to each other.

The ways of improving automatic matching procedures become more obvious when their methods are compared with the conventional stratigraphical techniques. It was seen in the Dartry Limestone example that was given in Section 12.2, that attempts with numerical methods failed even where it was relatively easy to establish a stratigraphic correlation in the field. Nevertheless, the methods used by the mapping geologist are quite complicated and, like so many almost instinctive procedures, can be analysed in only very general terms. Most lithostratigraphic mapping relies on building up a framework of marker horizons which are unusual lithologies within the area of study, as stated already. The choice of such units depends very much on geological knowledge which is not quantified very easily. A sandstone horizon in some deep-sea sediments will certainly be recognized as being different from, let us say, a sandstone in an estuarine series. Similarly, a marine shale within a coal-measure sequence will be rated differently from a similar horizon within, let us say, marine limestones. This rating does not depend on the rarity of the rock unit alone but also on the accumulated knowledge of the geologist who associates unique events in the history of the environment with such marker beds. A similar remarkable feature of the practising geologist, if he is to be compared with a computer, is his facility for making adjustments in the thickness scales of sedimentary sections. Not only is it quite possible for the trained stratigrapher to match two sections in which the rates of sedimentation were quite different but he knows also which lithologies are likely to change and furthermore, he knows where to expect such changes in thickness from his environmental studies.

Once the framework of the marker horizons has been completed, details can be correlated and the lithostratigraphic correlation can proceed to the highest level of a bed by bed correlation, in some areas. The last stages of such a detailed correlation can be appreciated only by geologists who have actually undertaken such a study, since there occurs a curious moment when one "knows" that the correlation is now correct. Clearly, this is an unsatisfactory criterion and decidedly not one that can be used in an argument with a perhaps, more sceptical worker. This alone justifies the search for more objective methods.

The above somewhat lyrical account of field stratigraphic procedures is necessary for demonstrating the shortcomings of the previously discussed automatic correlation methods. In the Dartry Limestone example of Section 12.2, it was concluded from an examination of the significance of the correlation coefficient, that it was, in fact, the data of the experiments which were unsuitable for finding an acceptable correlation. The variable which was used was a thickness index of bedding (see Section 7.5) which was chosen because it gives a good general indication of changes in the environment and is easy to measure. However, the data required for correlation include unique features of the sedimentary sequence and frequently, these can be found only by examining a large number of different parameters.

It can be argued that although the environmental history may repeat itself many times, it can never be exactly the same. Consequently, each lithological unit will become unique within a certain section if the petrographic description of a rock unit is made as complete as possible. Sneath (1967) has followed up this argument and he increased the number of descriptors for cross-association programmes, experimenting with as many as 36 different lithological characters which ranged from rock type, colour, texture and fabric to fossil content. Unfortunately, this pilot study has not yet been extended to the correlation of sections from different localities and if the method is to be used for practical problems, it will be necessary to investigate the economy of such a procedure fairly closely. The amount of work involved in collecting the data for a study of this type is quite considerable and obviously, one must ensure that no irrelevant information is collected. It seems important that different weights should be attached to the matches between the various descriptors. For example, the cross-association programme of Sackin et al. (1965) in its present state, can compare a sandstone with a limestone (which would be recorded as a mismatch). At the same time, the two beds could be of a light colour and weather into a brown colour. Both these circumstances would be recorded as two matches so that the two horizons have a matching coefficient of 2/3. A great deal more work will be needed to develop this procedure into a more acceptable system which has not yet been done.

If correlations are made simply by matching successive rock types, one avoids the difficulty of having to deal with a variable vertical scale but at the same time, one loses the information that is contained in the thickness measurements which is often indispensible. It is well known that in matching well log data, one looks largely for the vertical pattern with which peaks occur, rather than the absolute values of maxima and minima. It is probably true that if well log data happened to come in a form in which the vertical scale is completely eliminated, it would be impossible to correlate the two sections. On the other hand, one must find a method to allow for changing vertical scales if the thickness information is to be incorporated into the mechanical procedures of matching. One can make a rough distinction between two types of thickness changes. Firstly, there is a systematic trend-like change which occurs relatively predictably in most sedimentary basins and secondly, there are the less systematic thickness variations which will incorporate stochastic elements, almost certainly. The separation into a trend and a stochastic component of the horizontal variation is comparable with the similar separation that is made in vertical records but the ultimate aim of stratigraphy in attempting to combine these elements, leads to very great difficulties.

The existence of regional trends in sedimentary basins is well recognized. Generally, maximum deposition occurs towards the centre of the basin and the rock units become thinner as the shore lines are approached and they terminate when erosion exceeds deposition. Of course, there are many modifications which could be made to this simple model but sections can be found which differ in their overall thickness within all sedimentary basins. When one correlates such sections against a standard section, one must as it were, either expand or shrink the sections to make them comparable with the standard. The geometry of such a procedure was investigated by Haites (1963), who pointed out that the correlation of two sections of unequal thickness involves the projection of one section on to the standard. The main principles can be made clear by Fig. 12.6, in which three sections from the middle Glencar Limestone have been correlated by means of the nine sedimentary cycles which occur in this interval. The base of each cycle is given

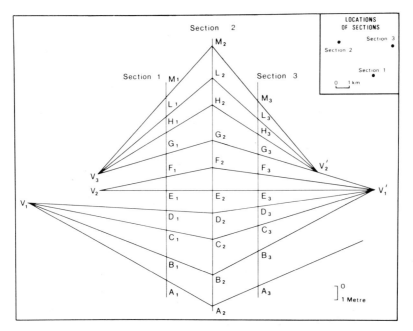

Fig. 12.6. Perspective correlation of three sections in the Glencar Limestone.

by a point A, B, ... on the line of a section, whereby the sections are, as a rule, vertical and parallel to each other.

It may be seen by comparing section 1 with section 2, that the lines which link the points $A_1 A_2 ... D_1 D_2$ intersect at a single point or vertex and that it follows that points A to D constitute a system which is known as perspectivity. Calling the thickness of the first cycle a, that of the second cycle b, and so on, one can write the condition under which the projective rays meet in a single centre of perspectivity or vertex by the equality of the cross-ratio:

$$\frac{(a + b)(c + b)}{b(a + b + c)} = \frac{(a_1 + b_1)(c_1 + b_1)}{b_1(a_1 + b_1 + c_1)} \qquad [12.8]$$

The formula can be proved by simple geometry (see Haites, 1963). Since the two lines (sections) which contain the points are parallel to each other, the perspectivity also implies the simple relationship:

$$\frac{a}{a_1} = \frac{b}{b_1} = \frac{c}{c_1} \qquad [12.9]$$

The ratios given in [12.9] clearly determine the horizontally measured distance of the vertex from the lines of the section. Since the quantities given in [12.8] are ratios, the condition for forming a vertex is not affected by the absolute scale that is used for plotting the stratigraphic sections. Haites makes use of this property and deliberately scales one section down in order to obtain better vertices. This is unnecessary in the present example, since the thickness variation is sufficient to lead to good intersections.

Fig. 12.6 illustrates a number of points which arise from the study of perspective correlation. Clearly, the points A_1 to D_1 of section 1 and the points B_3 to G_3 of section 3 can be correlated with section 2, simply by stretching the sections, according to the ratios given in [12.8]. Both sections 1 and 3 form a new vertex for the points G to M. Since the new centre of perspective is closer to the line of the sections, the sedimentation rates in the westernmost section must have increased in comparison with the rates for sections 1 and 3. This increase occurs earlier between the sections 2 and 1 than it does between the sections 2 and 3 because the vertices V_2 and V_3 have almost identical distances from the lines of the sections, implying that the ratio e_1/e_2 is the same as the ratio of g_1/g_2, and so on. It may be noted also that the first cycle in section 3 is abnormally thick and would indeed cause a shift from the right-hand to the left-hand side. A somewhat less drastic thickening can be seen also in cycle 4 of section 1, between the points D_1 and E_1, causing displacement of vertex V_2. Field studies show that these irregular thickness variations are caused by thickness variations of the shale and it is known that this clastic component was brought into the basin from the north to the northeast (Schwarzacher, 1968). The Glencar Limestone example, which involves sections of only 15—20 m in length and which come from localities which are very close to each other, shows that the relative stretching and shrinking of sections must follow a very complex pattern, the interpretation of which will require a greater amount of data. If it is further taken into account that the limestone—shale formations which have been used in this example probably represent a facies of low lateral variation, compared with other environments, it becomes understandable why so little progress has been made in three-dimensional stratigraphic studies.

12.5 THE CORRELATION OF SECTIONS WITH DIFFERENT VERTICAL SCALES

If it is suspected that the vertical scale of a section, as it is determined by sedimentation rates, varies systematically from one locality to the next, attempts can be made to compensate for this variation. Some of the methods which are suitable for such correlations were discussed in Sections 9.7 and 9.10, where it was seen that the removal of a lithological trend involves shrinking or expanding the vertical scale in order to adjust for differing sedimentation rates. There is then a very much better chance that the trend-free sections will provide good matches when cross-correlation methods are employed. Experiments were carried out using the Carboniferous limestone data of NW Ireland with a stretch factor that was set inversely proportional to the shale percentage and this yielded very high cross-correlations. Unfortunately this procedure accentuated the cyclicity of the sediments to such a great extent that the method was useless for stratigraphic correlation purposes but it does seem possible that the method might be applied usefully to sedimentary sequences which show less cyclicity than these particular sections.

An important contribution to the problem of correlating sections which have variable vertical scales is given by Neidell (1969) who used the so-called ambiguity and similarity functions for comparing two sections. The ambiguity function was first used in the processing of radar signals. If one wishes to calculate the correlation between an outgoing and a returning signal which is reflected from a moving target, one can achieve this by stretching the time or distance scale between the two series that are to be correlated. The ambiguity function $A(h, \alpha)$ is therefore a very appropriate tool for correlating two sections with different thickness scales. The function is really a covariance between two variables x_i and $y_{i(\alpha)}$ which is calculated for various lags h, and it can be written as:

$$A(h, \alpha) = \frac{1}{N} \sum_{-(N-1)/2}^{(N-1)/2} x_{i+h} \, y_{i(\alpha)} \qquad [12.10]$$

N is the number of points and is taken as an odd number. The peculiar notation $y_{i(\alpha)}$ means that the elements of the series y_i have been stretched by a factor α and then have been resampled at the same constant interval as x_i. The practical calculation of ambiguity functions poses several difficulties which arise largely from the interpolation of the stretched series. Since it is assumed that the stratigraphic observations are discrete samples of a continuous variable, they must contain inaccuracies already. The sample interval determines the frequency above which fluctuations cannot be recognized from the record and this Nyquist frequency as it is called in spectral analysis, was considered in Section 8.5. Resampling of the stretched data, introduces an additional bias which must be kept as small as possible. The best interpolation formula preserves the significant Fourier components of the stretched data and Neidell therefore uses an interpolation based on the function:

$$\phi t = \frac{\sin \pi t/\Delta t}{\pi t/\Delta t} \qquad [12.11]$$

where Δt is the sampling interval.

It was mentioned in Section 8.5 that fluctuations above the Nyquist frequency cause aliasing of the data. This phenomenon is usually kept at a minimum by choosing small enough sample intervals. However, if a section is

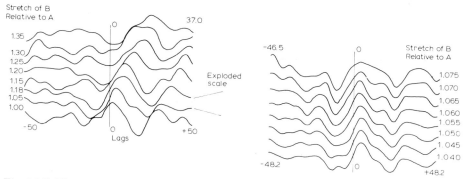

Fig. 12.7. The ambiguity function (after Neidell, 1969).

stretched artificially, then the Nyquist frequency is reduced and the treated section contains much more aliasing than the untreated section, against which it is compared. Filtering techniques can be used to compensate for this effect and Neidell (1969) gives a complete computer programme for the calculation of ambiguity functions. The programme performs the stretching, interpolation, filtering and correlation by a number of subroutines. The results are displayed as cross-covariance diagrams like the ones that are shown in Fig. 12.7, each curve corresponding to a given stretch factor. The results in this figure were calculated by Neidell and give the correlation between two magnetic profiles crossing the South Atlantic ridge, whereby section *B* has been stretched relative to section *A*. A final difficulty which must be mentioned, is the normalization of the ambiguity function, a problem which has not yet been solved apparently, and this must be kept in mind when interpreting curves like the ones given in Fig. 12.7. Because the ambiguity functions were not normalized, the comparison of the peaks must be made within the correlations referring to other stretches. Correlation peaks can be rated by measuring their width and their sharpness of resolution and by noting the percentage by which the maximum exceeds any secondary correlation. In considering Fig. 12.7, one would choose in all probability, the correlation curve which corresponds to a stretch of 1.065 as giving the best maximum. In addition, it may be noted that the ambiguity function appears to be very sensitive to small changes in the stretch factor and Neidell came to the conclusion that stretches of 0.1% can be measured by using this method.

The ambiguity function, as it has been discussed so far, is applicable only to correlation problems in which one section shows a constant stretch when compared with a second section. This is only rarely the case, as has been seen already, and even the short stratigraphical sections which were correlated in Section 12.4 contained at least two different stretch rates within a thickness of 15—20 m. In order to deal with variable stretch rates, Neidell proposed a similarity function S (h, α, K) which is an ambiguity function which is limited to a given range $K < N$. The similarity function can be written as:

$$S(h,\, \alpha,\, K) = \frac{1}{K} \sum_{-(K-1)/2}^{(K-1)/2} x_{i+h}\, Y_{i(\alpha)} \qquad\qquad [12.12]$$

and the calculation method is the same as the one that was used for [12.10].
There are no reports as yet, on the practical applications of the similarity
functions but it seems likely that difficulties may arise in choosing the range
of K. On the one hand, K must be large enough to yield a significant
covariance estimate and this often exceeds the available data, as was seen
previously in Section 12.3. On the other hand, K must be small enough to
resolve the changes of the vertical scale which are often very rapid. Only
experiments with real data can show whether a suitable compromise can be
found.
   A somewhat different approach to the correlation problem was attempted
by Dienes (1974) who gave only a very sketchy account of his methods,
unfortunately. Dienes uses a search method to find similarity between sections, whereby the variability of the sedimentary parameters is restricted by
some a-priori sedimentation model. Two special cases are discussed and the
following assumptions are made for the first situation. It is accepted that a
sedimentary parameter such as lithology for example, is a steady function
$f(x,z)$ that depends on a second steady function $g(x,\tau)$ which is the environmental history of the area. In these expressions, x is a geographical coordinate, z is the vertical thickness axis and τ is a time parameter. It is assumed
also that the following limitations exist between two localities x_1 and x_2:

$$|g(x_1\tau) - g(x_2\tau)| < E \qquad\qquad [12.13]$$

and: $|v(x_1\tau) - v(x_2\tau)| < \sigma$

$$v_{max} > v(x,\, \tau) > v_{min} > 0 \qquad\qquad [12.14]$$

The function $v\,(x\tau)$ stands for the sedimentation velocity at a certain locality
and for a given time period. These limitations enable one to give boundaries
to the range of possible lithological and thickness variations, which are expressed in terms of $\Delta\tau$ which is the permissible time error. Thus, if the
lithology is given in a standard section at position z one can search in the
second section for any lithology which is sufficiently similar to satisfy
[12.13] and one can search over a thickness range which satisfies [12.14].
An additional condition which is built into the algorithm prevents the viola-
tion of the law of superposition and similar lithologies must be ordered in
such a way that none of the correlation lines cross each other. The correlation that is achieved by this method is a true time-correlation which depends

of course on the existence of a time—thickness relationship which because $f(x,z)$ and $g(x,\tau)$ are assumed to be capable of differentiation, must be a steady deterministic function.

A second situation that is treated by Dienes (1974) is a model which incorporates a dependence between the lithology and the rate of sedimentation. For example, if it is accepted once again that:

$$f(x, z) = g(x, p)$$ [12.15]

where p represents a definite time interval comparable to the range K in Neidell's similarity function, then one can write:

$$z(x, p) = \int_0^p r[g(x, \tau)\, d\tau + u(x)]$$ [12.16]

The function r determining the rate of sedimentation depends indirectly on the lithology. This dependence could be approximated by several methods and Dienes estimates it from the thickness variations as they are observed in two adjacent localities. The model is similar in principle to the one that was used previously in the removal of stratigraphic trends.

Both the algorithm based on the limited variation model and the algorithm that is based on the lithology—sedimentation rate relationship have been tested on actual stratigraphic problems by Dienes, with good results. Unfortunately he gives very few details about the computational work that is involved.

The methods which have been discussed in this section are all promising and capable of further development. At the same time, it must be admitted that there is not yet any automatic procedure which can replace satisfactorily the correlation work that can be carried out by the trained stratigrapher.

12.6 STRATIGRAPHIC TIME CORRELATIONS

It has been stated before that stratigraphy involves three types of correlation: lithostratigraphic, biostratigraphic and chronostratigraphic correlations. The latter involves the recognition of time planes within the segments that are studied and this is obviously the most difficult problem in stratigraphic mapping. It is quite possible for example, that an economic geologist who finds valuable ore in a sediment is not interested in time correlation but is satisfied with establishing the geometrical shape of the lithosome that has his special interest. Nevertheless, information about both time and the lithology are essential for the theoretical reconstruction of different environments. Knowing the precise shape of a lithosome without any reference to planes of

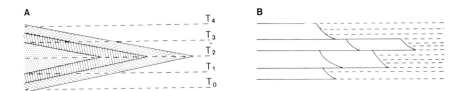

Fig. 12.8. The formation of a clastic wedge.

time is just as useless in this context, as knowing a precise but empty time-reference system.

It has been the prerogative of biostratigraphy to provide the framework for approximate time correlations for many decades. The method of absolute dating by measuring the products of radioactive decay is not sufficiently accurate for the detailed requirements of the mapping stratigrapher and as yet, it is uncertain whether techniques can be developed to make more use of the history of magnetic reversals which may be recorded in the stratigraphic column. The use of time correlations that are based on lithostratigraphic evidence has been criticized and continues to be considered quite unreliable by many geologists and therefore not worth attempting. This distrust of sedimentological methods is based partly on preconceived ideas and partly on real difficulties, some of which will be discussed in the following paragraphs.

Possibly, the most common objection to the use of lithostratigraphic methods is the fact that the boundaries of most sedimentary rock bodies are not parallel with time but intersect the time horizons at various angles. This situation is often illustrated by the classical transgression—regression model which is shown in Fig. 12.8A. A clastic wedge of shallow water deposits pushes into a set of marine sediments which represent deep-water conditions. The time planes in the diagram are horizontal and the lithological boundaries are inclined and so they show the time transgressive nature of the lithosomes. Clearly, one could redraw the figure and show the lithological boundaries as if they were horizontal, in which case the time planes would have to be inclined so that they transgressed the sedimentary units.

A more critical study of the model that is presented in Fig. 12.8A shows that it is highly inadequate. In particular, the inclination of the lithological boundaries against the time planes is quite exaggerated. Model A implies that sedimentation was continuous during the regression and transgression and that it was uninterrupted by erosion. Also, it implies that the sedimentation rates were the same for every type of lithology that was deposited. Both of these assumptions are contrary to what is observed in reality. Model B has been drawn in Fig. 12.8 to show the formation of a clastic wedge in somewhat more realistic terms. In this diagram, the advancing clastic sediments are shown as a number of current bedded units which have been deposited much more rapidly than the marine beds on the right-hand side. The time planes within the clastic facies are shown by solid lines and within the deeper marine facies, by broken lines. The latter are spaced relatively closely to indicate the lower sedimentation rates of the marine sediments. The discrepancy between the sedimentation rates implies that the clastic units must contain surfaces of prolonged non-deposition and therefore some of the time planes in these units represent a stratigraphic hiatus. Non-deposition is indicated by the merging of the time planes when coming from the marine environment towards the foreset beds of the delta. The overall effect of introducing variable sedimentation rates is an increased parallelism between the time planes and the lithological boundaries. It was argued in Section 2.3 that very often intervals of non-sedimentation lead to the formation of bedding planes which are recognizable features. Model B would lead one to expect that although time planes cross lithological boundaries as with model A, this should be noticed during careful mapping. In spite of being rather crude, model A in Fig. 12.8 permits one to make a rough estimate of the error which can be made in substituting a lithostratigraphic correlation for a time correlation.

It was mentioned in Section 2.3 that bedding planes can be recognized as sedimentary surfaces and therefore as true time planes by petrographic examination but it was also pointed out that in practice, the tracing of bedding planes can be very difficult. Since a bedding plane can hardly ever be examined in detail over very many miles, it is possible that some time transgression occurs. For example, Shaw (1964) states correctly that a bedding plane must represent a change of environment. Unless this change occurs simultaneously throughout the entire basin, the bedding plane must represent a boundary between two environments which has migrated and which is therefore essentially diachronous. This part of Shaw's argument seems reasonable but the further conclusion that the sharpness of a bedding plane reflects the abruptness with which one environment grades into another laterally is quite hypothetical. It seems much more likely that the "sharpness" of bedding planes is caused by the period of non-deposition and possibly erosion which such planes usually represent. It is also misleading to ignore the speed with which the bedding planes must have spread and there is every reason for assuming that this was rapid, compared with the migration of facies boundaries. Rough estimates of lateral speeds of sedimentary transport were given in Table 2.5 and these were shown to be very rapid in comparison with the vertical sedimentation speeds. However, direct evidence for the rapid development of bedding planes is provided by the fact that these planes intersect facies boundaries, almost invariably. If sedimentation slows down or ceases completely at a bedding plane there is an increased chance of it becoming a true time plane, as has been discussed already.

One can make rough estimates of the error which arises from substituting a lithostratigraphic for a chronostratigraphic correlation, by returning to the relatively crude model of cyclic sedimentation that is shown in Fig. 12.8A. If one accepts that in most sedimentation processes the sedimentation surface and therefore the time plane is horizontal, then the error can be determined from the geometry of the lithosome. For example, in the case of a clastic wedge like the one that is shown in Fig. 12.8, the maximum inclination of the time plane is determined by the dips of the upper and lower surfaces of the wedge. If the sedimentation is cyclical and is determined by shifting shorelines, then one might go further in limiting the possible inclination of the time planes by the following arguments. If transgression and regression occurred simultaneously throughout the depositional area, the time planes cannot transgress more than one hemicycle. Transgression through a full cycle would imply that regression occurred in one part of the basin whilst transgression occurred in another. This may be seen clearly from a consideration of Figs. 12.9A and B. The transgression of the time plane through a further hemicycle would imply that there are three different sets of clastic wedges which operate strictly in phase, although they originate from different areas of the basin. This is shown in Fig. 12.9C. It cannot be denied that these alternative models are possible or that situations like these may arise in delta environments. However, in shallow epicontinental seas, models of type C and even of type B, do introduce complications which must be justified by additional geological evidence. For the sake of simplicity, it seems reasonable to accept the simple transgression—regression model. The following considerations give some idea of the order of magnitude of such errors. Some Pennsylvanian cyclothems of the Midwestern United States of America can be traced with confidence over a distance of 500 km, at least. These cyclothems could be regarded as hemicycles and if one assumes a cycle to be 15 m thick, then this would mean that the maximum inclination of the time plane is 3 cm/km. If one makes the very generous allowance of $100 \cdot 10^3$ years per cycle, then the time error made by correlating lithostratigraphically

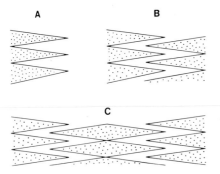

Fig. 12.9. Multiple transgressions in a basin.

over a distance of 1 km, would be a hundred years, on average. The error seems small when it is recalled that this is a maximum error which should arise only under the most unfavourable conditions and it is quite acceptable for most purposes. The rough calculation also permits one to estimate the minimum speed with which the transgression occurred and by basing this on the same average sedimentation rates, one obtains a value of 10^6 B (μmm per year), which compares reasonably well with the values that are given in Table 2.5.

Naturally, the rates with which the lithological boundaries migrate in a horizontal direction are only relevant to problems involving the correlation of sediments which originated by processes which are similar to those presented in the model of shifting facies belts, shown in Fig. 12.9. Under such circumstances, a vertical rock sequence must reflect the horizontal arrangement of different facies and one can predict one of these from a knowledge of the other. The equivalence of the two sequences is known as Walther's law and it is found to hold in an approximate way for many deposits which were laid down in shallow shelf seas. However, this "law" must not be interpreted too literally and one soon runs into difficulties if one attempts to draw a facies map from a detailed stratigraphic section. An obvious problem is presented by the horizons which represent erosional breaks in a sequence and which remove part of the evidence about the original lateral distribution pattern. A much more important consideration is the variety of environmental conditions which contribute towards sedimentation, some of which may migrate slowly, others more rapidly, and some may be effective almost instantaneously over large areas of the depositional basin. The latter include all the factors which control the vertical settling of sedimentary materials such as volcanic ash, plankton-derived sediments, and possibly precipitated carbonates and evaporites. It has been noted already that many bedding planes are regarded as features which, if they are not formed instantaneously, move comparatively rapidly through the environment. The development of small-scale laminations and minor cycles within major cycles are probably similarly rapid. Indeed, the occurrence of complex cycles as described in Section 11.6, can be explained only by facies migrations which must have occurred at different rates and, as before, it would be difficult to explain both the major and the minor cycles by the same mechanism of transgression and regression that is implied by Walther's law.

Assessing the reliability of chronostratigraphic correlations that are based on lithological evidence, depends very much on the character of the environment. It must be largely a question of geological judgement. Correlations will be difficult in deltaic areas, where some of the lithosomes have complicated shapes. There seems to be very little reason why the use of lithostratigraphic correlations for basinal depositions with parallel and horizontal bedding should not give results which are as accurate, if not better than biostratigraphic correlations. Both types of correlation are considered to be approximations for time correlations. In most cases, the distribution of sedimentary material is as rapid as that of the organisms which eventually fossilize. Biostratigraphic methods have distinct advantages when one passes from the consideration of inter-basinal correlation to the wider problem of intrabasinal correlation. A more extensive study of cyclic sedimentation and of complex cycles in particular as they are defined in Section 11.6, may enable correlations to be made between lithologies that might have been formed in different environments. It appears that much more detailed mapping work is necessary before the lithostratigraphic methods can be either condemned completely, or accepted generally.

Chapter 13

Some problems of three-dimensional stratigraphic analysis

13.1 INTRODUCTION

A complete stratigraphical analysis must involve the study of three-dimensional rock bodies within a reference system of time, as was emphasized in the preceding chapter. Given ideal mapping conditions, it ought to be possible to obtain the shape and distribution of the lithostratigraphic units, at least. A lithological parameter, let us say r, can then be expressed as a function of the three Cartesian coordinates x, y, and z and it may be written as $r = r\ (x,y,z)$. Matheron (1965) calls the variable r a "regionalized" variable. This is intended to be a neutral term which indicates that the value of r at any given point, is the outcome of natural phenomena (Matheron, 1967). The regionalized variable can be generated by some stochastic process but it does not necessarily have to be so.

Functions of the type $r\ (x,y,z)$ are usually very complicated and erratic in geology and they are consequently very difficult to describe by straightforward mathematical expressions. For this reason, one must resort to sampling techniques and these latter invariably involve a random element when the function r is determined. Quite apart from the "sampling error", there is a distinct possibility that the actual shape of an observed rock body is the outcome of a random process.

The model concept can be employed once again, in order to simplify three-dimensional stratigraphic analysis and it is clear that two types of models are required. Firstly, there must be a "process model" which deals with the dynamic system of the environmental factors which will determine eventually the type of sediment which is produced. Secondly, it may be necessary to have in addition, some simplification of the observed configuration of the lithostratigraphic units and this can be achieved by substituting a model for the geometrical shape of the rock body. It follows that the "process model" can be either some mathematical expression, or it can be any set of rules which portray the actual process. This model is used to generate the "geometrical model" which itself may be a simplified version of a rock distribution that has been observed in the field. The term "process model" was introduced by Sloss (1962) and was contrasted with the term "response model" which refers to the resulting configuration of sediments. A response model might be thought to be almost the same as a geometrical model but the geometrical model is always independent of time, whereas the sedimentary response model is a feature that develops with time. Thus time-dependent response models are feasible. Indeed, the process model is time-dependent by definition and it often interacts with the response model and feed-back relationships can exist between the two (Krumbein and Sloss, 1963).

Three-dimensional stratigraphic analysis can be simplified by thinking of a rock body as if it consisted of a series of two-dimensional sections. This treatment can be applied also to the different types of models which have just been discussed. For example, the geometrical model can be thought of as being a series of superimposed maps and if each map represents an iso-chronous surface, this can be made the basis for a dynamic process model. In a similar way, maps of this kind can be used for plotting any sedimentary parameter and a time-dependent response model can be constructed by choosing the vertical spacings between the maps as a time axis. Let us assume that the parameter which is plotted on a system of maps like this, represents the thickness z that was deposited during unit time. Each map then represents a record of the function $z_t = z(x,y)_t$. Geographical variation can be ignored in dealing with single sections, making it possible to concentrate on the simpler function $z = z(t)$, in the earlier models. When it is realized that the variable z is a function of both time and areal variation, the analysis becomes considerably more complicated and one of the major problems in the three-dimensional approach is to assess the extent to which the two functional dependencies can be separated.

13.2 THE AREAL VARIATION OF SEDIMENTS

That conditions of sedimentation and consequently the rocks which are formed in an environment change with their geographical distribution, is self-evident. The facies variation is very often systematic in its nature and it is related directly to such geographical factors as the distance from land, the water depth, or a combination of both. Indeed, palaeogeographic maps of an environment can only be reconstructed because these relationships are known. Deterministic models can be developed for some facies variations. For instance, the thickness of an individual bed which has been produced by a unidirectional current pattern is known to decrease exponentially with the distance from its source. Similar dependencies exist for other lithological parameters like grain size or roundness, for example. Theoretical models which attempt to explain these observations, can be based on the simple differential equation that was introduced in Section 3.5. The diffusion process that was introduced in the same section, provides an alternative way of obtaining a facies distribution model. One may assume, for example, that the source of the sediment is a topographically high area and that the sediment will spread from the source in accordance with [3.14], which gives the thickness of the sediment for each time increment.

Sedimentation processes incorporate various stochastic elements, by their very nature and almost any process will be associated with a certain number of random fluctuations, regardless of whether it involves sedimentation from transporting currents or settling or organic precipitation. If these variations occur independently during deposition in different localities, they will generate non-correlated noise in the areal distribution of the sediments. There must be many additional processes which occur during sedimentation and which can lead to correlated random terms. For example, a stream bed may be covered with pebbles which are scattered at random on a sandy surface. The pebbles will cause disturbances in the flow pattern which lead to the well known scour marks being made behind the pebbles and possibly, to increased deposition in front of them. Thus the random disturbances will affect adjacent areas and introduce correlated random deviations into the areal distribution of the sediments. Very similar effects can be expected to arise from wash-outs, sand bars and ripple marks. Indeed, the latter can be regarded as a case of oscillating areal distribution, where sediment transport leads to geological cycles in a horizontal direction.

The formation of correlated random disturbances is by no means restricted to the sediments that are produced from unidirectional currents alone. The ecologist, for example, is quite familiar with the occurrence of high-density patches in the distribution of plants and animals. The clustering together of various plant and animal communities may have a number of causes, and the mode of reproduction is a common one. The distribution of organisms can have a profound effect on sedimentation and can lead to areal variation which, once again, is highly autocorrelated. Unfortunately, no detailed studies have been made yet to investigate the areal correlation structure of sediments or to relate this to the primary processes of deposition. It is obvious that the practical difficulties which arise in making the observations (Section 12.1), have been a major obstacle but if they could be overcome by using suitable exposures, it would afford a very promising field of research.

The various types of areal variation which have been discussed in this section, suggest that one could construct models for the regional facies variations which are very similar to the models that are used for describing the time series properties of single sections. Indeed, Agterberg (1967, 1970) suggests the following general model:

$$\text{data} = \text{trend} + \text{signal} + \text{noise} \qquad [13.1]$$

The term "trend" refers to deterministic functions such as polynomials and it is assumed that this part of the areal distribution is caused by one of the deterministic processes which were discussed at the beginning of this section. The signal is assumed to be stationary in the wide sense and is characterized best by its autocorrelation. The noise is an uncorrelated random disturbance.

In accepting the Agterberg model ([13.1]), one might hope that the estimation of areal variation could proceed by using methods which are similar to those used in the estimation of the lithological variation in time-series analysis. Unfortunately, a number of difficulties arise which have not been met with in previous problems and an attempt will be made to show how these can be overcome, at least partially.

13.3 THE ESTIMATION OF AREAL TREND

The study of areal trends can be considered first of all. This is approached traditionally by fitting polynomials to the lithological data, using the methods of regression analysis. Thus the method is similar to some of the trend-fitting procedures which have been used previously in time-series analysis, but with one important difference. In time-series analysis, trends like the linear or the exponential trend can be explained by simple physical processes, making it possible to assess the effects which the removal of such trends would have on the data. Although similar, simple processes like the exponential decrease of grain size with increasing distance from the source area, can operate in areal variation, the geographical configuration of depositional basins can lead to very complicated distribution patterns which cannot be approximated by low-order polynomials. For example, if a basin's shore line is taken to be the source area for the sediments, then this is an essential boundary condition which will determine the distribution of the sediments. Clearly, the sedimentation patterns can be approximated by low-order polynomials only if the basin shapes are geometrically simple ones like circular or ellipsoidal basins.

One has only two options in dealing with the complex facies patterns which occur more usually. One can either abandon the idea that a simple relationship exists between the geographical coordinates and the lithological variation, or one can restrict the investigation to a sub-area which is small enough to be uninfluenced by the boundary conditions of the basin shape. For example, one might study the lithological variation along a series of sections or narrow strips which are taken at right angles to the shore line, in order to find the relationship between sedimentation and the distance from the land. In this case, one has to assume that the individual sections are independent of each other and that lateral relationships can be ignored. This situation can be compared with a study of sediment transport which restricts itself to laminar flow where the particles are allowed to move in definite planes only and where turbulence terms are omitted. Models like these can give valuable information about current velocities and sediment transport but they are unavoidably incomplete. In a similar way, low-order polynomial trends can be used to bring out the essential facies variation but they cannot always by interpreted in a physical sense.

An alternative approach to the analysis of areal trend was initiated by Krige (1966) who designed a method which can produce a trend surface map which is not based necessarily on the assumption of a simple functional relationship existing between the lithological variation and its geographical distribution. Indeed, it is implicit in Krige's approach that the method could be applied equally successfully to data that are generated by a random or by a deterministic process. The method was developed primarily as an aid in the evaluation of ore reserves. The amounts of rare minerals like gold fluctuate widely within a given rock body so that one has to introduce a certain amount of smoothing in order to make predictions or interpolations. This is achieved by making use of a weighted moving average, using methods which are similar to the ones which are used in time-series analysis. The area is subdivided into a number of blocks, measuring 30 m^2 for example. An estimate of the mean mineral content $\overline{X}_0$ is made at a point which is centred in one of these blocks. The estimate is based on all the sampling data X_1 within this block and those for a number of surrounding blocks (X_i, $i = 1, 2,...n$). A linear prediction equation is used and one can write for the estimate of X_0:

$$\hat{X}_0 = \sum_{i=1}^{n} \hat{a}_i X_i + \hat{a}_m \overline{X} \qquad [13.2]$$

The weights a_i can be determined in areas for which accurate estimates of X_0 exist. This done, one can apply [13.2] to areas for which only the X_i's are known. The term $a_m \overline{X}$ is a correction for the regional mean $\overline{X}$ and the constant regression is given by:

$$\hat{a}_m = 1 - \sum_{i=1}^{m} \hat{a}_i \qquad [13.3]$$

An alternative method of finding the weight factors a_i can be used when the two-dimensional autocorrelation of an area is known (Agterberg, 1970). Consider a number of points, $p_1, p_2,...p_n$ which surround a point p_0 and for which a measured lithological variable X_i with an assumed zero mean is known. The value at p_0 can be estimated by:

$$\hat{X}_0 = \sum_{i=1}^{n} a_i X_i \qquad [13.4]$$

The least-square estimates for the coefficients a_i lead to the solution:

$$\begin{bmatrix} a_1 \\ a_2 \\ \cdot \\ \cdot \\ \cdot \\ a_n \end{bmatrix} = \begin{bmatrix} 1 & r_{1,2} & r_{1,3} \cdots r_{1,n} \\ r_{2,1} & 1 & r_{2,3} & \cdot \\ \cdot & & & \cdot \\ \cdot & & & \cdot \\ r_{n,1} & r_{n,2} & \cdots\cdots 1 \end{bmatrix}^{-1} \begin{bmatrix} r_{0,1} \\ r_{0,2} \\ \cdot \\ \cdot \\ \cdot \\ r_{0,n} \end{bmatrix} \qquad [13.5]$$

which is the two-dimensional equivalent of the Yule-Walker relation ([7.31]) as it is used in time analysis. If the autocorrelation function is isotropic in the plane, then a value r_{ij} will depend only on the distance ζ_{ij} between two data points. Such distances can be determined from the geographical coordinates x,y and:

$$\zeta_{ij} = \sqrt{(X_i - X_j)^2 + (y_i - y_j)^2} \qquad [13.6]$$

The distances ζ_{ij} are calculated between each of the surrounding points p_1, $p_2...p_n$ and r_{ij} is determined from the known autocorrelation function r_ζ. The coefficients a_i can then be determined from [13.5] and they may be used for smoothing in the manner which is indicated by [13.2].

The term "kriging", has been used by some geologists to describe this type of smoothing operation, although it is not entirely clear why such an exotic terminology should be thought necessary. Equation [13.2] constitutes a linear filter which can be applied to the data and as such, it is a very flexible method. It should prove very useful in sedimentological problems where the analytical expressions of the polynomial trend surface are too rigid. Nevertheless, its main application will continue to be in the study of areal random processes and these will be discussed in the next section.

13.4 STATIONARY RANDOM PROCESSES IN THE AREAL DISTRIBUTION

Agterberg's general model which has just been introduced, refers to the component which is generated by a stationary random process as the "signal". This does not imply necessarily that this component is the most important one in a regional study. Agterberg (1970) makes it quite clear that the three components depend on the sampling scheme that is being used and if a small area is sampled in detail, the "signal" may very well become part of the trend. The situation is rather similar to the one that is found in time-series analysis and perhaps, it is advantageous to replace the term "signal" by the more neutral term of "oscillating component".

The analogy between geographical variation and time-series models is not a complete one, however. Time is a unidirectional parameter by definition and this means that the state of a variable $f(t_0)$ only can be determined by past events $f(t_{-1}, t_{-2}...)$ and that the future values of $f(t_1, t_2...)$ can have no physical influence on the value at t_0. This is not true necessarily, for the regional variables and for example, $f(x_0)$ can be determined by $f(...x_{-2}, x_{-1}, x_1, x_2...)$ that is, it can be determined by points which are in the past as well as in the future, if distance is regarded as being the equivalent of time. There are exceptions, of course. In the study of sediment transport problems, where $f(x)$ may represent a variable which is measured along the non-reversible down-stream direction, the distance function becomes equivalent to a time function. Whittle (1954) calls such relationships "unilateral", whereas the normal areal variation which may be observed along a straight line is "bilateral". Agterberg (1970) gave a summary of Whittle's approach to multidimensional autocorrelation and the following discussion is based essentially on this.

One may consider first the simplest bilateral correlation scheme which can operate along a straight line. Assume that a point p_0 lies half way between points p_1 and p_2. The values X_1 and X_2 are known and X_0 is to be predicted. It seems reasonable to give the values X_1 and X_2 equal weights, $a_1 = a_2$ and the prediction for X_0 is therefore:

$$\hat{X}_0 = \hat{a}_1 X_1 + \hat{a}_2 X_2 \qquad [13.7]$$

Using [13.5], one obtains:

$$\begin{bmatrix} a_1 \\ a_2 \end{bmatrix} = \begin{bmatrix} 1 & r_2 \\ r_2 & 1 \end{bmatrix}^{-1} \begin{bmatrix} r_1 \\ r_1 \end{bmatrix} \qquad [13.8]$$

where r_1 is the correlation coefficient for the distance $p_0 p_1 = p_0 p_2$ and r_2 is the correlation coefficient for the distance $p_0 p_2 = p_0 p_1$. It follows that:

$$\hat{a}_1 = \hat{a}_2 = \frac{r_1}{1 + r_2} \qquad [13.9]$$

For a series of discrete data, a value of x_k which is half way between X_{k-1} and X_{k+1} can be obtained from the model:

$$X_k = a(X_{k-1} + X_{k+1}) + \epsilon_k \qquad [13.10]$$

It can be proved that the term ϵ_k is an uncorrelated random value only if the series X_k ($k = 1, 2,...$) is a first-order Markov chain. Equation [13.10] is thus the equivalent expression to the unilateral time-series process:

$$X_k = CX_{k-1} + \epsilon_k \qquad [13.11]$$

and since it has the Markov property, one can write its autocorrelation function as:

$$r_\zeta = r_1^{|\zeta|} \qquad [13.12]$$

Equation [13.9] can be simplified by using [13.12], to give:

$$a = \frac{r_1}{1 + r_1^2} \qquad [13.13]$$

and it may be noted that the equivalent expression for the unilateral process is $C = r_1$, as can be seen from [7.27].

A continuous autocorrelation function can be formulated by letting the sample interval tend towards zero. From [13.10], one can derive a stochastic differential equation:

$$\frac{d^2 X}{d\zeta^2} \alpha^2 X_\zeta = \epsilon_\zeta \qquad [13.14]$$

where $\alpha^2 = (1 - 2a)/a$ and this leads to the autocorrelation function:

$$r_\zeta = (\alpha|\zeta| + 1) e^{-\alpha|\zeta|} \qquad [13.15]$$

The equivalent expression for the unilateral process has been noted already and it is:

$$r_\tau = e^{-\beta|\tau|} \qquad [13.16]$$

and the two expressions in [13.15] and [13.16] are clearly not the same. The Markov property of the bilateral process can be defined as:

$$P(X_\zeta|X_\delta; \delta \neq \zeta) = \lim_{\Delta\zeta \to 0} P(X_\zeta|X_{\zeta-\Delta\zeta}, X_{\zeta+\Delta\zeta}) \qquad [13.17]$$

whereas the unilateral process has been defined as:

$$P(X_t|X_\tau; \tau < t) = \lim_{\Delta t \to 0} P(X_t|X_{t-\Delta t}) \qquad [13.18]$$

The differential equation which was given in [13.14] can be generalized for the p-dimensional space and Whittle (1954) gives a general solution for the autocorrelation functions which arise from such processes. He considers particular solutions for the two-dimensional as well as for the three-dimensional process. The autocorrelation function of the two-dimensional process turns out to be:

$$r_\zeta = \alpha\zeta \, Y_1(\alpha\zeta) \qquad\qquad [13.19]$$

where $Y_1(\alpha\zeta)$ is a modified Hankel function, meaning a Bessel function of the second kind. The autocorrelation for the three-dimensional process gives the familiar exponential:

$$r_\zeta = e^{-\alpha|\zeta|} \qquad\qquad [13.20]$$

Whittle's results are of considerable importance in the interpretation of geological data. They suggest that if one makes the transition from the discrete regional processes to the continuous random processes, the exponential autocorrelation function (which is a condition for the Markov property) can be found only in three-dimensional processes. Empirical autocorrelation functions which were derived from geological data, have been sampled along straight line transects. For example, samples were taken at regular intervals and analysed for ore content in studies by Krige (1962) and Agterberg (1965, 1967). Such data were then treated exactly like a time series and a correlogram was calculated. It was found that the autocorrelation function of all the examples could be approximated by an exponential expression that is comparable with [13.20]. One must conclude in applying Whittle's results, that the random processes which were responsible for the mineral distribution must have operated in three dimensions. This is a reasonable assumption to make in the particular case of the mineral distribution. It should be noted that if the autocorrelation structure is exponential in the three-dimensional space, then it will also be exponential in any lower dimension, such as along the direction of transects which were used in these studies, for example. However, if one were to restrict the physical process model to a lower dimension then it could not provide an explanation of the observed distributions, according to Whittle's results. This restriction must be kept in mind when one proceeds to analyse areal variation by a series of independent transects, as was recommended in Section 13.3. A one-dimensional model could be quite inadequate to explain the observed correlation structure of a lithological variable.

Unfortunately, no quantitative work on the areal correlation structure of either sedimentological or stratigraphical data is available yet. This is in spite of it being likely that many of the features which are observed on bedding planes could have originated from processes which were essentially two-dimensional. Of the available data, those which come from ecological observations on vegetation patchworks are the most relevant. For example, Pielou (1964) showed that if an area which is populated by a single species that occurs in clusters, is classified into two states involving plants being either present or absent, then a transect which is taken in any direction and records the states that are encountered can be represented by a two-state Markov chain. It is very likely that the process which is responsible for the distribution of the plants is essentially two-dimensional and this appears to contradict Whittle's results. However, Switzer (1965) showed that there exists at least one discrete random process in the plane, which has the property that the states which are encountered along a straight-line transect have the properties of a Markov chain. Switzer's process will be referred to again in the next section.

The estimation of the areal autocorrelation is straightforward, providing that the data points are situated on a regular sampling grid. Under such circumstances, correlograms can be calculated along straight-line transects for any direction in the plane. Naturally, if the areal distribution of the correlation structure is isotropic, a single correlogram is all that needs to be calculated. The correlogram can be represented graphically by contours in an x,y coordinate system, the origin of which represents the zero lag position. The isotropic case, for example, can be shown by circular contours which normally decline in value from the centre of the circle, which has the maximum correlation coefficient of 1. Most anisotropic correlation structures will yield ellipsoidal contour lines.

Difficulties arise in the calculation of correlation functions when the data are obtained from irregularly spaced sample points. The problem of estimating the autocorrelation under such circumstances, was discussed by Agterberg (1970) who showed that at least an approximation of the autocorrelation structure can be obtained.

13.5 THE LITHOLOGICAL VARIATION IN STRATIGRAPHIC PROFILES

The preceding sections dealt with problems of three-dimensional rock bodies which were visualized as being a sequence of superimposed maps. In addition to this, a lithosome can be studied in a series of vertical slices which are the familiar stratigraphical profiles, of course. If one were to deal only with the geometrical distribution of rock types, there would be no difference between these two methods. However, the map represents only a short moment in the environmental history which is represented by a single time horizon. The profile, on the other hand, records geographical variation along its horizontal axis and its vertical axis measures the thickness of the sediments and is also a time axis, at least by implication.

   Whether the vertical axis should be divided into equal thickness or equal
time intervals, is a matter of choice. The profiles which result from these two
alternatives are different and they cannot be converted into each other unless both of the scales are given. From a purely practical point of view, it is
found that thickness profiles with added time lines are much more easily
read than the time profiles with added thickness information. However, the
latter are useful in the study of facies variation, where the thickness of the
strata can be ignored to some extent and time-scale profiles have been used
in simulation studies as well.

   The type of information that is needed for constructing a thickness profile
with added time lines, may be seen from Fig. 13.1. The two graphs on the
left-hand side of the figure give firstly, bed by bed, the length of time for
which a certain sedimentation process lasted and secondly, the rate with
which this process proceeded in each bed. The thickness of the beds is found
by multiplying the sedimentation rate by the time taken for the formation
of that particular bed. To keep the example as simple as possible, the sedimentation rates in Fig. 13.1 are chosen so that each bed attains a constant
thickness. Next, the time lines are constructed by multiplying the index of
the time line by the appropriate sedimentation rate for each locality. It may
be noted that the second time line enters bed 2 twice, for instance, and it is
the sedimentation rates for this particular bed which are relevant to finding
the position of the time line. The assumption of absolutely continuous sedimentation has been made in constructing the thickness profile that is shown
in the upper right-hand side of Fig. 13.1. This is a most unlikely situation, as
has been discussed at some length in Chapter 2, because petrographic studies
suggest that a break of deposition occurs after a bed has been formed and

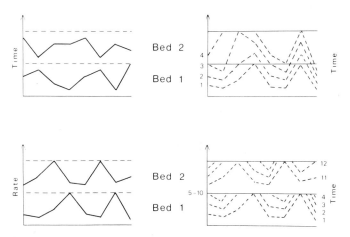

Fig. 13.1. The construction of a profile with time lines. Broken lines on the right show
time stages.

discontinuous sedimentation appears to be a more natural model. This more realistic situation is shown in the lower right-hand side of Fig. 13.1. Time lines have been constructed in this figure under the assumption that at least five time units elapse after the completion of bed 1, before the second bed is deposited. Under these circumstances, the time lines merge with the bedding plane which becomes an isochronous surface by this process, eventually. It is clear from the method of constructing the thickness profile and its time lines, using the data that are given on the left-hand side of Fig. 13.1, that this process is not reversible. As in all the previous examples, the thickness of the sediments is given by the product of the sedimentation rate and the period of time over which the sedimentation occurred. It is natural that in constructing a process model, one is interested mainly in the sedimentation rates and in the duration of the sedimentation process, both variables being dependent on the geographical position. Any model attempting to explain an observed profile must make some assumptions about the sedimentation rates, the length of time for which a particular sedimentation process lasted and the regional distribution of these two factors. One can think of various geological situations where either the sedimentation rates or the duration of the sedimentation processes are the main interest. In high-energy fluviatile environments, for example, one might think of sedimentation as consisting of a patchwork of areas where deposition occurs over different and variable periods of time, but with approximately constant rates for each facies. In basinal sedimentation, on the other hand, the time-dependent factors may be constant over large areas but the sedimentation rates may vary regionally either at random, or following some systematic trend. Different models must take such geological peculiarities into account.

The following specific example has been chosen to illustrate some other points that arise from the design of models which generate geological profiles. Assume that some random-walk-like mechanism exists to produce exponentially distributed sedimentation steps. This situation was discussed in Section 6.3 and may arise when sedimentation is steady, apart from irregular and exponentially distributed interruptions which are followed by a certain time of non-deposition. In this situation, the thickness of the sediment that is deposited in unit time can be expressed by a Poisson process with, let us say, density ρ. Let it be assumed further that at regular time intervals, sedimentation changes occur so that some marker horizons such as bedding planes, are formed. The incidents producing the bedding planes are also distributed exponentially with density $\lambda(\lambda < \rho)$. This sedimentation model was considered in Sections 6.2 and 11.4 and it was seen from [11.11] that the thicknesses of the beds which are measured in a single section are exponentially distributed. The model may be expanded into two dimensions by considering a string of localities that are arranged along a straight line and by assuming that each locality generates its own section. If sedimentation proceeds quite independently in each locality, then both the sedimentation

A   B

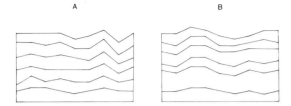

Fig. 13.2. Two artificial profiles based on independent random sedimentation (A) and synchronous random sedimentation (B).

steps occurring with density ρ and the incidents which lead to bedding planes occurring with density λ, will happen everywhere at different times so that they have only their distribution laws in common. This would result in the type of profile which is shown in Fig. 13.2A. The thickness distribution of beds which are measured in individual sections, is the distribution of [11.11] and the thickness distribution which is obtained by measuring the same bed in different localities, is identical to the vertical distribution of single sections.

This independence of different localities is very unlikely from a geological point of view since it implies that there are no regional dependencies within such an environment. A much more realistic model can be obtained by assuming that the incidents producing the bedding planes occur almost simultaneously over the distance covered by the profile. For example, one may assume that the time events which subdivide the section are caused by the influx of clastic material into a predominantly carbonate environment and that this spreading is relatively rapid, in comparison with the vertical sedimentation rates. Modifying the model in this way does not alter the bed-thickness distribution as it is measured in individual sections, but it does change the thickness distribution of an individual bed that is measured in different localities. One and the same bed now represents a time interval which is constant in every locality and therefore its thickness must be determined by the compound Poisson distribution that was introduced in [6.26] and [6.27]. The effect of bedding planes occurring simultaneously, may be seen in the short simulated section which is illustrated in Fig. 13.2B. The difference between the first model (A) and the second model (B) is obvious. When sedimentation is independent (model A), the beds cannot be correlated by matching the thicknesses of beds from different exposures but this can be done for model B. When model A operates, all the mean values which are measured for an identical bed in different localities will be the same. When model B operates, the averages which are derived from the different localities do permit the reconstruction of the function q_T (see [11.8]) and one can estimate at least, the relative time history of the sedimentation process. Use was made of this argument when discussing the example in Section 11.4.

It has been pointed out that model A is unrealistic because of its independence of regional variation. However, the same is true for model B to a lesser extent. The latter assumes that the minor fluctuations which can be described by parameter ρ are completely independent of the geographical location, whereas the incidents causing bedding planes were taken to operate with deterministic certainty wherever they occurred throughout the region. According to what has been discussed already, it seems more likely that sedimentation processes are regionally autocorrelated on a minor scale and that incidents such as the influx of clastic sediments are stochastically disturbed in their regional pattern. Realistic models for generating geological profiles should combine the stochastic processes that are responsible for the vertical sequence with the stochastic processes that are responsible for the lateral variation but double stochastic models lead to difficulties, as was seen in the previous section. A simple example will illustrate this.

The Markov-chain model has proved very popular in the description of vertical sections and so one might consider generating stratigraphic profiles by placing a great number of sections having the Markov property, next to each other. It can be seen immediately from Fig. 13.3 that such Markov chains can only exist next to each other if they are completely independent. The first section can be taken as an example. The state which occupies cell (2,1) should depend only on the content of the cell (1,1). Similarly, the state which occupies cell (2,2) should depend only on the state of (1,2). Next, assume that the Markov property exists in the horizontal distribution of the states and that for example, the series (1,1), (1,2), (1,3) ... constitutes a Markov chain. This must lead to the conclusion that state (1,2) depends on (1,1) and that state (2,2) depends on both state (1,2) and state (2,1), which is a contradiction of the Markov property. Of course, this was brought out by Whittle's results which were considered in the previous section, where it was mentioned also that the only simple two-dimensional Markov process that is known so far, is the Switzer process. Switzer (1965) showed that if an area is subdivided by a series of random lines to produce a Poisson field with density $\lambda \mathrm{d}p\mathrm{d}\theta$, where p and θ are the coordinates which are used for defining the lines, then a number of convex cells result. If the states occupying the cells are chosen at random and independently, from a two-state system, then the resulting pattern that is shown in Fig. 13.4 has the property that any straight-line transect through this area constitutes a Markov chain. A model of this type is by no means as unrealistic as it may appear to be from the diagram which is given in Fig. 13.4, where the density of the random lines was kept relatively low. If the concept of sedimentary "quanta", which was used primarily to describe stratigraphic time units, is expanded into two dimensions, then it might be possible to identify such cells as real sedimentological units but more work will be needed before this can be attempted.

Although the Markov process provides a simple model for random variation, other schemes such as the higher-order autocorrelated processes, prove

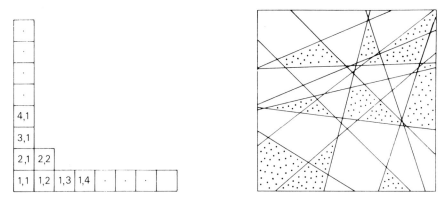

Fig. 13.3. The combination of a vertical and horizontal Markov chain.

Fig. 13.4. Realization of the Switzer process.

to be important in the study of vertical variation and these must be combined with the lateral variation processes which themselves might be of similar complexity. Not only are there no data for such studies but also there is no complete theory for two- and three-dimensional random processes. It is not surprising that the models which have been developed for generating geological profiles are as yet very simple and have not yielded very exciting results.

13.6 MODELS FOR SIMULATING GEOLOGICAL PROFILES

A model which is particularly instructive in connection with the problems which have just been discussed, was developed by Krumbein (1968). The model is based on the principle of transgression and regression "cycles", where it is assumed that a marker horizon such as a beach deposit, for example, moves backwards and forwards within the studied profile. The sedimentation process is studied by observing the position of the marker horizon at various fixed localities, such as the sections A, B, C ... in Fig. 13.5. One can say that the system is in state A at a given time, when the marker horizon is recorded in section A. It is assumed that sedimentation is steady and constant so that the thickness of a deposited bed can be equated directly with time. To give a full description of the strand-line movement, one must indicate whether a given state is approached during the trangressive or the regressive phase of the cycle. If it is assumed in Fig. 13.5 for example, that the sea is on the left-hand side and the land is on the right-hand side, then states A, B, ... are all part of the transgressive phase and states A', B' ... are all part of the regressive phase. One can assume further that the system operates at regular time intervals and this enables one to write a transition

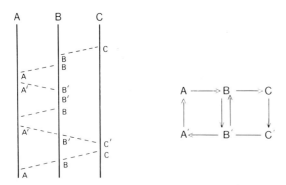

Fig. 13.5. Transgression and regression cycles simplified after Krumbein (1968).

matrix for the position of the marker horizon. For example, if the transgression—regression cycle is entirely deterministic, meaning that if the strand line migrates from A to B to C and returns to A, then the following cyclical matrix describes the process:

$$
\mathbf{P} =
\begin{array}{c c}
& \begin{array}{c c c c c c} A & B & C & C' & B' & A' \end{array} \\
\begin{array}{c} A \\ B \\ C \\ C' \\ B' \\ A' \end{array} &
\left[\begin{array}{c c c c c c}
0 & 1 & 0 & 0 & 0 & 0 \\
0 & 0 & 1 & 0 & 0 & 0 \\
0 & 0 & 0 & 1 & 0 & 0 \\
0 & 0 & 0 & 0 & 1 & 0 \\
0 & 0 & 0 & 0 & 0 & 1 \\
1 & 0 & 0 & 0 & 0 & 0
\end{array}\right]
\end{array}
\qquad [13.21]
$$

The deterministic model specifies that whenever one state is reached the marker horizon must move to the next state in the following time increment. A stochastic element can be introduced by allowing a certain probability that the strand line remains in the same position, for a variable amount time. In the deterministic model of [13.21], the full transgression must completed before a regression can occur and a further stochastic eleme. which was introduced by Krumbein, permits a reversal of the transgression cycle at any intermediate state. If only three sections are considered, as in the example above, then the reversals can occur not only at A and C but also at B. This is indicated in the diagram of the possible transitions which is shown on the right-hand side of Fig. 13.5. The process is Markovian and a typical transition matrix could be written as:

$$
\mathbf{P} = \begin{array}{c} \\ A \\ B \\ C \\ C' \\ B' \\ A' \end{array}
\begin{array}{cccccc}
A & B & C & C' & B' & A' \\
\left[\begin{array}{cccccc}
0.02 & 0.98 & 0 & 0 & 0 & 0 \\
0 & 0.02 & 0.78 & 0 & 0.20 & 0 \\
0 & 0 & 0.02 & 0.98 & 0 & 0 \\
0 & 0 & 0 & 0.02 & 0.98 & 0 \\
0 & 0.20 & 0 & 0 & 0.02 & 0.78 \\
0.98 & 0 & 0 & 0 & 0 & 0.02
\end{array} \right]
\end{array}
\qquad [13.22]
$$

The example assumes that the rates of transgression and regression movement are equal but it would be a simple matter to write the transition matrix in such a way that the transgression is more rapid than the regression. This would lead to more "natural looking" simulation results.

The lateral and vertical variation patterns which result from the profiles that are generated by this model, are of interest. The first important observation is about the lateral distribution of the lithologies. Since the model is based on the concept of migrating parallel facies belts, the order of these lithologies cannot be disturbed and consequently, the sequence of rock types along any time plane is entirely deterministic. For example, if one associates rock type 1 with marine sediments, rock type 2 with the littoral sands and rock type 3 with terrestrial sediments, then the sequence of the rock types always must be 1, 2, 3, in going from the left to the right-hand side along any horizontal plane. It follows that the entire profile can be constructed if the rock sequence is known for any of the single sections. Section B may be considered as an example and if the system is in state A, then rock type 3 will be deposited in that section. The states can be equated with the rock types in the following way. $A = A' = 3, B = B' = 2, C = C' = 1$. The matrix in [13.22] can be rearranged in order to bring out the rock sequence which is generated in section B (see [13.23] on the following page). This is still a Markov matrix, of course, but the 3×3 matrix referring to the succession of the lithologies is not Markovian and the stratigraphical section can be described by a Markov chain, only if it can be decided on lithological grounds, whether a sediment is formed during the transgressive or the regressive phase of a cycle. The stratigraphic sequence of undifferentiated lithologies can be modelled by a semi-Markov process and for this to be achieved, it is only necessary to find the distribution of the waiting times for each state. These may be obtained from [13.23]. It can be seen that the

		1	1	2	2	3	3
		C	C'	B	B'	A	A'
1	C	0.02	0.98	0	0	0	0
1	C'	0	0.02	0	0.98	0	0
2	B	0.78	0	0.02	0.20	0	0
2	B'	0	0	0.20	0.02	0	0.78
3	A	0	0	0.98	0	0.02	0
3	A'	0	0	0	0	0.98	0.02

$$S = 2 \qquad\qquad\qquad\qquad\qquad\qquad\qquad\qquad\qquad [13.23]$$

waiting time for state 1 for instance, is determined by the length of any chain which consists only of state C or C'. However, the probability of such an uninterrupted chain of C's having a length of n, is given by the first passage probability of state B to B', since these two states always terminate such chains. It follows from the arguments in Section 5.4, that one can write:

$$p_{11}^{(n)} = f_{BB'}^{(n)} = p_{BB'}^{(n)} - \sum_{v=1}^{n-1} f_{BB'}^{(n-v)}\, p_{BB'} \qquad\qquad [13.24]$$

Equivalent expressions can be found for all nine submatrices of [13.23] and after the appropriate calculations, it should be possible to find a matrix $\mathbf{P}(n)$ which provides a complete model for any section that is selected in this system.

An examination of Krumbein's model shows that an observed Markov sequence of lithologies is incompatible with the concept of transgressions and regressions, even if the transgression pattern follows a Markov process. The limitations of this model are caused largely by the strictly deterministic relationships existing between the vertical and the lateral variations, which have been discussed already as "Walther's law" in Section 12.6. If more flexible models are to be achieved, a search must be made for processes which not only permit random variations in the vertical direction of the profile but in its horizontal direction also.

Jacod and Joathon (1971, 1972) constructed some models which were intended to be the basis for simulation studies which were designed to investigate an actual stratigraphic problem as well as proving process models for some rather complex sedimentation environments. The basic assumptions in the Jacod-Joathon model are simple. Sedimentation is a non-decreasing function $Q_t(x)$ which depends on the time t and on the geographical location, x. The thickness of the sediment which is deposited during the time interval Δt

is proportional to:

$$\Delta Q_t(x) = Q_{t+\Delta t}(x) - Q_t(x) \qquad [13.25]$$

Sedimentation is a function of the water depth W_t and the amount of crustal sinking R_t and it can be expressed as:

$$dW_t(x) = dR_t - f[W_t(x)] \, dQ_t(x) \qquad [13.26]$$

which is equivalent to [3.21]. The function $f[W_t(x)]$ indicates the dependence of sedimentation on the water depth and this could be determined by the amount of sediment dispersion, as mentioned in Section 3.7. It is assumed in addition, that the environmental factors R_t and $Q_t(x)$ are random functions. The tectonic component R_t is represented by a random walk, whereby both the step size and the intervals between the steps are distributed exponentially with the parameter γ. In addition to the discontinuous tectonic movement, there may be an additional constant subsidence which proceeds with speed β.

The specification of the variable $Q_t(x)$ is rather more difficult. Not only is it dependent on time of course, but also on the geographical location x. In order to generate such a variable, Jacod and Joathon (1971) employ a stratagem which is based on the following scheme. The two horizontal axes of a Cartesian system are used to represent the geographical location while the vertical axis represents the time. The horizontal plane is scattered with Poisson-distributed points of a given density. Each point is in the centre of a disc-shaped body which is flat on its base and its upper surface is formed by a parabolic cap. Each time interval Δt is represented by a new map of Poisson-distributed bodies. When the scheme is viewed in a two-dimensional profile, the disc-shaped bodies appear in section as they are shown in Fig. 13.6 and the distribution of such cross-sections is again Poisson. Each of the randomly distributed bodies is used to define a function $G_i(x)$ which will determine the sedimentation $Q_t(x)$. The following rules are used. Take a point where $x = a$, in Fig. 13.6, then the perpendicular line which is erected in point a, will represent the time history of this location. It is assumed that when the line does not pass through a disc, $dQ_t(a) = dt$, but when it passes through disc number i, $dQ_t(a) = G_i(a)$. A typical realization of such a process is shown on the right-hand side of Fig. 13.6, where it is assumed that Q_t carries out a jump which is proportional to the intercept of the time path passing through any disc.

The scheme which has been outlined in the preceding paragraph can be modified in several ways and a typical application of the Jacod-Joathon model is to simulate the lenticular sedimentation of a high-energy environment. For example, one can specify that the functions $G_i(x)$ are associated with the deposition of sand, whereas clay may be deposited in between the

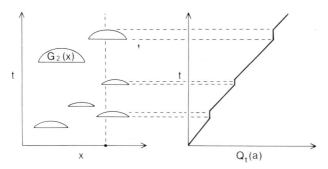

Fig. 13.6. The Jacod-Joathon model of sedimentation.

discrete bodies. If an environment like this is subjected to further irregular tectonic movement which affects the sedimentation rates according to [13.26], then one can simulate some very realistic profiles. Indeed, Jacod and Joathon (1972) have shown that by selecting a limited number of parameters and by comparing the simulation results with actual field measurements, the model can produce results which are in very good agreement with the observed structures. Simulated geological structures can be used in the study of practical applications like the gas or the ground-water flow in complex geological bodies and actual field measurements can be used to either confirm or improve the quality of the model. At the same time, it must be admitted that the work of Jacod and Joathon throws very little light on the genesis of random processes which operate regionally. It is assumed that the function $Q(x)$ is constant during a short time interval Δt and that it is determined only by the water depth of the basin at that moment. This state of average sedimentation is disturbed regionally by a number of randomly distributed centres which give rise to local lens formation. Sedimentation in localities which are close to each other is not independent because each lens has a finite size but the model does not allow for the interaction of overlapping lenses. Indeed, it is assumed that the maximum diameter of the lenses is smaller than the distance between the sample points. Therefore the model does not give any information which might be used to investigate why and how lenses form and our ignorance of this process cannot be disguised by the arbitrary assumption that the centres of such disturbances are distributed in an independent and random manner.

13.7 THE ENVIRONMENTAL MODEL

The environmental model was introduced originally in Section 1.3 as a system of factors which are linked together by functional relationships. Perhaps it is wishful thinking to attempt to replace this whole complex of physical activity by a system of mathematical expressions which simulate the environmental processes and which will generate a three-dimensional body of sediments, comparable to stratigraphic systems in nature. One brave attempt in this direction was made by Harbaugh (1966) who compiled a computer programme which can simulate the sedimentation processes occurring in a basin which contains various sediment-producing organisms and which receives clastic sediments from various directions. An attempt is made to show how organic growth, the distribution of sediments and the geometry of the basin interact to provide various sedimentary bodies. Although this model is vastly simplified, it can simulate quite complicated rock distributions.

The main lesson to be learned from Harbaugh's attempt is possibly two-fold. On the one hand, it can be demonstrated that even quite simple models can produce some very realistic three-dimensional sediment bodies. On the other hand, the model highlights once again how little is known about the sedimentation process. Most of the functional relationships, such as the dependence of organic development on external factors and the dependence of sedimentation on the water depth or the distance from the land, as well as other relationships, are known incompletely. All the initial conditions such as the geometry of the basin, the tectonic framework and the influx of the external sediment have to be guessed at before the model can operate. If the model incorporates yet more random processes to make it more realistic, then the number of unknown parameters increases correspondingly.

The question which arises is whether an attempt should be made to make the models ever more realistic and to develop well-fitting empirical models, as discussed in Section 3.4, or whether one should concentrate on the theory of environmental processes. It appears that the latter approach can provide the more meaningful explanations, but it is hampered seriously by the complexity of the processes being investigated. However, theoretical models can deal with simplified situations or situations concerned with definite aspects of sedimentation. For example, Krumbein's work which was considered in Section 13.4 showed that the relationship between lateral and vertical rock variation needs urgent investigation. Very little is known yet, about the random processes which affect the areal variation of sedimentological parameters and even less is known about the compatibility of various stochastic schemes in three dimensions. It is possible that quite simple theoretical models combined with detailed stratigraphical mapping and sedimentological research may help towards a better understanding of the space—time relationships in sediments. It is only after the basic problems of environmental interaction have been understood, that one can hope to approach complex operations like stratigraphic interpretation or stratigraphic correlation, from a quantitative point of view.

References

Agterberg, F.P., 1965. The technique of serial correlation applied to continuous series of element-concentration values in homogeneous rocks. *J. Geol.*, 73: 142—154.

Agterberg, F.P., 1966. The use of multivariate Markov schemes in geology. *J. Geol.*, 74: 764—785.

Agterberg, F.P., 1967. Mathematical models in ore evaluation. *J. Can. Oper. Res. Soc.*, 5: 144—158.

Agterberg, F.P., 1970. Autocorrelation functions in geology. In: D.F. Merriam (Editor), *Geostatistics, a Colloquium*. Plenum, New York, pp. 113—142.

Agterberg, F.P. and Banerjee, I., 1969. Stochastic model for the deposition of varves in glacial lake Barlow—Ojibway, Ontario, Canada. *Can. J. Earth Sci.*, 6: 625—652.

Alexander, W.H., 1961. *Elements of Mathematical Statistics*. Wiley, New York, 367 pp.

Allen, J.R.L., 1965. A review of the origin and characteristics of recent alluvial sediments. *Sedimentology*, 5: 89—191.

Anderson, R.Y. and Kirkland, D.W., 1960. Origin, varves and cycles of Jurassic Todilto Formation, New Mexico. *Am. Assoc. Pet. Geol. Bull.*, 77: 241—256.

Anderson, R.Y. and Koopmans, L.H., 1963. Harmonic analysis of varve time series. *J. Geophys. Res.*, 68: 877—893.

Anderson, R.Y. and Koopmans, L.H., 1968. Statistical analysis of the Rita Blanca varve time series. In: R.Y. Anderson and D.W. Kirkland (Editors), *Paleoecology of an Early Pleistocene Lake on the High Plains of Mexico. Geol. Soc. Am. Mem.*, 113: 59—75.

Anderson, T.S. and Goodman, L.A., 1957. Statistical inference about Markov chains. *Am. Math. Stat.*, 28: 89—110.

Barrel, J., 1917. Rhythms and the measurement of geologic time. *Bull. Geol. Soc. Am.*, 28: 745—904.

Bartlett, M.S., 1946. On the theoretical specifications and sampling properties of auto-correlated time-series. *Suppl. J.R. Stat. Soc.*, 8: 27—85.

Beerbower, I.R., 1964. Cyclothems and cyclic depositional mechanisms in alluvial plain sedimentation. *Kans. Geol. Surv., Bull.*, 169 (1): 31—42.

Bendat, I.S. and Piersol, A.G., 1966. *Measurements and Analysis of Random Data*. Wiley, New York, 390 pp.

Bharucha-Reid, A.T., 1960. *Elements of the Theory of Markov Processes and Their Applications*. McGraw Hill, New York, 468 pp.

Billingsley, P., 1961. Statistical methods in Markov chains. *Am. Math. Stat.*, 32: 12—40.

Bonham-Carter, G.F., 1972. Optimization criteria for mathematical models used in sedimentology. In: D.F. Merriam (Editor), *Mathematical Models of Sedimentary Processes*. Plenum, New York, 271 pp.

Bonham-Carter, G.F. and Sutherland, A.J., 1968. Mathematical model and FORTRAN IV program for computer simulation of deltaic sedimentation. *Kans. Geol. Surv., Comput. Contrib.*, 24: 1—56.

Bott, M.H.P., 1964. Formation of sedimentary basins by ductile flow of isostatic origin in the upper mantle. *Nature*, 201: 1082—1084.

Briggs, L.I. and Pollack, H.N., 1967. Digital model of evaporite sedimentation. *Science*, 155: 453—456.

Bubnoff, S., 1947. Rhythmen, Zyklen und Zeitrechnung in der Geologie. *Geol. Rundsch.*, 35: 6—22.

Bubnoff, S., 1950. Die Geschwindigkeit der Sedimentbildung und ihr endogener Antrieb. *Misc. Acad. Berolinensia.*, 1—32.

Burnaby, T.P., 1953. A suggested alternative to the correlation coefficient for testing the significance of agreement between pairs of time series and its application to geological data. *Nature*, 172: 210—211.

Campbell, V.C., 1967. Lamina, laminaset, bed and bedset. *Sedimentology*, 8: 7—26.

Carr, D.D., Horowitz, A., Hrabar, S.V., Ridge, K.F., Rooney, R., Straw, W.T., Webb, W. and Potter, P.E., 1966. Stratigraphic sections, bedding sequences and random processes. *Science*, 154: 1162—1164.

Carrs, B.W., 1967. In search of geological cycles using a technique from communication theory. *Kans. Geol. Surv., Comput. Contrib.*, 18: 51—56.

Carrs, B.W. and Neidell, N.S., 1966. A geological cyclicity detected by means of polarity coincidence correlation. *Nature*, 212: 136—137.

Churchill, R.V., 1958. *Operational Mathematics.* McGraw Hill, New York, 337 pp.

Cotter, D.C., 1968. *A study of Some Cyclic Characteristics of a Limestone—Shale Section in the Visean of N.W. Ireland.* Unpubl. thesis, Queen's University, Belfast, 222 pp.

Cox, D.R., 1962. *Renewal Theory.* Methuen, London, 142 pp.

Cox, D.R. and Lewis, P.A.W., 1966. *The Statistical Analysis of Series of Events.* Methuen, London, 285 pp.

Cox, D.R. and Miller, H.D., 1965. *The Theory of Stochastic Processes.* Wiley, New York, 398 pp.

Dacey, M.F. and Krumbein, W.C., 1970. Markovian models in stratigraphy. *J. Int. Assoc. Math. Geol.*, 2: 175—191.

Davis, H.T., 1933. *Tables of the Higher Mathematical Functions, 1.* Principia Press, Bloomington.

Davis, J.C. and Cocke, I.M., 1972. Interpretation of complex successions by substitutability analysis. In: D.F. Merriam (Editor), *Mathematical Models of Sedimentary Processes.* Plenum, New York, 271 pp.

Davis, J.C. and Sampson, R.J., 1967. FORTRAN II time trend package for the IBM 1620 computer. *Kans. Geol. Surv., Comput. Contrib.*, 19: 1—28.

De Raaf, J.F., Reading, H.G. and Walker, R.G., 1965. Cyclic sedimentation in the Lower West Phalien of North Devon, England. *Sedimentology*, 4: 1—52.

Dienes, I., 1974. General formulation of the correlation problem and its solution in two special situations. *Int. Assoc. Math. Geol.*, 6: 73—81.

Doveton, J.H., 1971. An application of Markov chain analysis to the Ayrshire coal measures succession. *Scott. J. Geol.*, 7: 11—27.

Duff, P.McL.D. and Walton, E.K., 1962. Statistical basis for cyclothems: a quantitative study of the sedimentary succession in the east Pennine coalfield. *Sedimentology*, 1: 235—255.

Duff, P.McL.D., Hallam, A. and Walton, E.K., 1967. *Cyclic Sedimentation.* Elsevier, Amsterdam, 280 pp.

Elias, M.K., 1964. Depth of Late Paleozoic sea in Kansas and its megacyclic sedimentation. *Kans. Geol. Surv., Bull.*, 169(1): 87—106.

Elliott, R.E., 1961. The stratigraphy of the Keuper series in southern Nottinghamshire. *Proc. Yorks. Geol. Soc.*, 33: 197—234.

Elliott, R.E., 1968. Facies, sedimentation successions and cyclothems in productive coal measures in the East Midlands, Great Britain. *Mercian Geol.*, 2: 351—371.

Emiliani, C., 1955. Pleistocene temperatures. *J. Geol.*, 63: 538—578.

Emiliani, C. and Geiss, I., 1957. On glaciations and their causes. *Geol. Rundsch.*, 46: 576—601.

Feller, W., 1957. *An Introduction to Probability Theory and its Applications, 1*. Wiley, New York, 461 pp.

Fiege, K., 1937. Untersuchungen über zyklische Sedimentation geosynklinaler und epikontinentaler Räume. *Abh. Preuss. Geol. Landesanst.*, 177: 1—128.

Fiege, K., 1952. Sedimentationszyklen und Epirogenese. *Z. Dtsch. Geol. Ges.*, 103: 17—22.

Fischer, A.G., 1964. The Lofer cyclothems of the Alpine Triassic. *Kans. Geol. Surv., Bull.*, 169 (1): 107—149.

Fischer, A.G., 1969. Geological time—distance rates: the Bubnoff unit. *Geol. Soc. Am. Bull.*, 80: 549—552.

Fournier, F., 1960. *Climat et érosion: la relation entre l'érosion du sol par l'eau et les précipitations atmosphériques*. Univ. Paris, 201 pp.

Fox, W.T., 1964. FORTRAN and FAP program for calculating and plotting time-trend curves using an IBM 7090 or 7094/1401 computer system. *Kans. Geol. Surv., Spec. Dist. Publ.*, 12: 1—24.

Gabriel, K.R. and Neumann, J., 1957. On a distribution of weather cycles by length. *Q. J. R. Meteorol. Soc.*, 83: 375—380.

Gingerich, P.D., 1969. Markov analysis of cyclic alluvial sediments. *J. Sediment. Petrol.*, 39: 330—332.

Granger, C.W.J. and Hatanaka, M., 1964. *Spectral Analysis of Economic Time Series*. Princeton University Press, Princeton, 299 pp.

Haites, T.B., 1963. Perspective correlation. *Bull. Am. Assoc. Pet. Geol.*, 47: 553—574.

Hannan, E.J., 1960. *Time Series Analysis*. Methuen, London, 152 pp.

Hannan, E.J., 1961. Testing for a jump in the spectral function. *J. R. Stat. Soc.*, B 23: 394—403.

Harbaugh, J.W., 1966. Mathematical simulation of marine sedimentation with IBM 7090/7094 computers. *Kans. Geol. Surv., Comput. Contrib.*, 1: 1—52.

Harbaugh, J.W. and Bonham-Carter, G.F., 1970. *Computer Simulation in Geology*. Wiley, New York, 575 pp.

Harbaugh, J.W. and Merriam, D.F., 1968. *Computer Applications in Stratigraphic Analysis*. Wiley, New York, 282 pp.

Hawkins, D.M. and Merriam, D.F., 1973. Optimal zonation of digitized sequential data. *J. Int. Assoc. Math. Geol.*, 5: 389—395.

Heirtzler, J.R., 1968. Sea-floor spreading. *Sci. Am.*, 219. 60—70.

Hjulström, F., 1939. Transportation of detritus by moving water. In: P.D. Trask (Editor), *Recent Marine Sediments. Soc. Econ. Paleontol. Mineral. Spec. Publ.*, 4: 5—31.

Holmes, A., 1965. *Principles of Physical Geology*. Nelson, London, 1288 pp.

Hull, E., 1862. On iso-diametric lines, as means of representing the distribution of clay and sandy strata as distinguished from calcareous strata, with special reference to the Carboniferous rocks of Britain. *Q. J. Geol. Soc. London*, 18: 127—146.

Jacod, J. and Joathon, P., 1971. Use of random genetic models in the study of sedimentary processes. *J. Int. Assoc. Math. Geol.*, 3: 265—279.

Jacod, J. and Joathon, P., 1972. Conditional simulation of sedimentary cycles in three dimensions. In: D.F. Merriam (Editor), *Mathematical Models of Sedimentary Processes*. Plenum, New York, 271 pp.

Jenkins, G.M., 1961. General considerations in the analysis of spectra. *Technometrics*, 3: 133—166.

Jessen, W., 1956. Das Ruhrkarbon (Namur Cob.—Westfal C) als Beispiel für extratellurisch verursachte Zyklizitätserscheinungen. *Geol. Jahrb.*, 71: 1—20.

Jopling, A.V., 1964. Interpreting the concept of the sedimentation unit. *J. Sediment. Petrol.*, 34: 165—172.

Jopling, A.V., 1966. Some applications of theory and experiment to the study of bedding genesis. *Sedimentology*, 7: 71—102.

Karlin, S., 1966. *A First Course in Stochastic Processes*. Academic Press, New York, 502 pp.

Kendall, M.G., 1946. *The Advanced Theory of Statistics, II*. Griffin, London, 521 pp.

Kendall, M.G. and Buckland, W.R., 1960. *A Dictionary of Statistical Terms*. Oliver and Boyd, Edinburgh.

Kendall, M.G. and Stuart, A., 1968. *The Advanced Theory of Statistics, 3*. Griffin, London, 2nd ed., 557 pp.

Kheiskanen, K.I., 1964. Some sedimentological dynamic features of the Middle and Upper Yatulian basin in central Karelia. *Sov. Geol.*, 1964 (12): 58.

Kolmogorov, A.N., 1951. Solution of a problem in probability theory connected with the problem of the mechanism of stratification. *Am. Math. Soc. Transl.*, No. 53: 171—177.

Koopmans, L.H., 1967. A comparison of coherence and correlation as measures of association for time or spacially indexed data. *Kans. Geol. Surv., Comput. Contrib.*, 18: 1—4.

Korn, H., 1938. Schichtung und absolute Zeit. *Neues Jahrb. Mineral., Abh.*, 74: 50—188.

Krige, D.G., 1962. Statistical application in mine valuation. *J. Inst. Min. Surv. S. Afr.*, 12: 45—95.

Krige, D.G., 1966. Two-dimensional weighted moving average trend surface for ore valuation. *Symp. Math. Stat. Comput. Appl. Ore Valuation, Johannesburg, 1966, Proc.*, pp. 13—33.

Krinsley, D.F. and Funnel, B.M., 1965. Environmental history of quartz sand grains from the Lower and Middle Pleistocene of Norfolk, England. *Q. J. Geol. Soc. London*, 121: 435—461.

Krumbein, W.C., 1967. FORTRAN IV computer programs for Markov chain experiments in geology. *Kans. Geol. Surv., Comput. Contrib.*, 13: 1—38.

Krumbein, W.C., 1968. FORTRAN IV computer program for simulation of transgression and regression with continuous-time Markov models. *Kans. Geol. Surv., Comput. Contrib.*, 26: 1—38.

Krumbein, W.C. and Dacey, M.F., 1969. Markov chains and embedded Markov chains in geology. *J. Int. Assoc. Math. Geol.*, 1: 79—96.

Krumbein, W.C. and Graybill, F.A., 1965. *An Introduction to Statistical Models in Geology*. McGraw Hill, New York, 475 pp.

Krumbein, W.C. and Scherer, W., 1970. Structuring observational data for Markov and semi-Markov models in geology. *Office U.S. Naval Res., Tech. Rep.*, 15, 59 pp.

Krumbein, W.C. and Sloss, L.L., 1963. *Stratigraphy and Sedimentation*. Freeman, San Francisco, 2nd ed., 660 pp.

Krynine, P.D., 1942. Differential sedimentation and its products during one complete geo-synclinal cycle. *An. Congr. Panam. Ing. Minas Geol., Santiago*, 1942: 536—561.

Kuenen, Ph.H., 1950. *Marine Geology*. Wiley, New York, 551 pp.

Kulinkvich, A.Y., Sokranov, N.N. and Churinova, I.M., 1966. Utilization of digital computers to distinguish boundaries of beds and identify sandstones from electric log data. *Int. Geol. Rev.*, 8: 416—420.

Lanczos, C., 1957. *Applied Analysis*. Pitman, London, 539 pp.

Lumsden, D.N., 1971. Markov chain analysis of carbonate rocks: applications limitations and implications as exemplified by the Pennsylvanian system in Southern Nevada. *Geol. Soc. Am. Bull.*, 82: 447—462.

Mann, C.J., 1967. Spectral-density analysis of stratigraphic data. *Kans. Geol. Surv., Comput. Contrib.*, 18: 41—45.

Mann, C.J., 1970. Randomness in nature. *Geol. Soc. Am. Bull.*, 81: 95—104.

Matheron, G., 1965. *Les variables régionalisés et leur estimation*. Masson, Paris, 306 pp.

Matheron, G., 1967. Kriging, or polynomial interpolation procedures. *Ec. Natl. Supér. Mines Trans.*, 70: 240—244.

Merriam, D.F., 1963. The geologic history of Kansas. *Bull. Kans. Geol. Surv.*, 162: 1—317.

Merriam, D.F., 1970. Comparison of British and American carboniferous rock sequences. *J. Inst. Assoc. Math. Geol.*, 2: 241—264.

Merriam, D.F. and Sneath, P.H., 1967. Comparison of cyclic rock sequences using cross association. In: C. Teichert and E. Yokelson (Editors), *Essays in Paleontology and Stratigraphy. Kans. Univ. Dep. Geol. Spec. Publ.*, 2: 523—538.

Mizutani, S. and Hattori, I., 1972. Stochastic analysis of bed-thickness distribution of sediments. *J. Int. Assoc. Math. Geol.*, 4: 123—146.

Moore, R.C., 1936. Stratigraphic classification of the Pennsylvanian rocks of Kansas. *Kans. Geol. Surv., Bull.*, 22: 1—256.

Moore, R.C., 1950. Late Paleozoic cyclic sedimentation in central United States. *Int. Geol. Congr., 18th, London, 1948, Rep.*, 4: 5—16.

Moore, R.C., Frye, J.C., Jeweth, J.M., Lee, W. and O'Connor, H.G., 1951. The Kansas rock column. *Kans. Geol. Surv., Bull.*, 89: 1—132.

Neidell, N.S., 1969. Ambiguity functions and the concept of geological correlation. *Kans. Geol. Surv., Comput. Contrib.*, 40: 19—29.

Öpik, E.J., 1967. Climatic changes. In: S.K. Runcorn et al. (Editors), *International Dictionary of Geophysics*. Pergamon, Oxford, 2 vols.

Otto, G.H., 1938. The sedimentation unit and its use in field sampling. *J. Geol.*, 46: 569—582.

Payne, T.G., 1942. Stratigraphical analysis and environmental reconstruction. *Bull. Am. Assoc. Pet. Geol.*, 26: 1697—1770.

Pearn, W.C., 1964. Finding the ideal cyclothem. *Kans. Geol. Surv., Bull.*, 169 (2): 399—413.

Pearson, K., 1948. *Tables for Statisticians and Biometricians, 1*. University Press, Cambridge, 143 pp.

Pettijohn, F.J., 1957. *Sedimentary Rocks*. Harper, New York, 2nd ed., 718 pp.

Pettijohn, F.J. and Potter, P.E., 1964. *Atlas and Glossary of Primary Sedimentary Structures*. Springer, Berlin, 370 pp.

Pielou, E.C., 1964. The spatial pattern of two-phase patchworks of vegetation. *Biometrics*, 20: 156—167.

Pollack, H.N., 1968. On the interpretation of state vectors and local transformation operators. *Kans. Geol. Surv., Comput. Contrib.*, 22: 43—45.

Potter, P.E. and Blakely, R.F., 1967. Generation of a synthetic vertical profile of a fluvial sandstone body. *J. Soc. Pet. Eng.*, 1967: 243—251.

Potter, P.E. and Siever, R., 1955. A comparative study of Upper Chester and Lower Pennsylvanian stratigraphic variability. *J. Geol.*, 63: 429—451.

Quenouille, M.H., 1947. A large sample test for the goodness of fit of autoregressive schemes. *J. R. Stat. Soc.*, 110: 123—129.

Quenouille, M.H., 1949. Approximate tests of correlation in time series. *J. R. Stat. Soc.*, B, 11: 68—84.

Quenouille, M.H., 1952. *Associated Measurements*. Butterworth, London, 242 pp.

Quenouille, M.H., 1957. *The Analysis of Multiple Time Series*. Griffin, London, 104 pp.

Read, W.A., 1969. Analysis and simulation of Namurian sediments in Central Scotland using a Markov-process model. *J. Int. Assoc. Math. Geol.*, 1: 199—219.

Read, W.A. and Merriam, D.F., 1972. A simple quantitative technique for comparing cyclically deposited successions. In: D.F. Merriam (Editor), *Mathematical Models of Sedimentary Processes*. Plenum, New York, 271 pp.

Rivlina, T.S., 1968 (1970). A stochastic model of stratification (the case of unlimited interstratal erosion). Engl. Transl. In: *Topics in Mathematical Geology*. Consultants Bureau, New York, 281 pp.

Robertson, T., 1952. Plant control in rhythmic sedimentation. *Congr. Av. Etudes Stratigr. Géol. Carbonifère, Heerlen, 1951, C. R.*, 3: 515—521.

Sackin, M.J., Sneath, P.H.A. and Merriam, D.F., 1965. ALGOL program for cross-association of non-numeric sequences using a medium-size computer. *Kans. Geol. Surv., Spec. Dist. Publ.*, 23, 36pp.

Sander, B., 1936. Beiträge zur Kenntnis der Anlagerungsgefüge. *Mineral. Petrogr. Mitt.*, 48: 27—139.

Sander, B., 1948. *Einführung in die Gefügekunde geologischer Körper, 1*. Springer, Wien, 215 pp.

Scheidegger, A.E., 1961. *Theoretical Geomorphology*. Springer, Berlin, 327 pp.

Scherer, W., 1968. *Applications of Markov chains to Cyclical Sedimentation in the Oficina Formation eastern Venezuela*. Unpubl. thesis, Northwestern Univ., Evanston, 93 pp.

Schwarzacher, W., 1946. Sedimentpetrographische Untersuchungen kalkalpiner Gesteine. *Jahrb. Geol. Bundesanst.*, 1947: 1—48.

Schwarzacher, W., 1947. Über die sedimentäre Rhytmik des Dachstein Kalkes von Lofer. *Verh. Geol. Bundesanst.*, 1947: 175—188.

Schwarzacher, W., 1954. Die Grossrhytmik des Dachstein Kalkes von Lofer. *Tschermaks Mineral. Petrogr. Mitt.*, 4: 44—54.

Schwarzacher, W., 1958. The stratification of the Great Scar Limestone in the Settle district of Yorkshire. *Geol. J.*, 2: 124—142.

Schwarzacher, W., 1964. An application of statistical time-series analysis of a limestone shale sequence. *J. Geol.*, 72: 195—213.

Schwarzacher, W., 1967. Some experiments to simulate the Pennsylvanian rock sequence of Kansas. *Kans. Geol. Surv., Comput. Contrib.*, 18: 5—14.

Schwarzacher, W., 1968. Experiments with variable sedimentation rates. *Kans. Geol. Surv., Comput. Contrib.*, 22: 19—21.

Schwarzacher, W., 1969. The use of Markov chains in the study of sedimentary cycles. *J. Int. Assoc. Math. Geol.*, 1: 17—39.

Schwarzacher, W., 1972. The semi-Markov process as a general sedimentation model. In: D.F. Merriam (Editor), *Mathematical Models of Sedimentary Processes*. Plenum, New York, 271 pp.

Selley, R.C., 1970. Studies of sequence in sediments using a simple mathematical device. *Q. J. Geol. Soc. London*, 125: 557—581.

Shaw, A.B., 1964. *Time in Stratigraphy*. McGraw Hill, New York, 365 pp.

Shrock, R.R., 1948. *Sequence in Layered Rocks*. McGraw Hill, New York, 507 pp.

Sloss, L.L., 1962. Stratigraphical models in exploration. *J. Sediment. Petrol.*, 32: 415—422.

Sloss, L.L., 1963. Sequences in the cratonic interior of North America. *Geol. Soc. Am. Bull.*, 74: 93—114.

Sloss, L.L., 1964. Tectonic cycles of the north American Craton. *Kans. Geol. Surv., Bull.*, 169 (2): 449—460.

Sloss, L.L., 1972. Synchrony of phanerozoic sedimentary—tectonic events of the north American craton and the Russian platform. *Int. Geol. Congr., 24th, Montreal 1972, Sect.*, 6: 24—32.

Sneath, P.H.A., 1967. Quality and quantity of available geologic information for studies in time. *Kans. Geol. Surv., Comput. Contrib.*, 18: 57—61.

Stacey, F.D., 1969. *Physics of the Earth*. Wiley, New York, 324 pp.

Stille, H., 1924. *Grundfragen der vergleichenden Tektonik*. Borntraeger, Berlin, 443 pp.

Stockmann, K.W., Ginsburg, R.N. and Shinn, E.A., 1967. The production of lime mud by Algae in south Florida. *J. Sediment. Petrol.*, 37: 633—648.

Suess, E., 1892. *Das Antlitz der Erde, 1*. Tempsky, Wien, 778 pp.

Switzer, P., 1965. A random set process in the plane with a Markovian property. *Am. Math. Stat.*, 36: 1859—1863.

Truemann, A.E., 1948. The relation of rhythmic sedimentation to crustal movements. *Sci. Progr. (London)*, 36: 193—205.

Twenhofel, W.H., 1939. *Principles of Sedimentation.* McGraw Hill, New York, 673 pp.

Umbgrove, J.H.F., 1947. *The pulse of the Earth.* Nijhoff, The Hague, 357 pp.

Urey, H.C., Lowenstam, H.A., Epstein, S. and McKinney, C.R., 1951. Measurements of paleotemperatures and temperatures of the upper cretaceous of England, Denmark and the southeastern United States. *Bull. Geol. Soc. Am.*, 62: 399—416.

Van Houten, F.B., 1964. Cyclic lacustrine sedimentation, Upper Triassic Lokatong Formation, central New Jersey and adjacent Pennsylvania. *Kans. Geol. Surv., Bull.*, 169 (2): 497—531.

Van Vleck, J.H. and Middleton, D., 1966. The spectrum of clipped noise. *Proc. Inst. Electr. Electron. Eng.*, 54: 2—19.

Vella, P., 1965. Sedimentary cycles, correlation and stratigraphic classification. *Trans. R. Soc. N. Z.*, 3: 1—9.

Vistelius, A.B., 1949. On the question of the mechanism of the formation of strata. *Dokl. Akad. Nauk S.S.S.R.*, 65: 191—194.

Vistelius, A.B., 1961a. *Materials for the Lithostratigraphy of the Productive Strata of Azerbaidzhau.* Akademia Nauk, Moscow, 157 pp.

Vistelius, A.B., 1961b. Sedimentation time-trend functions and their application for correlation of sedimentary deposits. *J. Geol.*, 69: 703—728.

Vistelius, A.B., 1967. *Studies in Mathematical Geology.* Consultants Bureau, New York, 294 pp.

Vistelius, A.B. and Faas, A.V., 1965. The mode of alternation of strata in certain sedimentary rock sections. *Dokl. Akad. Nauk S.S.S.R.*, 164: 40—42.

Vistelius, A.B. and Feigelson, T.S., 1965. On the theory of bed formation. *Dokl. Akad. Nauk S.S.S.R.*, 164: 158—160.

Vortisch, W., 1930. Ursache und Einteilung der Schichtung. *Jahrb. Geol. Bundesanst. Wien*, 80: 453—496.

Weller, J.M., 1964. Development of the concept and interpretation of cyclic sedimentation. *Kans. Geol. Surv., Bull.*, 169 (2): 607—621.

Wells, J.W., 1963. Coral growth and geochronometry. *Nature*, 197: 948—950.

Wheeler, H.E., 1958. Primary factors in biostratigraphy. *Bull. Am. Assoc. Pet. Geol.*, 42: 640—655.

Whittaker, E. and Robinson, G., 1960. *The Calculus of Observation.* Blackie, London, 397 pp.

Whittaker, E.T. and Watson, G.N., 1969. *A Course of Modern Analysis.* University Press, Cambridge, 608 pp.

Whittle, P., 1952. Test of fit in time series. *Biometrika*, 39: 309—318.

Whittle, P., 1954. On stationary processes in the plane. *Biometrika*, 41: 434—449.

Whittle, P., 1954. The statistical analysis of a sunspot series. *J. Astrophys.*, 119: 251—260.

Yule, G.V., 1927. On a method of investigating periodicities in disturbed series, with special reference to Wolfer's sunspot number. *Philos. Trans. R. Soc. London, Ser. A.*, 226: 267—298.

Index

te Due